Rich. d Owen .

Natur Edwards

I am
Yours [illegible]
J.W. Eschardt

Agréer mes civilités,
Efairmaie

Henry Stevens

H. Crewe

Hugh Reid

Howard Saine

Yours
very truly
F.O. Morris

[illegible]
ton Rowley

Yours Very Truly
Symington Grieve

Dr
Wilh. Blasius

H.E. Dresser

Rowland [illegible]
L. Reichenbach

[illegible]
Löbbecke

Yrs truly
2 A Harvie Brown

I [illegible]

yours sincerely
Rothschild

Paul Hahn

J.J. [illegible]

Hugh [illegible] Barclay

Alfred Roberts.

da Meyerman ather

J.G. [illegible]

[illegible]

J.W. Baedeker

Inspektor Reenley

THE GREAT AUK

BY

ERROL FULLER

A PETER N. NEVRAUMONT BOOK

HARRY N. ABRAMS, INC., PUBLISHERS

Library of Congress Cataloging-in-Publication Data
Fuller, Errol.
The great auk / by Errol Fuller.
p.   cm.
Includes bibliographical references (p.   ).
ISBN 0–8109–6391–4
1. Great auk. I. Title.
QL696.C42F85   1999
598.3'3—dc21   99–24632

First published in Great Britain in 1999
Copyright © Errol Fuller 1999
Distributed in 1999 by Harry N. Abrams, Incorporated, New York

This book was designed and published by Errol Fuller,
65 Springfield Road, Southborough, Kent, England

Colour separations by Sfera International, Milan, Italy
Printing and binding by Sfera International, Milan, Italy

British Library Cataloguing in Publication Data
A catalogue record for this book is available from the British Library.

Frontispiece: The Great Auk Imagined off Eldey Island by Ray Harris Ching

Endpapers: Great Auk signatures—a collection of autographs belonging to men with an
interest in Garefowls

Half title: A mandible discovered by Leslie Tuck on Funk Island and given to
Ralph S. Palmer around 1960. Dr. Palmer gave it to the author in October 1996

Pages following: Three views of Hoppa's Auk (bird no. 12)

A collection of eight eggs belonging to Dr. Jack Gibson. From left: (top row) eggs no. 15,
egg no. 16, egg no. 19, egg no. 21, (bottom row) egg no. 17, egg no. 18, egg no. 22 and
egg no. 20. All are approximately natural size. *Courtesy of Dr. Jack Gibson*

A Great Auk cigarette tin. This very rare tin contained 100 cigarettes. "Great Auk" was an
English brand manufactured for a few years around the time of World War I. It could only
be bought from three or four retail outlets. *Courtesy of Peter Southon*

For Rowan, Errol and Frankie

The...thrill is not to kill, but to let live.
J.O. Curwood

Animals, our brethren, in suffering, disease and death
Charles Darwin

Pingouin Brachyptère ♂

THE GREAT AUK

Contents

PROLOGUE

It is the morning of the third day of June in the year eighteen hundred and forty-four. A fishing boat has come close to the remote Icelandic island of Eldey. Despite the season, a thin mist hangs over the island and the sun shines only palely through the clouds, hardly managing to warm the stones. Three men, Sigurðr Islefsson, Jón Brandsson and Ketil Ketilsson, have struggled ashore from an eight-oared boat that waits, with the rest of the crew, for their return. As they clamber over the rocks and stones, hundreds of seabirds rise up and scuttle from their rough, lurching path, flapping their way to safety. One of the men gives a call and points through the swirling mist as it blows in off the sea. Two birds, larger than the others, struggle up and waddle clumsily and frantically away, moving as fast as their short legs will carry them. Unlike the other birds, these two cannot fly and desperately they make for the only sanctuary they know - the safety of the water. But the men are too fast! In a flash one of them reaches the nearest bird and bludgeons it to the ground where it flaps hopelessly until it is picked up and strangled. The other bird waddles on, panic and distress driving it forward, its little, useless wings helplessly outstretched. Almost, it has reached the water. Almost, but not quite! One of the Icelanders darts in front and cuts it off from the sea. Another comes up behind. As the bird turns just a yard or two from safety, the first man seizes it firmly and breaks its neck. The men roughly pick up the poor broken bodies and return across the strand to their boat.

Tradition has it that these two birds were the very last of their kind, the last Great Auks. Perhaps they were. But, more likely, a few others still clung to existence in the lonely places of the far north. What seems certain, however, is that very soon afterwards, somewhere in the deep waters of the North Atlantic or in a hollow upon a wind-lashed reef - perhaps sheltering from the fury of the waves - the last Great Auk died.

A Gallery of Auks (artists unknown unless stated) (Centre). Illustration from H. Davis's *The Natural History of Animals* (1902). (Top row, from left). Watercolour by W. Kuhnert from *The Harmsworth Natural History* (1911); engraving (source unknown); woodcut by T. Bewick; engraving by T. Wood from A. Newton's *Dictionary of Birds* (1893-6). (Second row). Engraving from S. Baird's *The Waterbirds of North America* (1884); engraving from J. Wood's *The Illustrated Natural History* (1872). (Third row). Drawing by A. & C. Lydon from R. Sharpe's *Sketchbook of British Birds* (1898); drawing by D. M. Henry. (Bottom row). Engraving from *The Field* (March 27 1875); engraving by W. Freeman from A. Mangin's *Mysteries of the Ocean* (1874); engraving from Knight's *Pictorial Museum of Animated Nature* (c.1850); Watercolour by F. Frohawk from A. Butler's *British Birds with Nests and Eggs* (1896-8).

THE GREAT AUK

THE GREAT AUK

-Whilst thee the shores and sounding seas
Wash far away, where'er thy bones are hurl'd;
Whether beyond the stormy Hebrides,
Where thou, perhaps, under the whelming tide,
Visit'st the bottom of the monstrous world.

John Milton (*Lycidas*)

The Great Auk has always been peculiarly fascinating. Quite why is difficult to say.

Certainly, it has much to do with the fact that the bird is extinct. But there are many other extinct species and the truth is that most of them arouse little interest. In any case the Great Auk drew men to it long before it vanished.

Probably, it has something to do with the bird's strange-sounding name. Great Auk - like Dodo, Kookaburra or Bustard - is so curious and appealing to the ear that, once heard, it is unlikely to be forgotten. Even the Great Auk's other names - Garefowl, Penguin of the North, Riesenalk, Apponath - are all attractive in their way. But a name is - surely - just a name.

Perhaps, the fascination is connected with the contradictions surrounding the species: strikingly but soberly coloured, visually familiar - due to superficial resemblance to the Penguins - yet so inherently mysterious, easy to catch but so difficult to find, helpless on land, agile in the sea, this was a creature that abandoned its natural element for one that was far darker and more sinister. The Great Auk was unable to perform the one function that so characterises a bird, yet it could 'fly' through the water with a marvellous ease.

Even the supposed date of its extinction has a certain contradictory balance. The year 1844 is just recent enough to make the bird seem accessible, yet long enough ago to make it truly elusive. And this elusive quality was as real to those who actually knew the bird as it is to those who come long after. To the men and women who, 300 years ago, frequented the coasts, reefs and skerries of the far north, the Great Auk was, perhaps, a familiar enough sight - but only for a few weeks of each year. It came, suddenly in May - from where they knew not - climbed ashore for a few weeks then vanished again just as suddenly as it had arrived.

The notorious encounters of some of the last few individuals with man are so remarkable and so touching that they have passed - almost - into legend. There was the captive bird allowed to 'sport' in the sea with a cord attached to its leg; predictably, it escaped - at the mouth of the River Clyde. There was the poor little captive on St Kilda cruelly executed as a witch when a storm blew up that it was suspected of having caused; and

there is the chilling account of the killing of what were, perhaps, the last two individuals on the grim Icelandic island of Eldey.

But the tale of the Great Auk doesn't end with its extinction. The pursuit of the most meagre scraps of information, the recording of the most minute details of habit or appearance, the search for stuffed birds and preserved eggs have all obsessed certain men down through the years and the list of those who've spent much of their lives in such bizarre hunts is not a short one. The sums of money exchanged for specimens have often been extraordinary: £350 in 1895, £700 in 1936, £9,000 in 1971, $25,000 in 1974. These were all substantial amounts when judged by the standards of the times.

The Great Auk, large and plump, has tempted man since the dawn of history. Whatever the precise reasons for its hold on his imagination, it is a hold that has lasted, and even 150 years after the bird's apparent demise, the grip seems as powerful as ever.

Aquatint by J.J. Audubon and R. Havell from J.J. Audubon's *The Birds of America* (1827-38).

THE GREAT
AUK'S
APPEARANCE

THE GREAT AUK'S APPEARANCE

An ancient mariner... with a white waistcoat, a black 'beetle coat,' an ugly nose and a helpless pair of arms.

An anonymous correspondent to a London newspaper (1902)

ALCA impennis. *Great Auk.* Bill marked with feveral furrows; the upper mandible in part covered with black velvet-like feathers: between each eye and the bill a large white fpot: the reft of the head, neck, back, tail and wings, a glolfy black: tips of the fecondaries white: under part of the body white: legs black: length three feet. The wings too fmall for flight; the length from the tip of the longeft quill to the firft joint, being only four inches and a quarter.

Found only on our northern coaft: obferved never to wander beyond foundings: breeds on the ifle of *St. Kilda:* lays only one egg varioufly fpotted.

Hand-coloured engraving by P. Mazell from J. Walcott's *Synopsis of British Birds* (1789).

Hand-coloured lithograph by J. G. Keulemans from H.E. Dresser's *A History of the Birds of Europe* (1871-96).

The Great Auk, black and white, hugely beaked and shaped rather like a Penguin, could hardly be confused with any living bird. Its upright stance separates it from the vast majority of birds, its enormous beak marks it out from any Penguin and its overall size - around 30 inches (75 cm) - removes it from its superficially similar relative the Razorbill (*Alca torda*).

The sexes were, apparently, identical (although it is possible that slight external differences existed) but the plumage was subject to some seasonal change.

J.G.Keulemans lith.

Hanhart imp.

The Great Auk in close-up. (This page, from the top). Side view of the head; the head seen from above; the feet. (Facing page). The head seen from the front; the tail.

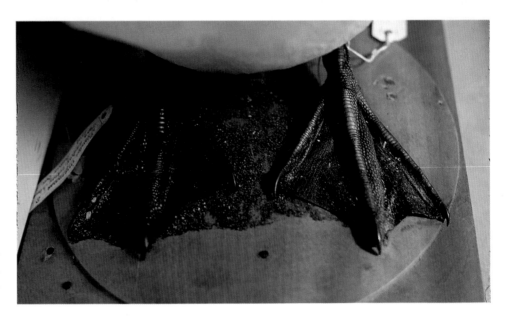

In winter the birds were at sea and seldom seen so it is the summer plumage that is most familiar. In this feathering the upperparts and the head were black, apart from a large white patch before each eye. The upper throat and sides of the neck were also black but here the feathers were usually tinged with chocolate brown. The wings were black with the tips of the secondary feathers white. The lower parts from throat to under tail coverts were white with most individuals showing a distinctive fringe of grey feathers on the flanks below the wings.

It is possible to give more detail to this description. The Great Auk's head was dominated by the huge, grotesque beak, a striking appendage quite unlike the bill of any Penguin and an object that sets these birds apart as profoundly different creatures. Yet the exact appearance of this beak in life is uncertain. Its colour in the living bird was never precisely or unequivocally reported, and preserved specimens don't necessarily provide a reliable guide. Because the colours on a bird's bill are subject to fading or other alteration after death, taxidermists and preparators regularly paint this part- often accurately, sometimes imaginatively. There is little doubt that the Great Auk's beak was dark in colour - blackish- with varying numbers of grooves (as many as twelve as few as three) on each side of the mandible, a variation usually explained as an indicator of age. In many of the surviving mounted specimens the grooves are marked with white but it remains uncertain how much (if any) of this colour is natural and how much is simply paint. Several nineteenth century commentators, men who handled specimens when they were comparatively fresh (most notably J.H. Gurney, 1869), were convinced that at least some of the pigment was natural. On the other hand, a number of early descriptions make no mention of any white on the beak. Perhaps this too was a variable feature. There is a certain natural elegance in the notion that - during the breeding season - the bill carried some white striping (as it does in the Razorbill) but the truth is that no-one really knows whether it did or not.

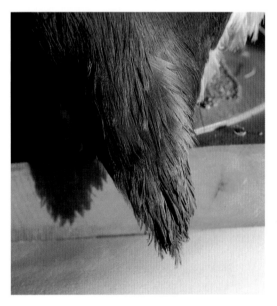

The colour of the rather small eye is another feature that went largely undescribed although, fortunately, there are at least two unequivocal records of it. During the early 1820's, a Scottish professor by the name of Fleming was able to observe a captive bird and described the iris as chestnut with the margin of each eyelid black. Sigriður Thorláksdótter, an Icelandic fishwife who in 1831 handled 23 freshly killed birds and saw one alive, said the eyes were hazel.

Many of the Icelandic fishermen who captured birds on the reefs and skerries off south-west Iceland mentioned a peculiar film or membrane that covered the eyes as the birds emerged from the water. Doubtless this is the feature known to bird anatomists as the nictitating membrane. The Icelanders called it the *blaðka* and some of them described it as blue and white in colour. Probably, the words 'blue and white' result from translation difficulties (the men's observations were noted down during conversations with two Englishmen, Alfred Newton and John Wolley); perhaps a light blue was meant, although grey would be a more usual colour for this particular feature.

According to Sigriður Thorláksdótter, the inside of the mouth was brightly coloured; she recalled it as 'light red.' Professor Fleming described it as orange, but perhaps the discrepancy is again due to linguistic difficul-

Average measurements	
Length	75cm - 80cm
Culmen	85mm
Bill depth	40mm
Wing	160mm
Tail	75mm
Tarsus	58mm
Middle toe	70mm

The underwing of Audubon's Auk (bird no. 20)- a photograph taken when it was re-stuffed during the 1920's.

Hand-coloured lithograph from N. Kjærbölling's *Ornithologia Danica - Danmark's Fugle* (1851-6). The picture shows clearly the brown throat.

ties. The possibility of some sort of reddish colour about the head may explain one of the mysteries of Great Auk literature. A certain Martin Martin wrote a late seventeenth century account of Great Auks on St Kilda, an account that perhaps resulted from a direct encounter with living birds. In an otherwise straightforward description he mentioned that the birds were 'red around the eyes,' a remark that has perplexed generations of Garefowl scholars. John Smith (1879) was the first to suggest that Martin, perhaps watching from a distance, saw birds opening and closing their beaks. The gape would reach behind and below the eyes and the bright colour perhaps misled Mr. Martin. Another early account suggests that the colour of the mouth's interior was yellow, however. In an unpublished manuscript kept today at the University Museum, Copenhagen, Otto Fabricius, author of *Fauna Groenlandica* (1780), is quite specific on this point. What, then, can be made of these differing accounts? It seems safest to assume that the colour was in reality a yellowy orange and that Sigriður Thorláksdótter's description of 'light red' rests on a linguistic misunderstanding. What Martin Martin saw remains a mystery.

On each side of the face, immediately behind the upper mandible, was a large lozenge-shaped patch of velvety textured white feathers that just touched the leading edge of the eye. Together with the dramatic outline of the beak and the small eye, this gives to the head a completely distinctive appearance.

Many of the preserved specimens (but not all) show a rich brown that suffuses the blackish feathering of the sides of the face and neck. Whether this colouring is a mark of older birds or has some sexual significance is not known.

The wings were tiny in proportion to the size of the body yet they remained entirely wing-like in character having no similarity to the

Penguin's flippery appendages; the stunted primary feathers were thin and pointed. The colour was black, although the secondaries were rather strikingly tipped with white.

The underparts from throat to tail were white but most preserved specimens show a distinctive fringe of longer feathers on the flanks. These feathers are coloured with a subtly pale shade of grey - almost lilac - and extend in a band from just below the wing right down to the tail. A few existing specimens lack this grey fringe, but what this means is unclear. Walter Rothschild (1907) believed this lilac grey shade to be a female characteristic but his reasons for this supposition are not known. Interestingly, Peter Lyngs (see *British Birds*, 1996, 89, 10, p.453) has noted a similar feature on Razorbills in the Baltic. Some 20% of a colony he investigated were marked with grey but he was unable to relate the colour to sex or to age. He makes the curious point that (as is almost the case with the Great Auk) he has never seen this Razorbill characteristic depicted or mentioned elsewhere.

The tail, consisting of fourteen feathers, was short but pointed, black above and white below.

The webbed feet were black and placed far back on the body. The tarsus was short. There were three toes - with a hind toe wanting - each with a rather small claw.

The ear apertures were very small, covered with feathers and stood in line with the opening of the beak at the back end of each cheek. The nostrils were at the rear of the beak and were elongated in shape running into a deep furrow. The tongue, according to Fabricius, was small, flat and shaped like a sword's blade.

In winter the birds underwent a plumage change similar to that experienced by their relatives the Guillemot (*Uria aalge*) and the Razorbill. The black of the foreneck and chin was replaced by white and the lozenge-shaped white patch between eye and beak disappeared.

The appearance of the chick is unknown and so too are any juvenile changes in plumage.

The Cambridge Auk (bird no.5) - a photograph showing the fringe of grey feathers. *Courtesy of the Zoological Museum of Cambridge University.*

*THE GREAT
AUK'S PLACE
IN THE WORLD
OF BIRDS*

THE GREAT AUK'S PLACE IN THE WORLD OF BIRDS

Museum specimens - a Great and a Little Auk. Photograph by Norrie of Fraserburgh (c.1900).

The Great Auk and related birds. Watercolour by Archibald Thorburn, reproduced in his *British Birds* (1918).

As suggested by its name, the Great Auk is classified in the bird family known as the Auks and Murres or, more technically, as the Alcidae. This family consists of more than twenty species among which are Auks and Auklets, Murres and Murrelets, Puffins, Guillemots, Dovekies and Razorbills. They are rather dumpy, short-winged birds, some with astonishingly shaped and coloured beaks. Mostly, they are of small to medium size and the Great Auk, as might be expected, is by far the largest. It was also the only flightless member of the family, most of its relatives being characterised by a fast and direct flying style.

All of the Alcidae are tied to the water and are confined to the seas, coasts and islands of the northern hemisphere. Their living is chiefly earned by bobbing on the ocean's surface, then diving for fish and crustaceans; some species feed on plankton. Most kinds breed in colonies and some of them form colonies that are huge in size.

Although birds from the family occur in both the North Atlantic and the North Pacific, there are none in southern seas. Here, their ecological position is taken - for the most part - by Penguins. In some superficial ways the two groups might seem similar but they have no close relationship to one another. Any seeming similarities are due entirely to similarities in their way of living. The Auks are, in fact, much more closely related to the Gulls and Waders, with which families they share anatomical peculiarities.

Although, doubtless, many Auk species evolved and then died out during remote prehistoric epochs, the Great Auk is the only member of the family to have become extinct during recent times.

One of the species that failed to survive into historical times appears, however, to have been very closely related to *Alca impennis* and was perhaps a direct ancestor. It was described by Storrs Olson (1977) from fossil bones found in a phosphate mine in North Carolina. Olson considered these relics so close in form to bones of the Great Auk that he named the species *Pinguinis alfrednewtoni* to commemorate the name of the nineteenth century Cambridge professor who placed so much Garefowl lore on record. Cyril Walker, an osteological specialist at the Natural History Museum, London has identified bones that he believes are similar to those of Olson's from Lower Pleistocene deposits at Kallo in Belgium.

Some writers suggest that Great Auks from American waters constituted a different species to those from Europe. This idea is put forward on the grounds of apparent differences in bone measurements, American birds seeming slightly larger. The evidence is hardly conclusive and there is no evidence at all as to whether or not there was interaction between birds from the two areas.

Black Guillemot. (summer & winter).   Puffin. (summer & winter)
Common Guillemot.   Brünnich's Guillemot.
Razorbill (summer & winter)
Great Auk.   Little Auk. (summer & winter)

THE LIFE OF
THE GREAT AUK

THE LIFE OF THE GREAT AUK

The sea-fowls are, first, Gairfowl, being the stateliest, as well as the largest sort, and above the size of a solan goose [Gannet], of a black colour ... a large white spot under each [eye], a long broad bill; it stands stately, its whole body erected, its wings short, flies not at all, lays its egg upon the bare rock, which, if taken away, she lays no more for that year ... It comes without regard to any wind, appears the first of May and goes away about the middle of June.

Martin Martin (1698)

Their food was fish and crabs. They made no nest, but laid their single egg upon the bare rock. Nobody knows what their exact breeding season was, though they were found with eggs in June and young in July; nobody knows how long they incubated their single egg, or how long the young stayed on the breeding rock before it swam away. The Gare-fowl ... laid its egg on low rocky islands and skerries, on to which it could struggle and waddle at any state of the tide. All its known breeding places were lonely and remote; they were all in the ... colder waters of the North Atlantic, though not, as is popularly supposed, in the Arctic.

James Fisher (1945)

The life of the Great Auk begins with the egg. Large in size, pyriform in shape and beautifully marked with streaks and blotches, no two Garefowl eggs were patterned alike; they were never identical. Can anything be deduced from this curious fact? The answer - probably - is yes.

It suggests that the Great Auk, like its close relative the Guillemot (*Uria aalge*) - a bird also noted for the individual patterning of its eggs - was a creature that, in its heyday, packed itself tightly onto its breeding islands to nest virtually shoulder to shoulder with its companions. The unique appearance of each egg served, presumably, as a device to aid identification, a means by which each pair of Auks could single out their own egg. When Great Auks waddled ashore to breed, they clearly did not want to be alone.

Early written records certainly bear this out. The terrible mass slaughters that took place could hardly have occurred on islands that were only thinly populated; nor could they have been effected among birds showing more individual traits. There can be little doubt that Great Auks, like their living relatives, Guillemots and Razorbills (*Alca torda*), enjoyed - indeed needed - the company of many others of their kind.

The pattern of the species' decline to extinction reveals that this was a creature whose long-term survival could only be effected in large groups. Sven Bengston (1984) believed that Garefowl colonies tended to be rather

The Great Auk at Home. Oil painting by J.G. Keulemans (donated to The Natural History Museum, London by Walter Rothschild).

small and others have followed this belief - but it cannot be right. Certainly, the colonies were small towards the end but this is hardly a reflection of a genuine preference. In the species' heyday things were surely different.

Following the model of other communal bird species, the Great Auk's behavioural responses can be crudely outlined: when one bird hopped ashore, so did its companions; if one swam to the right, it was quickly followed by others; if it moved left, so did the rest. It is an ironic fact that much of our knowledge of the living Garefowl comes from nineteenth century records of lost and isolated birds observed in entirely untypical situa-

Guillemots on Funk Island- a photograph taken on July 14 1959 by James Fisher. *Reproduced by kind permission of Clemency Thorne Fisher.*

Two Great Auk eggs with the smaller eggs of other birds (natural size). The upper figure is copied from egg no. 45, the lower shows a generalised pattern and doesn't seem to represent any specific specimen. Hand-coloured lithograph from F. Bædeker's *Die Eier der Europäischen Vögel* (1855-63). *Courtesy of Peter Blest.*

tions. Emotionally and physically, the Great Auk needed to live in large groups in order to thrive- and the very fact and manner of its extinction suggests nothing less. Why else would it have died out? Had it felt comfortable existing in small groups or parties, the species could always have found lonely places in the far north in which to secretly breed, despite the fierceness of human persecution and despite the rather specific requirements in terms of landing places and breeding spots that its flightlessness imposed. Perhaps a similar pattern of decline is occurring with the Jackass Penguin (*Spheniscus demersus*), a creature whose numbers have fallen sharply in recent decades. Though still numerically a very populous bird, it may be reaching a point of no return. Unlike the Great Auk, however, the Jackass Penguin has survived into an era in which its survival is likely to be encouraged.

Just as the individual colouring and marking of the Great Auk's egg paints a picture of a rigidly communal bird, so too does its pyriform shape. The species' nearest relative, the Razorbill, lays what may be described as an altogether more egg-shaped egg; this it incubates in a lying-down position. The Guillemot, however, with an egg shaped suspiciously like the Garefowl's, adopts a more upright stance with the egg's pointed end directed inwards, conforming to the bird's posture. What does this erect posture - to which the shape of the egg appears to accommodate itself - mean? Simply that Guillemots, standing virtually shoulder to shoulder, can pack themselves more tightly onto the cliff ledges on which they nest. With Great Auks being - unlike their relatives - flightless, and therefore unable to reach otherwise suitable nesting places (they could only take up those they could waddle to), viable breeding spots were probably at a premium. The pyriform egg, indicating upright posture and optimum use of available space, was - presumably - a very necessary development.

There is, in fact, a written description that confirms how closely Great Auks packed together. During 1858 John Wolley, celebrated egg collector and Garefowl enthusiast, travelled to Iceland and interviewed many of the sailors and fishermen who'd hunted down the last of the Great Auks. One

Illustration by Jens Frimer Andersen for P. Lyngs *Gejrfuglen* (1994). *Reproduced by kind permnission of the artist.*

of them, Jón Brandsson, a man who in 1844 strangled one of the last pair, described an earlier garefowling expedition:

> All the birds were together, close under the rocks and sitting so that they were shoulder to shoulder. They sit more upright than the blackbirds [Guillemots] which are nearer the sea.

(Above, left). Engraving by F. Specht from R. Lydekker's *The Royal Natural History* (1895).

(Above, right). Painting by R. Didier reproduced in *La Terre et La Vie* (1934).

In nature, function ruthlessly equals design and the pyriform egg shape confers another advantage. It is generally believed that this particular form is a device evolved to cut down chances of eggs rolling away in critical situations. On a narrow cliff ledge or on a rocky, uneven surface probably exposed to high or gusting winds, the result of uncontrolled movement is obvious. If a pyriform egg is knocked or blown it is likely to move in a circle - albeit quite a large one - rather than entering into a free roll from which the consequences might prove disastrous.

Tim Birkhead (1993) pours a certain amount of scorn on the idea of egg shape conveying this kind of advantage and preventing breakage. He suggests, quite rightly, that it would often - perhaps usually - fail to save the egg from catastrophe. This rather misses the point about natural selection, however. A device - or trick - doesn't have to work every time in order to become a successful agent of selection. It simply has to succeed often enough to confer an advantage in the struggle for life. This advantage cannot, meaningfully, be quantified. Who can say whether an advantage factor of one percent or ninety-nine is necessary to produce - over aeons - an evolutionary modification? The point is that pyriform shape does make significant, and potentially catastrophic, egg rolling just a little less likely.

The Great Auk almost certainly laid just a single egg during the course of a normal season. This fact alone is enough to reveal that Garefowls were comparatively long-lived birds; their size is a factor that serves to confirm this.

The Great Auk underwater. Computer generated image by Jack Griggs for the American Bird Conservancy's Field Guide, *All the Birds of North America* (1997). *Reproduced by kind permission of Jack Griggs and Harper Collins Publishers.*

Standing around 30 inches (75 cm) tall and weighing perhaps 5 kg, the Great Auk, with its stark black and white colouring, erect posture and formidable beak, was an imposing creature despite the grotesquely disproportionate little wings that rendered it quite incapable of flight.

-Through the air, that is; for underwater flight the wings and proportions of the body were beautifully designed. And it was in the water where the bird passed most of its time, by far the greater part of the year being spent at sea.

Only during its short summer breeding season did it struggle ashore- and then it paid a high price for its superb adaptation to water. Its little legs, set far back on the body, didn't allow the easiest, or quickest, movement on land. Because it was uncomfortable and vulnerable when out of the water its stay on land was as brief as possible. Whether it actually mated at sea like the Puffin (*Fratercula arctica*) or on land like its other close relatives is unknown, as is every other aspect of its courtship. It may be supposed that the yellowy orange colouring of the interior of the Great Auk's mouth represented a sexual stimulant- which supposition lies at the limit of our understanding of Garefowl sexual behaviour.

That the Great Auk's sojourn on land was markedly short is best shown in an early account of St Kilda, the remote island off the Scottish coast that was once a home to the species. Written by Martin Martin, the account was published in 1698 and, according to its preface, is based on:

> Nothing ... but what [the author] asserts for truth, either
> upon his own particular Knowledge, or from the constant and
> harmonious Testimony given him by the inhabitants, People so
> plain and so little inclined to impose upon Mankind, that perhaps
> no place in the World at this Day knows Instances like these of
> primitive Honour and Simplicity; a People abhorring lying,
> Tricks, and Artifices, as they do the most poisonous Plants,
> or devouring animals.

Martin states - basing his statement on the accounts given by these people - that Great Auks arrived at the start of May and were gone again by the middle of June. Due to technical adjustments to the calendar made during the eighteenth century, these dates should be advanced to mid May and late June but they indicate, nevertheless, that the yearly Auk visit was of remarkably short duration.

The dates may be reliable or they may not. They hardly tally with a traditional Icelandic date - St John's Day - for raiding one of the several breeding stations that went by the name of Geirfuglasker. This particular 'Geirfuglasker' was off the east coast and Icelanders sailed to their island around June 24th in confident expectation of finding eggs. The Auks were unlikely to hatch and vanish within a week. Nor do the St Kilda dates agree with a day in early August of 1808 when a privateer named *The Salamine* arrived at another (and rather better known) Geirfuglasker. On this occasion the crew, apparently, went ashore to wreak havoc and destruction among the nesting birds. Such discrepancies are perhaps caused by misunderstandings that cannot now be unravelled. Certainly, Vilhjálmur Hákonarsson, one of the fishermen interviewed by John Wolley in 1858, supplied dates that agree quite closely with those given by the St. Kildans. He said:

> The Garefowl laid at the end of May; the young hatched about
> Midsummer's Day. About July 6th, the birds left the rock

It can be added - for what it is worth - that Wolley considered this Vilhjálmur a man who told, 'the truth as best he could.' It can also be mentioned that he participated in some of the last raids to the island of Eldey, raids that were undoubtedly successful in harvesting birds and eggs.

Chromolithograph by A. Thorburn from Lord Lilford's *Coloured Figures of the Birds of the British Isles* (1885-98).

The title page to Martin Martin's book.

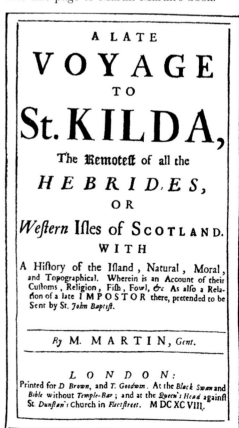

What cannot be in doubt is the fact that the birds made off as soon as they possibly could. An anonymous eighteenth century letter now kept at the National Library of Scotland (Advocates Library Ms 21.1.5 ff 183-5), referring again to St Kilda says:

> They lay their eggs a little above the sea mark on rocks of easy access; they carry off their young soon to feed them at sea.

We may be sure that Garefowls laid their eggs soon after swimming to shore. Several accounts - just like the last - speak of the laying occurring just beyond the reach of the waves. Depending on the situation, however, the birds sometimes penetrated a short distance inland or clambered onto the lower cliffs, sometimes laying on the bare rock, sometimes on accumulations of guano. Having produced their egg, the parents settled down to hatch it.

Whatever the actual count of days spent at the breeding station, their number was, of course, governed by the twin factors of incubation period and chick-rearing time. In the absence of properly compiled records, the only sensible assumptions that can be made lie in comparisons with the Garefowl's closest extant relatives, the Guillemot and the Razorbill. These two species have incubation periods of 32 and 35 days respectively; as the

Great Auks and their chick. Watercolour by D. Quinn painted for the cover of T. Birkhead's *Great Auk Islands* (1993). *Reproduced by kind permission of David Quinn and Tim Birkhead.*

Great Auk is considerably larger than either, it is reasonable to assume that the hatching of its eggs took a few days longer.

As far as chick-rearing is concerned, the Guillemot takes 21 days and the Razorbill 17; once again the Great Auk might be expected to be a little slower. In such considerations the effect of the species' flightlessness cannot be meaningfully assessed, and any conclusions reached are rendered all the more speculative by the strange fact that the Auk family is remarkably diverse in the development rate of chicks. Young Puffins, for instance, take seven weeks to reach a stage at which they are fit to leave their breeding colony, yet several Murrelet species that inhabit the North Pacific rear chicks capable of leaving after just two days.

Notwithstanding this curious diversity, the safest comparisons remain with the Razorbill and Guillemot. Once the chick has emerged from the egg it is attended constantly by at least one parent who will keep it warm and guard it from predators. Because of the intense feeding requirements and the pressure of bringing food back to an ever growing dependant, it is desirable for the young one to be encouraged to sea as rapidly as possible, before its increasing demands for food defeat the parent birds' ability to cope. Guillemot and Razorbill chicks leave the colony when they reach about a quarter of the adult body weight. With the Great Auk being flightless, however, the laborious task of conveying food back from the sea to the breeding colony was, perhaps, particularly onerous, a factor that may have completely altered such a ratio.

Speculation on the likely development rate of the Great Auk's chick is the only option for there is little direct evidence concerning it. Even its appearance can only be imagined. It might seem surprising that chicks were never taken for collections - along with the skins of adults and their eggs - but, if they were, none now survive. Various stories have been retailed in which young birds are mentioned, but these are hardly illuminating. When, for instance, the privateer *The Salamine* visited the Geirfuglasker in 1808 young birds were, reportedly, trodden down during the wholesale slaughter that took place. On the other side of the Atlantic, John James Audubon heard that fishermen often took young birds for bait. In an early French account of North American coasts Nicholas Denys (1672) stated that on the Grand Banks of Newfoundland, Great Auks carried their young on their backs. Apart, perhaps, from the last, which might suggest that the young left their colony at a particularly early age, these kinds of passing remarks are of little help in building up a meaningful picture.

The most detailed account of chicks comes in the writing of Otto Fabricius (1780) who stayed at Frederikshåb, south-west Greenland between the years 1768 and 1774. Fabricius claimed that, during the month of August, he saw - and caught - chicks, still with grey down among their feathers, off Greenland's west coast. The Danish Garefowl scholar Japetus Steenstrup, widely regarded as the 'father' of Great Auk research, utterly distrusted this information and Symington Grieve, author of *The Great Auk or Garefowl* (1885), deferring to the older man's opinion, expressed the belief that Fabricius was perhaps confused by individuals belonging to another species. As a consequence, the report remained largely disregarded until quite recently when Morten

Otto Fabricius (1744-1822).

Meldgaard (1988) resurrected it, pointing out that in other respects the eighteenth century commentator has proved a most reliable observer and writer. Among Fabricius's claims is one that he actually dissected a chick and discovered that it had eaten Roseroot (*Sedum rhodiola*) and other shore plants.

As well as cutting up dead baby birds, Fabricius also dissected adults. In doing so he provided the only precise information about what wild Garefowls actually ate. Obviously they were fish eaters - probably taking shoaling pelagic fish as well as diving to catch those on the sea floor - but Fabricius was able to list the actual species he found inside the birds' bodies: the Shorthorn Sculpin (*Cottus scorpius*) and the Lumpsucker (*Cyclopterus lumpus*). A captive Great Auk taken off the coast of Ireland during 1834 was fed on Sprats (*Sprattus sprattus*) and Olaus Wormius (1655) claimed that he fed a pet bird on Herrings (*Clupea harengus*).

Recent research using elaborate techniques to extract and analyse bone matter and soil fragments (*see* Olsen, Swift and Mokhiber, 1979 *and also* Hobson and Montevecchi, 1991) has attempted to define the exact feeding requirements of the species. Elaborate technology does not necessarily equal clear results, however, and some of the conclusions arrived at are problematical. Hobson and Montevecchi, on the basis of bone analysis, came up with some particularly interesting suggestions. Not only do they suspect that Great Auks ate crustaceans as well as fish, they concluded that the chicks were perhaps fed on plankton - probably brought back to land in the adult's exceptionally capacious gullet. Bill Bourne, author of a compact history of the species (1993), suggests that the structure of the throat was, perhaps, sufficiently robust to allow feeding of the young by regurgitation.

The capacious gullet was first subjected to academic scrutiny by the celebrated Newcastle taxidermist John Hancock during the nineteenth century. Hancock showed a particular interest in the throats and stomachs of birds and re-stuffed his own Garefowl specimen to reveal the true capability of the feature. Long before the time of Hancock, however, the capacity of the Great Auk's stomach was noticed by the people of St Kilda. Another manuscript in the Advocates Library of the National Library of Scotland (Ms 33.3.20 fol.21) refers to this Scottish island and also to the Garefowl. Known as the Robert Sibbald manuscript, it contains the following curious piece of information:

> The Garefowl ... is bigger than any goose ... Among the other commodities they export out of the island this is none of the meanest. They take the fat of ... fowls that frequent the island and stuff the stomach of this fowl with it, which they preserve by hanging it near the chimney, where it is dried with the smoke ... they sell it as a remedy... for aches and pains.

The gullet and stomach may have been especially capacious but it is the bill, clearly designed for the catching, carrying and slicing of fish, that is the Great Auk's most striking feature. An exaggerated version of the beak of the Razorbill - a species whose name speaks for itself - this was a powerful and formidable weapon. Grotesque and unwieldy it may appear but in reality it was a spear-like and streamlined agent of destruction.

Drawing by G.E. Lodge for D.A. Bannerman's and G.E. Lodge's *The Birds of the British Isles* (1953-63) *Courtesy of John Metcalf.*

Three pictures from the north-east of England by, or showing the influence of, John Hancock. (Above, left). Lithograph by J. Hancock from *A Catalogue of the Birds of Northumberland and Durham* (1874).
(Above, centre). Drawing by J. Duncan for the frontispiece to the large paper edition of his *Birds of the British Isles* (1900).
(Above, right). Oil painting by an unknown hand (thought to be J. Hancock or D. Embleton). *Courtesy of the Hancock Museum, Newcastle upon Tyne.*

The Razorbill - closest living relative of the Great Auk. Watercolour by Ray Harris Ching. *Reproduced by kind permission of the artist.*

(Above). Hand-coloured wood engraving from F. O. Morris's *A History of British Birds* (1856-62).

(Above, right). Great Auks at sea. Illustration by Jens Frimer Andersen for P. Lyngs *Gejrfuglen* (1994). *Reproduced by kind permission of the artisit.*

Great Auks at sea. An illllustration in J. Sellar's *The English Pilot* (1706).

Apart from engaging in the regular task of finding food, the activities of the Great Auk when it left the land can only be guessed at. Although (at least some of) the places where the birds bred are well enough known, where they went for the rest of the year is less certain. They passed the winter at sea and, presumably, used prevailing ocean currents in their migrations. One of the best recorded observations of winter birds seems to confirm this. The same Otto Fabricius who left notes about chicks in down, left a record of his sightings off south-west Greenland of winter Auks that had probably ridden the East Greenland Current. He watched them arrive each season in September, remain some distance off-shore, then leave again around January.

It is sometimes claimed that archaeological evidence reveals that Great Auks spent their winters far from their breeding colonies, in much the same way as their living relatives do today. This is not necessarily true. The archaeological evidence simply shows where Garefowls were caught and eaten; it does not reveal where they came from. Quite possibly, there were colonies in many places and the unfortunate creatures whose remains are found in prehistoric kitchen middens need not necessarily have wandered far. The Great Auk's flightlessness and size may have imposed constraints upon it that do not apply to its smaller relatives. On the other hand it may have wandered as much as they do. We just do not know.

Many birds, probably those that bred on islands close to Newfoundland, wintered on the Grand Banks or maybe farther south off the coast of New England. J. Sellar in *The English Pilot* (1706) maintained that the presence of Great Auks was a clear sign of shoaling water and that the Newfoundland Banks had been reached. Older shipping records suggest that a vessel needed to be within a hundred leagues of land to encounter them and, later, Pennant (1761-6) confirmed this:

> This bird is observed never to wander beyond soundings; and according to its appearance they direct their measures, being then assured that land is not very remote.

The eighteenth century fishermen and fowlers of Iceland were quite used to seeing Great Auks in the sea. So commonly did they encounter them that, according to Alfred Newton (1861), such occasions were scarcely worthy of note. As the years passed and the numbers of Auks diminished, so too did its use to sailors and records of the birds at sea become more sporadic.

Although the Great Auk has often been imagined as an Arctic bird, this is a misconception. Certainly, it was a creature of the far north, but it lived below the Arctic Circle right across the Atlantic. The misunderstanding probably arose because such words as *Iceland*, *Greenland* or *Labrador* are synonymous in many minds with *Arctic* or *North Pole*. It seems equally likely that any assumption of the species being polar in his habits owes something to the inevitable comparisons with Penguins. There is only one seemingly realistic record - from Disko Bay, West Greenland - of Great Auks occurring within the Arctic Circle, and even this record can be disputed.

From their known breeding stations - islands off of the Newfoundland coast, islands off southern Iceland, St Kilda and the Orkneys - Great Auks roamed the seas of the north with a range that stretched far and wide. This range was not dissimilar to that of the Razorbill but it didn't reach so far into northern latitudes; nor did it stretch so far to the south. Leaving aside

(Above, top). Coloured engraving from Sir William Jardine's *The Naturalists' Library - British Birds* (1838-43).

(Above). Life-size figure carved in wood by Peter Welsh of Porthtowan, Cornwall. *Collection of Kenny Everett.*

Hand-coloured lithograph from C.B. Cory's *The Beautiful and Curious Birds of the World* (1880-3).

the disputed Disko record, birds were seen as far north as the coasts of southern Greenland, mid-Norway and northern Labrador. To the south they straggled even farther afield. At one time they were certainly familiar birds around the coasts of the British Isles and several disputed early nineteenth century records indicate that they may have frequented the shores of northern France.

Remains from a much earlier period show that individuals reached Gibraltar and even Italy. Presumably, the advance of the great ice sheets pressed the Garefowl farther to the south than has been the case more recently. On the other side of the Atlantic, birds ranged down past New England and there is archaeological evidence to show that they penetrated as far south as Florida.

Rather like the Gannet (*Sula bassana*), the Great Auk - at least in recent historical times - showed a preference for breeding in colonies at just a few select localities, leaving many seemingly suitable sites unoccupied. Whether this was because of specific feeding requirements or whether it simply reflects the need of the species to escape human attention is uncertain. Perhaps it results from a combination of both circumstances with the influence of man being the weightiest of determining factors.

Like other birds and other Auks, the Garefowl underwent a moult. Salomonsen (1945) was of the opinion that they shed their 'flight' (hardly an appropriate word in the case of *Alca impennis*) feathers one at a time. This would be quite unlike the situation in the extant larger alcid species. These (like the ducks, but unlike many other birds) lose all flight feathers simultaneously rendering them incapable of flight for a short time. This simultaneous feather loss probably makes only a small alteration in swimming ability while ensuring that the flightless period is kept to a minimum. In the case of the Great Auk, with its much reduced wing area, diving ability might have been seriously impaired with this kind of moult; the length of the period of flightlessness was, of course, irrelevant.

The Great Auk (just like the Penguins) had, by abandoning the means of flight, made streamlining - in a way entirely suited for passage through water - perfectly possible. This very process of streamlining meant, however, that mobility on land would, inevitably, be restricted. The bird's legs, set far back to facilitate swimming, allowed only a waddle on land and the small wings, so efficient as paddles, could be used as little more than balancing devices. Perhaps the best idea of their disproportionate size is conveyed by comparison with those of the Razorbill. The Great Auk was a bird perhaps six times as heavy and twice as big in terms of linear measurement, yet its wings were almost identical in size to those of its otherwise much smaller relative. When the demand for stuffed Great Auks caused taxidermists to 'manufacture' fake ones, the making of the wings might have posed the biggest problem. It didn't. Taxidermists simply used wings taken from dead Razorbills! The wings, then, although useless for getting the bird airborne, still looked very much like wings. They had by no means acquired the flipper-like appearance that so characterises the wings of the Penguins. Committed to the water the Great Auk may have been, but not quite so totally as these remarkable southern counterparts.

Illustration by Johannes Larsen (1921). *Courtesy of the Johannes Larsen Museet, Denmark.*

Yet Garefowls certainly had to contend with problems - in the shape of terrestrial predators - that most Penguins do not. The presence in northern latitudes of so formidable a predator as the Polar Bear may have restricted the more northerly breeding sites available to them- but foremost among all enemies was, of course, man. Unlike the Penguins, Great Auks had the misfortune to live relatively closely to areas where human beings had developed the technological ability to tamper with them. Had Garefowls been birds of the south and the Penguins birds of the north, perhaps their status today would be reversed.

The Great Auk's abandonment of the air, the one factor that above all others makes comparison with Penguins so tempting, brought with it most of the elements that contributed to the species' downfall. Not only was it unable to escape its worst enemy when on land, its flightlessness created another, subtler, problem. After its wintering at sea was over and the time to return to land had come, the Great Auk's choices were, unhappily, limited. The number of localities with sufficient space to accommodate large colonies of birds yet also allow each individual easy access was restricted. For a large colony to survive it needed to be at a site inaccessible - preferably unknown - to man. It needed a rich and plentiful food supply close at hand. Added to this, the site had to be such that the birds could simply hop from the surf and waddle ashore as nimbly as they were able. Anything less spelled disaster.

When the call came and the long-lived, slow to multiply Great Auk dragged itself from the water to complete its yearly cycle and lay again its large and beautiful egg, its grip on life hung by the most slender thread.

(Facing page, top). Illustration in F.J. Fitzinger's *Bilder-Atlas (Wissenschaftlich-Populären Naturgeschichte der Vögel)*, 1862-4.

(Facing page, bottom). Oil painting by R. Horsfall in the collection of the Smithsonian Institution. Washington.

THE
LIVING BIRD

THE LIVING BIRD

When fed in confinement it holds up its head, expressing its anxiety by shaking the head and neck, and uttering a gurgling noise. It dives and swims under water, even with a long cord attached to its feet, with incredible swiftness.

John Fleming (1828)

During the summer of 1821 a Great Auk was caught alive on the island of St Kilda. Two men and two boys saw it from a boat as it sat on a low cliff ledge facing the sea. The men were landed- one at each end of the ledge- and they made their way stealthily towards the Auk. Meanwhile, the boys rowed the boat back to a spot directly beneath the bird's resting place, going as quietly as possible. This was a creature that could hear very well. As the men closed in the Great Auk became alarmed and jumped off its rock towards the sea - straight into the arms of one of the waiting boys .... who held it fast!

Happily, the bird was kept alive and, somehow, it came into the possession of a Mr. McLellan, the tacksman at St Kilda and also at the Isle of Glass, one of the Northern Hebrides.

A little later in the year *The Regent*, yacht of *The Commissioners of the Northern Lighthouses*, called at Glass during a tour of inspection. On the eve of the yacht's departure - the 18th of August - Mr. McLellan came on board and gave his bird to Mr. Stevenson, the Engineer for the Board of Commissioners, who, together with the Reverend John Fleming, intended to keep it until it died and then present it to the University Museum at Edinburgh.

Fleming published several descriptions of this individual, the best of them in *The Edinburgh Philosophical Journal* for 1824:

The bird was emaciated, and had the appearance of being sickly, but in the course of a few days became sprightly, having been plentifully supplied with fresh fish, and permitted occasionally to sport in the water with a cord fastened to one of its legs to prevent escape. Even in this state of restraint it performed the motions of diving and swimming under water with a rapidity that set all pursuit from a boat at defiance. A few white feathers were at this time making their appearance on the sides of its neck and throat, which increased considerably during the following week, and left no doubt that, like its congeners, the blackness of the throat feathers of summer is exchanged for white during the winter season.

Fleming and Stevenson apparently left *The Regent* at the Mull of Kintyre, departing from the boat in order to proceed to Glasgow. After their depar-

ture, the Great Auk escaped. Near the entrance to the Firth of Clyde, while it was taking its usual bath in the sea with the cord attached to its leg, it somehow slipped away. Grieve (1885) believed that it perhaps died soon afterwards, there being some evidence that an Auk's body was cast ashore at Gourock at about this time (*see* Gray, 1871).

The few remaining records of living Garefowls are largely of this kind. They are portraits of isolated birds, captured in unfamiliar situations or circumstances of distress, then removed to an environment that was completely alien, but an environment where the bird - surprisingly enough - was able to survive for a few weeks - or even months - in what seems to have been a cheerful enough state.

There are hardly any properly written accounts detailing first-hand observations of the birds in their natural, wild condition. Although seen frequently enough, encounters with Great Auks failed to inspire masterly literary efforts. Sailors and fishermen intent on the destruction of birds and eggs do not usually leave behind them ornithologically useful - or attractive - observations. Man slaughtered these birds in their thousands yet the scraps of information left behind by those who actually encountered them are few.

Hand-coloured lithograph by Edward Lear from J. Gould's *The Birds of Europe* (1832-7).

53

When John Wolley, the naturalist and Garefowl researcher, toured Iceland in 1858 and collected together the eye-witness accounts of Icelandic fisherfolk, he recorded them for posterity in his notebooks kept today at the Cambridge University Library. According to Wolley's translation of his interview with Jón Brandsson:

> The wings are kept close to the sides when the bird is at rest, but a little out (so that light shows under it) when it begins to run.

Another fisherman, Erlandar Gudmundson, indicated that when a bird was at sea:

> He always had his beak pointed upwards and was always moving his head from side to side.

The same Erlandar also had something to say about a curious tradition among Icelandic fishermen that the birds were blind - at least when they first came on land. Describing a living individual - and presumably referring to the anatomical feature known as the nictitating membrane- he insisted:

> The bird looked about as if from under blinkers. It could see with great difficulty until the water lifted the lid from its eyes.

The often cited adventure of William Bullock, jeweller and silversmith of Sheffield, Liverpool and London, provides - despite the fragmentary written record of it - perhaps the most vivid account of a free Garefowl in its natural environment. In one of the early years of the nineteenth century Bullock travelled to the Orkney Islands in pursuit of the strange dream that has infected many men - the possession of a stuffed Garefowl. He didn't get the bird he wanted (at least not on the occasion of his visit; he was sent

(Below, left). Watercolour by William Lewin for the special edition of his *Birds of Great Britain* (1789-94). This edition consisted of 60 copies and the 324 plates in each were hand painted by the author. No two copies are, therefore, entirely alike.

(Below, right). Hand-coloured engraving by William Lewin for the standard edition of his *Birds of Great Britain* (1795-1801).

a specimen later) but spent several hours in a rowing boat in hot pursuit of an individual that was too quick, too evasive and altogether too much the master of its watery world for the party of hunters that dogged it.

According to George Montague (1813) who recorded Bullock's experience just a year or so after it happened:

> The...male Mr. Bullock had the pleasure of chasing for several hours in a six-oared boat, but without being able to kill him, for although he frequently got near him, so expert was the bird in its natural element, that it appeared impossible to shoot him. The rapidity with which he pursued his course under water was almost incredible.

The most comprehensive report of the living Garefowl in its natural condition comes from the pen of Alfred Newton (1861) but, clear though it is,

The Great Auk imagined off Eldey Island. Oil painting by Ray Harris Ching. *Reproduced by kind permission of the artist.*

it does not stem from the Cambridge professor's own observations. It is simply a compilation of the various reminiscences that he'd heard from Icelandic fishermen when he toured Iceland at the same time as his friend John Wolley:

> They swam with their heads much lifted up, but their necks drawn in; they never tried to flap along the water, but dived as soon as alarmed. On the rocks they sat more upright than either Guillemots or Razor-bills, and their station was further removed from the sea. They were easily frightened by noise, but not by what they saw. They sometimes uttered a few low croaks. They have never been known to defend their eggs but would bite fiercely if they had the chance when caught. They walk or run with little, short steps, and go straight like a man. One has been known to drop down some two fathoms off the rock into the water.

As far as the free Garefowl is concerned, Newton's words round off a review of Great Auk literature. There is nothing else! For additional descriptions it is necessary to go back to the captives.

The earliest written account of the close observation of a living Garefowl comes from Olaus Wormius (1655) who acquired an individual from the Faroe Islands. He kept it at Copenhagen for several months and believed it to be a fairly young bird. Wormius reported that it could swallow a whole herring and, if it was hungry, would take three in quick succession.

A similar pet bird is mentioned by J. Wallis (1769):

> The Penguin, a curious and uncommon bird was taken alive a few years ago in the island of Farn [the Farne Islands lie off the coast of Northumberland] and presented to the late John William Bacon, Esq. of Etherston, with whom it grew so tame and familiar that it would follow him with its body erect to be fed.

This individual may even be identical with one of the stuffed birds now at the Hancock Museum, Newcastle-upon-Tyne, although there is no positive evidence of this and the likelihood is only slim.

In 1831, when Icelandic fishermen had just started raiding the island of Eldey with the intention of killing Great Auks, they managed to destroy 23 individuals. A twenty fourth bird they took away alive, wishing to see if the merchants they dealt with would send it - still living - to foreign countries. When the raiding party got back to the mainland one of the men wrapped the bird in a cloth and carried it - apparently for fun - under his arm to the village of Keflavik. The poor thing struggled a little at times and tried to bite but the seaman had taken the precaution of tying its beak with string. At Keflavik and Reykjavik the merchants seem to have shown a disappointing lack of imagination for the bird was, it seems, killed by a man named Thomsen. It is supposed to have made no sound as it died.

The capture of a second bird at St Kilda around the year 1840 had the extraordinary - and awful - consequence of the bird being beaten to death after having been kept alive for several days; its captors thought it might be a witch! During its confinement the poor creature was imprisoned in a bothy some way up the rock known as Stac-an-Armin, and the men had plenty of time to watch it. It made a great noise not unlike that of a Gannet

Watercolour by Allan Brooks.

(*Sula bassana*), although louder, and also made a violent clacking sound when closing its beak. Evidently it didn't perform this action very much, keeping its mouth open:

Very long and often, as if it would never shut its bill again.

The unhappy bird showed this tendency to keep its mouth open most particularly when anyone came near it - in the circumstances a perfectly sensible action. Unfortunately for the Auk a storm blew up and sealed its fate. The men, thinking their captive a master of wind and weather, considered such an act of thaumatergery deserving of a capital sentence and they cruelly beat the Great Auk to death with sticks.

There is one other recorded instance of a fairly lengthy encounter with a living Garefowl. This concerns the capture of a young female near the entrance to Waterford Harbour, Ireland during 1834. The bird seemed to be starving; certainly it showed little fear of humans for it was enticed to the side of a fisherman's boat with a handful of sprats that were either thrown or held out to it. Eventually it died and its stuffed skin still exists in the Museum of Trinity College, Dublin, but during the few months that it lived, it seems to have become quite tame even though one of its owners (it passed through several hands) thought it, 'rather fierce.' Strangely, in addition to eating freshwater fish - which it apparently preferred to salt-water varieties - the Auk became quite partial to potatoes mashed in milk.

Dr. Burkitt (*see* Gurney, 1868), the man who saved the beleaguered little creature for posterity, charmingly described the living bird in words that return us almost exactly to our starting point. It had:

A habit of frequently shaking its head in a peculiar manner, more especially when any particularly favourite food was presented to it.

EXTERMINATION

EXTERMINATION

> God made the innocence of so poor a creature to become such
> an admirable instrument for the sustenation of man.
> Richard Whitbourne (1622)

There need be no doubt that man's fierce persecution of the Great Auk was the direct cause of its extinction, that man hunted this bird out of existence.

Some writers choose to side-step this idea and suggest that such words don't paint an entirely accurate picture, that the Penguin of the North was in decline long before human intervention and that its future looked bleak no matter what. They insist that *Alca impennis* was a doomed species.

Maybe it was.

There is nothing particularly unlikely about such an idea. The fossil record shows that a succession of flightless seabirds - built roughly along the lines of the Great Auk - have developed over the last 100 million years only to face ultimate extinction. And there is substantial archaeological evidence to show that the Garefowl's range has been shrinking for a long time.

What remains undeniably obvious, however, is the fact that these birds were eliminated wherever - and more or less as soon as - man could get at them. A species in natural decline it may (or may not) have been, but it is certain that Garefowls would be living today if man had left them alone.

The pattern of the bird's decline reflects very surely the advance of man's technical ability. Wherever Auks could be reached, they would inevitably be destroyed. Sadly for the Garefowl, *Homo sapiens'* increasing efficiency in the logistics of destruction was entirely matched by his desire to completely fulfil his potential and to exterminate in horrifically large - and seemingly totally unnecessary - numbers.

The Great Auk's flightlessness, and its vulnerability when it came ashore, more than facilitated man's dark intentions. The argument that the species was over-specialised and that its loss of the ability to become air-borne rendered it hopelessly unfit for survival, is perfectly true- but only in terms of the bird's interaction with man. Without his malevolent presence, the Garefowl's inability to fly would have been a comparatively unimportant issue in the battle for life. Penguins survive well enough without the power of flight, but they have one important advantage. They inhabit areas where man, by and large, did not develop the technology to get at them, at least not until his inclination to do so was no longer matched by his inclination to destroy. This isn't to say that Penguins weren't butchered, it is simply a recognition that Penguin butchery wasn't the intensely sustained persecution that Garefowl hunting was. Another suggestion put forward to establish man's ultimate innocence is the notion that this was never a particularly numerous species and that it lived mostly in small colonies which, by virtue of their size, maintained a fragile balance with survival. There are

A Last Stand. Oil painting by Errol Fuller.

all sorts of very good reasons for supposing the opposite - that this was a bird that liked the company of large numbers of its own kind. The notion that it didn't is largely based on a misinterpretation of written records. To interpret these correctly it should be remembered that the records coming down to us all date from periods long after man had made his first serious inroads into the numbers of birds and had already laid waste to the more accessible breeding areas.

The example of Funk Island, where Garefowls gathered in their thousands, is surely enough to show the truth.

The question of numbers introduces another issue. Is it really possible to hunt a species right out of existence when it occurs over such a vast area? As a rule hunting affects individuals, not species. Habitat destruction and the breaking of ecological balances are the common causes of bird extinctions. It is usually only in cases like that of the Dodo (*Raphus cucullatus*)- where the individuals are conspicuous and restricted to comparatively small islands or areas- that an unequivocal finger can be pointed at hunting as the sole cause of extinction. What is different, then, about the Great Auk?

Great Auk and Penguin. Coloured plate from W. Kirby's *Natural History of the Animal Kingdom* (1889).

Here is a bird with the potential to inhabit all the waters, islands and coasts of the North Atlantic. How could hunters have actually found out the last few and eliminated them? This is really the crux of the matter. Had the Great Auk enjoyed living in small, discreet parties, a few could always have found out-of-the-way places in which to perpetuate their race. But for a species that could only flourish in large, conspicuous groups, such subterfuge was clearly impossible. The Great Auk was hunted out of existence in much the same sense as the Passenger Pigeon (*Ectopistes migratorius*). Once its numbers had fallen below a certain level, no matter that the number was still very large, the species was not viable and was doomed.

Whether the two birds killed in the famous encounter on Eldey during June of 1844 were really, as is so often supposed, the last of their kind is beside the point. With creatures like the Great Auk or the Passenger Pigeon it isn't necessary to hunt them down to the last few individuals to ensure extinction. The Great Auk was finished as a species long before 1844.

Three Funk Island skeletons. This photograph (c.1900) was probably taken at the Museum of Comparative Zoology, Cambridge, Mass.

The earliest account of Garefowl massacres is given by Jacques Cartier, the French explorer who sailed from St. Malo to Newfoundland in 1534. Translated by H.P. Biggar in 1924, his account of what was presumably Funk Island reads as follows:

> We ... sailed... as far as the Isle of Birds ... Some of these... are as large as geese ... black and white with a beak like a crow's. They are always in the water, not being able to fly in the air, inasmuch as they have only small wings... with which... they move as quickly along the water as the other birds fly through the air. And these birds are so fat it is marvellous. In less than half an hour we filled two boats full of them, as if they had been stones, so that besides them which we did not eat fresh, every ship did powder and salt 5 or 6 barrels full of them.

A small beginning perhaps but a good sample of what was to come. Two years later a Mr. Hore and 'divers other gentlemen,' undertook a voyage to Newfoundland. An account of their expedition is given by Hakluyt (1552-1616):

> From the time of their setting out from Grauesend, they were very long at sea, aboue two months ... vntil they came to the Island of Penguin ... whereon they went and found it full of great foules white and grey, as big as geese, and they saw infinite numbers of their egges. They draue a great number of the foules into their boates vpon their sayles, and tooke up many of their egges, the foules they flead and their skinnes were very like hony combes full of holes being flead off: they dressed and eate them and found them to be very good and nourishing meat.

By 1622 Richard Whitbourne was describing how Garefowls were herded, hundreds at a time, down gangplanks into boats waiting to receive them. Quite how this was effected he does not make clear but the fate of the poor birds is quite apparent. As he so poetically says:

> Such an admirable instrument for the sustenation of man.

It is evident that Great Auks had, for centuries, been systematically exterminated at all their more accessible breeding sites. Naturally, no records remain of these grim skirmishes in man's battle for survival. How many of these breeding stations there were and where they were situated is information that is lost forever. The records that remain relate only to the Great Auk's last strongholds - a few comparatively remote and isolated islands.

Of these, Funk Island was probably the most important. It is estimated (*see* Nettleship and Birkhead, 1985) that the first Europeans to come across this small piece of land just off the Newfoundland coast may have found a colony of up to 10,000 pairs. Such a convenient larder of fresh meat was bound to be plundered, especially by ships that had just made the long voyage across the Atlantic. For something like 300 years the larder held up, despite the fearful onslaught on the birds. Eaten or used for bait, boiled down for their fat or stripped of their feathers, the Great Auks of Funk Island were exploited in every way that human ingenuity could devise. To facilitate removal of their feathers individuals were scalded in vats of boil-

A settlement in West Greenland in the mid-eighteenth century. The bird in the foreground, raising its small wings and about to be harpooned, is thought to be a Great Auk. Engraving by Johanne Fosie from Hans Egede's *Det Gamle Grønlands Nye Perlustration* (1741).

Catching the Great Auk on the coasts of Newfoundland. Engraving by F.W. Keyl and E. Evans from G. Kearley's *Links in the Chain* (circa 1880). *Courtesy of Peter Blest.*

ing water and the fires to heat the cauldrons were fuelled by the fat from already butchered birds. Great stone pounds were built (signs of which still exist on the island) into which the little victims could be driven and kept alive until such time as their flesh or feathers were needed. According to Aaron Thomas (1794, *but in* Murray, 1968):

> If you come for their Feathers you do not give yourself the trouble of killing them, but lay hold of one and pluck the best of the Feathers.

You then turn the poor Penguin adrift, with his skin half naked and torn off, to perish at his leasure.

Whether or not Thomas, a rating aboard H.M.S. *Boston*, participated in these horrible activities is not quite clear for - with a certain understatement - he adds:

This is not a very humane method but it is the common practize.

Finally, he catalogues the horrors of the island with an ambivalence of manner that is difficult to reconcile with someone who is totally brutish:

A nineteenth century photograph of a Funk Island mummy.

(Facing page, top left). Engraving from G. Shaw's *General Zoology* (1826).

(Facing page, top right). Hand-coloured engraving by F. Martinet from E.L. Daubenton's *Planches Enluminees* (1765-81).

(Facing page, bottom). Engraving by P. Paillou from T. Pennant's *British Zoology* (1776).

While you abide on this island you are in the constant practize of horrid cruelties for you not only Skin them Alive, but you burn them Alive also to cook their Bodies with. You take a kettle with you into which you put a Penguin or two, you kindle a fire under it, and this fire is absolutely made of the unfortunate Penguins themselves. Their bodys being oily soon produce a Flame; there is no wood on the island.

Recently it has become fashionable to express a degree of scepticism over such remarks. W.A. Montevecchi and L.M. Tuck (1987), for instance,

GREAT AUK.

Funk Island. *This photograph by W.A. Montevecchi is reproduced with his kind permission.*

Engraving from T. Parkin's pamphlet *The Great Auk or Garefowl* (1894).

regard them as distorted and exaggerated. That these commentators find it difficult to envisage - and believe in - the infliction of such cruelties is, perhaps, entirely to their credit. The sheer wickedness of human beings is not a subject on which it is healthy to dwell, and all the more so because it is so commonplace.

Aaron Thomas dwelled on it a great deal, however. The vile practises he described had been going on for many, many years by the time he witnessed them. Whenever sailors and ships were in the vicinity of Funk Island at the appropriate season such barbarous activities were performed. The wonder is that the Great Auks held out here for so long.

Hunted originally for their meat, the exploitation escalated over the years into a general persecution for the other commodities the birds could provide; feathers, fat and oil. For what reason seamen needed to light great fires on Funk Island and keep them burning for weeks on end is not really known, but it is certain that they did and certain also that the fat and oil from Garefowls served - at least in part - to fuel these great infernos. When, long after the Great Auks and the fires were gone, Funk began to be investigated by those curious about the fate of the birds, vast quantities of bones were found (a surprisingly large number of skeletons exist in museums around the world made up from assortments of bones found on the island) and Newton (1865) reported that a 'Yankee speculator' removed great quantities of top soil and mould from the rocks. So rich was this - composed in large part of the rotted-down remains of dead birds - that it was sent to Boston to be sold as agricultural manure.

William Bourne (1993) makes the interesting point that had Garefowls been the island's sole tenants, without sharing it with hoards of smaller seabirds, they might have clung to existence. Reduced to comparatively small numbers, their own exploitation might eventually have become uneconomic with the remnants of the colony being left in peace. Bourne's point is that the large number of smaller birds provided sufficient inducement for hunters to keep on coming and, when they did, the Great Auks along with the Gannets - being the largest and potentially the most profitable - naturally became the prime targets. But bird species are unlikely to be offered such convenient isolation as Bourne would wish and the notion is, in any case, at complete variance with the sequence of events that occurred, later on, off the coast of Iceland.

As the eighteenth century drew to its close, so too did the Great Auk's tenure of Funk Island. The stock of birds, once almost unimaginably plentiful, was finally depleted. From here on, the Great Auk would be a stranger to the coasts of Newfoundland. The last scenes in man's ruthless extirpation of the Garefowl would be played out on the other side of the Atlantic.

The bird skerries off the south-west corner of Iceland provided the very last stronghold of the Great Auk. Here, the birds were perhaps a little more fortunate than those on Funk Island simply because the implications of the geography were different. Apart from the islands being *en route* to the Icelandic capital, Reykjavik, there was no pressing reason for ships from the great populous centres of Europe to pass them. Comparatively speaking, they were out of the way. One of the terrible ironies of Funk

EXTERMINATION

Island was that it lay at the far end (from Europe) of a transatlantic crossing, and the sailors who'd endured the journey were hungry. In the beginning it wasn't necessarily cruelty or greed for profits that drove seafaring men to slaughter Garefowls; it was simply empty bellies. The Icelandic islands and skerries were left, therefore, largely alone. Besides, they were dangerous.

It was the Icelanders themselves, with their comparatively primitive and unsophisticated equipment and correspondingly basic needs, who presented the threat to the Great Auk. That they'd been catching the birds for a very long time is beyond doubt. By the time that reliable records began to be kept the Garefowl had retreated to those reefs and skerries that were most inaccessible to humans. Although Icelanders were in the habit of raiding the famous Geirfuglasker in mid-summer for both eggs and birds (and had been for centuries) there were many years when they didn't go at all, judging the tides and winds too dangerous for such missions. The cost of visiting the Icelandic islands was potentially very high. For those who came only in small fishing boats, the currents and tides were incredibly treacherous and a volcanic *röst* added considerably to the hazards.

After consulting a manuscript in the Library at Reykjavik - a document relating specifically to the Geirfuglasker - Alfred Newton (1861) believed he had good grounds for supposing that during the seventeenth and eighteenth centuries there were long periods - 75 years or more - during which no-one visited the reef. According to Friedrich Faber (1827) a landing could only be attempted in calm weather and even then it was necessary for a man to spring from a boat onto the rock with a rope tied around his middle so that he could eventually be dragged off again through the ever boiling waves. Probably, only a very few men showed both the inclination and the daring to go and if a leader of such men died or was otherwise indisposed he was not, perhaps, easily replaced. Despite the destruction that was undoubtedly wrought, Garefowl populations were given a certain amount of breathing space, time to replenish the ranks of the fallen. In much the same way Beothuk Indians (now as extinct as the birds they hunted) harvested the Great Auks of Funk Island before the coming of the European.

By the early 1800's, however, the Great Auks that came every summer to the Geirfuglasker were, it seems, just about the only breeding birds left.

For just how long this dwindling little band could have withstood the intermittent egging and birding raids of the Icelanders, there is no way of estimating. The answer is irrelevant anyway for the situation was fast changing. A series of incidents occurred that were to prove catastrophic.

One of the knock-on effects of Britain's war with Napoleon was the conflict that broke out between England and Denmark. This resolved itself into a number of ugly battles and skirmishes. One of the latter involved a 22 gun privateer named *The Salamine*. Carrying British colours she sailed to Thorshavn, capital of the Faroe Islands. After pillaging the town, the crew press-ganged a man by the name of Peter Hansen and carried him off to be their guide as they progressed towards Iceland. When the privateer reached Reykjavik the crew promptly sacked the town but then, before leaving Icelandic waters, the vessel somehow chanced on the Geirfuglasker; unfortunately for the Great Auks, this visit

coincided not only with a period of fine weather but also with the season when the birds were ashore. On either the 7th or the 8th of August (this date seems surprisingly late in the year but it is usually accepted) the sailors spent a whole day on the reef happily killing the birds, trampling down eggs and young and generally wreaking havoc. After this unpleasant little excursion *The Salamine* sailed away and on the way home deposited Hansen back on the Faroes. The effect of this ugly incident wasn't quite over, however, for it was to have an unfortunate consequence some five years later.

A serious famine developed on the Faroes, largely as a result of the Danish government's decree that any intercourse with the British was to be punished by death. By 1813 this isolation from trade contacts had so affected the islands that the Faroese were in a desperate state and a ship was sent to Iceland for supplies. Its pilot was none other than Peter Hansen, a man who now knew all about the Geirfuglasker and its store of birds. Just as unfortunate for the Great Auks was the timing of the expedition - the

(Above) Watercolour by John A. Ruthven. *Reproduced by kind permission of the artist.*

(Facing page). Chromolithograph by J.G. Keulemans from W. Rothschild's *Extinct Birds* (1907).

month of July - and the state of the sea - unusually calm. So gentle were the waves that when the vessel, the schooner *Færöe*, came off Cape Reykjanes it was becalmed. A boat was lowered and some of the crew rowed off to either the Geirfuglasker or the nearby Geirfugladrángr where, as Hansen expected, they found birds. The Faroese killed as many as they could and loaded their boat until it was full, leaving behind the bodies of those they couldn't carry scattered on the rocks. Their intention was to return for these, but they never did. A wind sprang up and Hansen took the decision to sail immediately for Reykjavik. On arrival the Faroese had 24 birds on board as well as others already salted down.

These kind of losses were, of course, paltry compared to those that occurred on Funk Island during previous centuries but this Icelandic colony was a much smaller one and, as far as we know, it was the last.

Following Hansen's expedition and for the next seventeen years the Great Auks were left in comparative peace; a few birds were killed and, no doubt, some eggs were taken. Then, the colony was devastated by an event that was nothing to do with man. He was to take full, and typical, advantage of the opportunity it presented, however.

Early in the year 1830, probably at the start of March, there began a series of volcanic disturbances that lasted for perhaps twelve months.

During this period the Geirfuglasker disappeared beneath the waves. Several birds were, apparently, washed ashore in a distressed state during these eruptions (among them, perhaps, the Garefowl whose stuffed skin is now in Nantes), but the major consequence for the Great Auk as a species was that it lost its home. With the Geirfuglasker gone, the surviving individuals chose to come ashore on nearby Fire Island - the island of Eldey.

Wild and forbidding, Eldey might have provided an ideal refuge but it held one big disadvantage - and it was the same disadvantage that down the centuries had precluded the Great Auk from one nesting site after another. It was accessible to man. The unpredictable currents and surf that once surrounded the Geirfuglasker, and so often prevented human approach, were nowhere near so fearsome or threatening.

No-one knows quite why a certain Brandur Guðmundsson decided to visit Eldey during the summer of 1830. There is no record of Garefowls having occurred there before; perhaps he was after smaller birds or maybe he was thwarted by the disappearance of the Geirfuglasker and decided to look elsewhere before going home. However it was, his journey uncovered the fact that the surviving birds were using the island as their new nesting site; the fate of the species was sealed. Guðmundsson and his companions took several individuals then came back and took several more. The next year 24 were brought away including a live bird about which little is known except for the fact that its death, from one cause or another, soon followed. Such small harvests - although so devastating to the beleaguered little band of Garefowls - hardly justified the arduous, and still quite dangerous, journey. But the Great Auk wasn't, of course, the only quarry; the fishing boats came back loaded with smaller seabirds to make the expedition worthwhile.

Around about this time another factor came into play. No longer were the birds and their eggs required simply to supplement the diet of the local fishing people. The Great Auk was in demand as a specimen. So too were its eggs. Prices were offered that must have seemed surprising to the poor fisherfolk who were actually catching the birds. Whenever it was possible, the Great Auk was actively sought out, no longer as just the largest and most desirable of the seabirds but now as the very justification and purpose of the hunting trips.

While the Icelanders systematically hunted down the last few individuals remaining vulnerable to them, a few birds straggled elsewhere.

In 1821 the celebrated incident occurred when a Garefowl was caught at St Kilda and, for once, kept alive. Allowed to sport in the water with a length of cord attached to its leg, it eventually escaped. Another bird was captured - near the entrance to Waterford Harbour, Ireland - and it too was allowed to live although, sadly, it succumbed after just a few months of captivity. A third bird caught alive was treated in an altogether more barbarous manner.

This third bird was another St Kilda capture. The islands of this remote little archipelago had formerly provided a nesting place for the species but by the nineteenth century the breeding colony was long gone and the visit of a Garefowl was a very rare occurrence. To most of the islanders, particularly the younger ones, the bird was totally unfamiliar.

Men of St Kilda (c.1880). This photograph by Norman MacLeod for George Washington Wilson probably shows some of the men who took part in the events of 1840.

Stac-an-Armin. A photograph from J.H. Gurney's *The Gannet* (1913).

Stac-an-Armin photographed by James Fisher. *Reproduced by kind permission of Clemency Thorne Fisher.*

EXTERMINATION

Main Street, St Kilda. Finlay MacQueen outside no.5.

The tale that comes down to us of the capture of a Garefowl in or about the year 1840 is a strange and a dark one. Although the year is not known with certainty, the month was July. Five men, wandering on the great rock of an island known as Stac-an-Armin, caught a large and plump bird asleep on a ledge. Nearly 50 years afterwards Harvey-Brown and Buckley (1889) described the event:

> Three of them were Lauchlán McKinnon, about 30 years of age - and now, or till recently, alive - his father-in-law, and the elder Donald McQueen - both now dead ... It was Malcolm McDonald who actually laid hold of the bird, and held it by the neck with his two hands, till others came up and tied its legs.

Having caught their bird on its ledge about halfway up the stac (which in this particular place sloped up relatively easily) the men took it to their bothy where they confined it for three days. Apparently:

> It used to make a great noise ... It opened its mouth when anyone came near it [and] nearly cut the rope with its bill. A storm arose, and that, together with the size of the bird and the noise it made, caused them to think it was a witch. It was killed on the third day after it was caught, and McKinnon declares they were beating it for an hour with two large stones before it was dead; he was the most frightened of all the men, and advised the killing of it.

After its death the men flung the bird's body behind their bothy.

What, perhaps, gives this story an additionally awful dimension is the fact that the people involved are so real, so starkly identified. Mary Bones (1993) was even able to supply details about their otherwise mundane lives in her paper on the Great Auk and St Kilda. Lauchlán McKinnon (1808-95) lived at No.1 Main Street and was married three times. Donald McQueen

THE ISLAND OF BORERAY, ST. KILDA. 6206. G.W.W.

Borerey, one of the islands making up the St Kilda group. Photograph by Norman MacLeod for George Washington Wilson, taken from the summit of Stac-an-Armin.

(c1807-80) was one of his brothers-in-law and lived at No.10. Malcolm McDonald (1800-46) of No.8 was the father of McKinnon's third wife.

As Mary Bones (1993) points out, it is easy enough to see how the bird's unhappy end came about. These were island men, reared in lonely isolation and brought up in a world steeped in superstitions from which they were protected by a powerful, yet faltering, Christianity. When cut off from their homes on the island of Stac-an-Armin by foul weather, they might easily fall back on the old ways and beliefs and come to see the strange bird with which they'd cooped themselves up as the sinister cause of all their troubles. But even so...

This poor, abused creature, separated from others of its kind, was not the first Garefowl to meet a cruel and barbarous death at the hand of man, but it was to be among the last.

In June of 1844 the final scene in the grisly tragedy was played out. Pursued for their meat, their fat, their stomachs, their oil, their feathers and now, at the end, for their price as specimens, the last birds had taken refuge on the island of Eldey and this was the stage for man's final encounter with them. The death of a pair killed here during that summer marked the end of man's relationship with the living Garefowl.

The survival of the Great Auk was incompatible with the coming of technological man and from the moment that humans took to the sea in

76

boats the species was doomed. From his cosy nest in Cambridge, Alfred Newton (1865) - perhaps the subtlest of all Garefowl scholars - wrote the words that best sum up the Great Auk's downfall:

> The merciless hand of man, armed, perhaps, with only the rudest of weapons has driven the Garefowl, first from the shores of Denmark and then from those of Scotland. At a later period it has been successively banished from the Orkneys, the Færöes and St. Kilda. Then too, a casual but natural event has accelerated its fate. The eruption of a submarine volcano on the east coast of Iceland by laying low one of its chief abodes has contributed effectually to its destruction. But worse than all this has been the blow which, on the discovery of America, came upon the portion of the race inhabiting the Newfoundland islets, when it was brought suddenly face to face with a powerful and hitherto unknown enemy, and where the result has been what invariably happens, when a simple tribe of savages used only to the primeval customs of its forefathers is all at once confronted with invaders of the highest type of civilisation- "The place thereof knoweth it no more."

1844

1844

From a historical point of view the year 1844 isn't particularly memorable. Fox Talbot began publishing the first photographic book, *The Pencil of Nature*, there was a Potato Famine in Ireland, a supposedly inhabited island in the South Pacific - Tuanahe - may have sunk into the ocean (on the other hand it may never have existed), baked bean promoter Henry Heinz was born and the last known pair of Great Auks were done to death on a remote island in the North Atlantic.

Whether these birds were actually - as is so often believed - the last of their kind is hardly important. They stand as a powerful symbol for the extermination of all the others. Nor, perhaps, does the exact nature of their deaths matter. After all, dead is dead. And as for the tale of the two plump little corpses, their story simply illuminates the precise commercial reason for these two particular deaths. It is a reason that bears only peripherally- albeit so dramatically- on the underlying causes and history of the death of the species.

That the tale of these two birds can be recorded in sensational detail is due entirely to the efforts of two Englishmen, John Wolley and Alfred Newton, who travelled together to Iceland in the 1850's where they met - and interrogated - many of those who actually took part in the incident. Although Wolley died at the early age of 36, he left for posterity the painstakingly compiled notebooks (now preserved at the Library of Cambridge University) that he labelled 'Garefowl Books.' In these he detailed all that he'd managed to discover while in Iceland. His companion on the trip, Alfred Newton, published an abstract of these findings in *The Ibis* for 1861, shortly after Wolley's premature death but most of Wolley's information remains overlooked; down the years, few people have looked at it.

The deaths of the two birds of 1844 form a climax to a tragedy that moved forward with a terrible inevitability from the moment that the Geirfuglasker vanished beneath the waves. When the remaining Great Auks made the nearby island of Eldey their new summer quarters they left themselves horribly vulnerable. Between the years 1830 and 1841 Great Auks were regularly taken from the island by parties of Icelandic fishermen. Most summers, early in June, boats set off from the adjacent mainland and returned with dead birds; yet with each passing year the haul got smaller and smaller. For 1831, the evidence indicates that 24 individuals were taken; in either 1840 or 1841 just three were obtained. For three years no-one bothered to make the voyage out to the lonely island or, if they did, they found nothing. Then, during 1844, a merchant, Herr Carl Siemsen (a dealer through whose hands no less than 21 birds and 9 eggs reputedly passed on their way to collectors and museums), commissioned a raiding expedition with the sole intention of obtaining Auks.

The pen of Alfred Newton (1861) vividly describes the situation and the end to which the 'last' of the Great Auks were brought:

The party consisted of fourteen men: two of these are dead, but with all the remaining twelve we conversed. They were all commanded ... by Vilhjálmur [Hákonarsson], and started in an eight-oared boat from Kyrkjuvogr, one evening between the 2nd and 5th of June 1844. The next morning early they arrived off Eldey. In form the island is a precipitous stack, perpendicular nearly all round ... a shelf ... slopes up from the sea ... until it is terminated abruptly by the steep cliff ... At the foot of this inclined plane is the only landing place; and further up, out of the reach of the waves, is the spot where the Gare-fowls had their home. In this expedition but three men ascended: Jón Brandsson ... Sigurðr Islefsson and Ketil Ketilsson. A fourth, who was called upon to assist, refused, so dangerous did the landing seem. As [these] men clambered up, they saw two Gare-fowls sitting among the number-less other rock-birds ... and at once gave chase. The Gare-fowls showed not the slightest disposition to repel the invaders, but immediately ran along under the high cliff, their heads erect, their little wings somewhat extended. They uttered no cry of alarm, and moved, with their short steps, about as quickly as a man could walk. Jón with outstretched arms drove one into a corner, where he soon had it fast. Sigurðr and Ketil pursued the second, and the former seized it close to the edge of the rock, here risen to a precipice some fathoms high, the water being directly below it. Ketil then returned to the sloping shelf whence the birds had started, and saw an egg lying on the lava slab, which he knew to be a Gare-fowl's. He took it up but, finding it was broken, put it down again ... All this took

The Island of Eldey - a photograph taken by Hjálmar Bárðarson for his book *Birds of Iceland* (1986). *Reproduced by kind permission of Hjálmar Bárðarson.*

place in much less time than it takes to tell it. They hurried down again, for the wind was rising. The birds were strangled and cast into the boat, and the two youngest men followed. Old Jón, however, hesitated about getting in, until his foreman threatened to lay hold of him with the boat hook; at last a rope was thrown to him, and he was pulled in through the surf.

This, then, is Alfred Newton's summary of the accounts of those who actually participated in the raid, accounts given little more than a decade after the events themselves occurred.

Two of the main players, Sigurðr Islefsson and Ketil Ketilsson, gave their own individual versions of events and Wolley's hand-written notes detailing his interview with each man remain, a century and a half later, on the pages of his 'Garefowl Books.' According to Sigurðr:

The rocks were covered with blackbirds [Guillemots] and there were the Geirfugles ... They walked slowly. Jón Brandsson crept up with his arms open. The bird that Jón got went into a corner but [mine] was going to the edge of the cliff. [I] caught it close to the edge - a

The ledge on Eldey where the 'last' two Great Auks were killed. A photograph taken by Hjálmar Bárðarson for his book *Birds of Iceland* (1986). *Reproduced by kind permission of Hjálmar Bárðarson.*

Last crew which got Geirfugl — Ofeigur

✓ Vilhjalmur Hakonarson	formalen p.39	
✓ Jon Brandsson	p.47	41
Olafr Jonson	? a young man now dead (said Dr.?) p.115	
✓ Jon Ejolfson	now in Fuglevik p.249	32
✓ Eirikr Olafsson p.115		32
Jon Ketilsson	was never to Eldey as himself says p.140.	
✓ Ketil Ketilsson p.53		21
✓ Jon Biarnasson p.104		41
✓ Gudmr Hakonarson p.80-82		26?
✓ Jon Thorsteinsson	now in Garda, Skagastrdunn — pp.239,240	33
✓ Sigurdur Islefsson p.104		25
✓ Jon Gunnarsson pp.34...+138		56
✓ Gunnar Haldorsson p.108		19
✓ Ofeigur Hendriksson p.113		38
Biorn Thordarson p.235	now i Gardlins i Leyra, was certainly not	
✓ Svein Porvardsson	dead. v. p.241	39?

A page from the 'Garefowl Books' of John Wolley showing his list of the crew that sailed to Eldey in June 1844. *Courtesy of the University Library, Cambridge.*

precipice many fathoms deep. The black birds were flying off. I took him by the neck and he flapped his wings. He made no cry. I strangled him.

On being more closely questioned, Sigurðr offered the following additional details:

My bird ran [from where it was first seen] about 20 fathoms [40 metres]. It walked like a man ... but moved its feet quickly. It began to move as the people came on the rock and as the blackbirds [Guillemots] began to fly. Its wings lay close to its sides - not hanging out. As it was held by the neck it hung its wings out a little.

More than one Icelandic fisherman reported that when Garefowls died they simply raised their wings and sighed.

Ketil Ketilsson's account confirms the general impression given by Sigurðr:

[We] ran together after the birds, but as [we] got near the edge of the precipice [my] head failed me and I stopped; Sigurðr went on and seized the bird ... This one must have run quicker than a man can walk ... It held itself quite straight up, with its wings close to its sides, and as it ran it made no noise or cry.

The bird that Jón Brandsson ran after did not get nearly so far though it went in much the same direction. Wolley went on to record that Ketil

The eyes of the 'last' Great Auks preserved in spirits along with other internal parts. Photograph by Geert Brovad. *Courtesy of the Zoologisk Museum, Copenhagen.*

believed it was this other bird that was sitting on the egg:

> My informant, as soon as he was thrown out of the chase, went to look at the place from which the bird had started and there he saw an egg lying on the ... lava slab which he took up ... It was cracked or broken on the side upon which it was lying. Ketil laid it down again where he had found it.

The morning after the boat's return to its home bay of Kyrkjuvogr, the expedition leader, Vilhjálmur Hákanarsson set out to deliver the two trophies to Herr Siemsen in Reykjavik. On the way he met - by curious coincidence - the son of Peter Hansen, pilot of Faroe, a man by the name of Christian. An unexpected bargain was struck and Christian Hansen came away with the two dead Garefowls for a price of 80 rigsbank dollars (about £9).

Herr Siemsen never received the birds he'd hoped to get; they were sold instead to a Reykjavik apothecary by the name of Möller who thought them sufficiently important to commission a portrait. The picture was painted by a French artist named Vivien, who was staying in Iceland at the time, and for a number of years it hung in the apothecary's house. It was seen by Wolley and Newton but is now lost. Once the portrait was finished, Herr Möller began the process of skinning the birds and soon afterwards they were sent to Denmark. Möller also saved the skinless bodies and these too he sent - presumably in spirits - to Denmark.

What then happened to the skins - for which such effort was expended- is, surprisingly, something of a mystery. Somehow, amid all the frantic Garefowl research of the nineteenth century, they were lost track of. Several of the surviving stuffed specimens, notably those in Kiel, Bremen and Oldenburgh were tentatively identified with them. The most likely candidates, however, are the birds now in Los Angeles and in Brussels. Both these specimens were in the hands of a little known Copenhagen dealer by the name of Israel in the year 1845. Where Israel's birds came from - if they weren't the birds of 1844 - is an unanswerable question.

The ultimate destination of the two skinless corpses is well known; they were acquired by the Royal Museum, Copenhagen. Preserved in spirits, the internal organs still exist at the Univesitetets Zoologiske Museum. They are the remains of the bodies of a male and a female.

(Photograph on third page following). Fabergé Great Auk made of rock crystal with cabochon-ruby eyes, satin finished and set in silver; height 3 ins (73mm). *The Royal Collection © Her Majesty Queen Elizabeth II.*

(Photograph on fourth page following). A pair of silver pepperettes by L. Neresheimer and Co., stamped with importer's mark, *Chester, 1904;* each weighs 9 ounces and is 5 ins. (130 mm) high. *Reproduced by kind permission of Peter Petrou Antiques, London.*

AUKS AND MEN -
THE CULT OF
THE GAREFOWL

AUKS AND MEN - THE CULT OF THE GAREFOWL

As on the death of an ancient hero myths gathered around his memory as quickly as clouds round the setting sun.

Alfred Newton (1896)

To those who eked out a meagre living among the bleak isles and skerries of the far north, to sailors and fishermen anxious to replenish supplies before venturing into the icy waters of the Arctic, this bird had always been irresistible. But as the Garefowl passed into myth among these peoples of the north, the Great Auk steadily acquired legendary status with a more sophisticated public. Nineteenth century city dwellers, who'd never glimpsed grey Atlantic seas, became aware that the bird once existed and perhaps did still. It fast became a cult creature and to a limited extent it retains this position even today. With the exception of the Dinosaurs, the Dodo and, perhaps, the Mammoth, no creature carries a name so synonymous with the word extinction.

Nowhere did the process of its elevation from the rôle of curious bird to that of icon of extinction find a clearer or more elaborate expression than in Victorian Britain - but this was by no means just a British phenomenon. Right across Europe and also in the United States, a strange passion for all things relating to Great Auks spread. The bird was no longer an edible commodity but its appeal to human beings was not about to end.

No general natural history book would be complete without an entry describing - and preferably also figuring - the Great Auk. No editor of a newspaper, magazine or journal would lightly pass up an opportunity for running an article about the bird itself or about its relics. Steadily the Great Auk crept into literature and art. Pictures of the bird were relentlessly copied and re-copied as they were published and re-published over and over again. All kinds of decorative and souvenir items were made. The celebrated jeweller Carl Fabergé made small replicas, one of which can be found today in the Queen's collection. Stuffed birds and preserved egg shells were sold for high prices - sums beyond the dreams of ordinary individuals - and fake specimens were made, sometimes just for fun, sometimes with a deliberate intention to deceive. The species was featured on postage stamps and cigarette cards, with even a cigarette brand bearing its name. Great Auks had caught the popular imagination in a way that was hardly predictable.

The first burst of awareness and enthusiasm came from academics and natural historians and it was their initial devotion to the cause that inspired interest among the general public. In the decades immediately following the Great Auk's disappearance, men like John Wolley, Alfred Newton, George

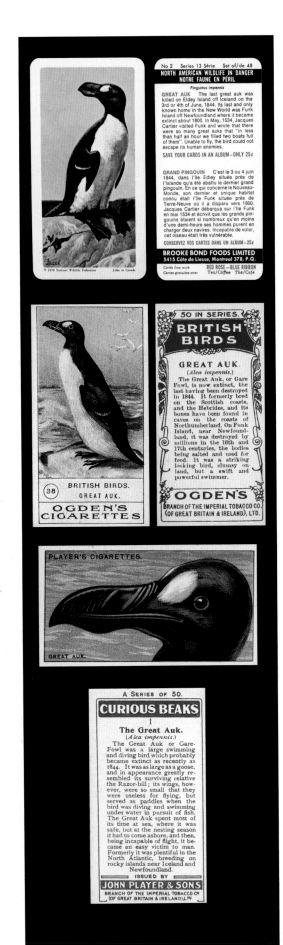

Dawson Rowley and John Hancock began to make determined attempts to piece together the story of the Garefowl and later in the century a younger generation - Symington Grieve, Edward Bidwell and Thomas Parkin among them - took over the work. In Europe their counterparts were scholars like Wilhelm Blasius, Japetus Steenstrup, Victor Fatio, and H. Duchaussoy while in the United States F.A. Lucas and S.F. Baird became keen enthusiasts. Always the pursuit of knowledge went hand in hand with the acquisition of specimens. Alfred Newton, for instance, who - along with Wilhelm Blasius - was perhaps the most thorough of all Garefowl scholars, owned several eggs, plotted to get several more, and enjoyed lifelong daily access to Cambridge University's stuffed specimen. Blasius owned a stuffed bird but never got an egg.

The names of the dealers who supplied such trophies are as inextricably linked with the unravelling of the story of the Great Auk as are the names of the scholars. Men like Mechlenburg, Apothecary of Flensburg, Israel of Copenhagen, Brandt and Salmin of Hamburg or the Frank family of Leipzig, Amsterdam and London were instrumental in establishing the craze for the Great Auk. The important London dealers, Leadbeater, John Gould and Rowland Ward - not to mention the famous old auction house of Stevens' - all played a significant part.

Then there were the obsessed collectors: Champley, Hewitt, Thayer, Massey, Middlebrook, Garvagh, Jourdain are all names tightly tied to the tale of the Great Auk. Together with many others, these men helped save Auk remains from oblivion and added to the rich fund of Garefowl lore.

Although knowledge of the species is largely a knowledge gathered together after the bird's extinction, the wonder and curiosity that the Great Auk inspired is something rooted deeply in the past.

Cave paintings in the Grotte Cosquer, southern France (assuming they do indeed, as is generally supposed, show Great Auks) testify to primitive man's emotional and spiritual involvement with the birds; these paintings are judged to be around 20,000 years old. More recently, although still long, long ago, Garefowl beaks were used in elaborate funeral rites the significance of which can only be guessed at. Seven graves in a cemetery, perhaps 4,000 years old, discovered at Port au Choix, Newfoundland contained beaks and in one grave two hundred of them had been used to entirely cover a human body. Possibly they were good luck charms or perhaps they symbolised prowess at fish catching. Maybe they adorned a decorated coat or sheet that has long since rotted away (similar garments are still made by the Aleuts using the feathers and beaks of smaller Auk species). Perhaps the beaks were placed in the graves in the rôle of attendants that would accompany the dead to another world.

Other prehistoric rituals and acts of magic connected with primitive man's involvement with the Great Auk have vanished into time, but coming much closer to today, the Garefowl was linked to a ritual of a rather different kind. One of the earliest attempts at bird conservation by means of the law was directed at safeguarding the Great Auk. Considering the terrible events that occurred on Funk Island, there is a certain irony attached to efforts made to restrict egg collecting there. The punishment meted out to wrongdoers was a dire one. In 1793 several men suffered severe public floggings for infringements against this restriction and

G. Eduards ad viv. delin .   J. M. Seligman excudit .   Joh. Sebast. Leitner sculps.
Cum.Priv. Sac.Caes.Majestatis.
Penguin Arcticus .   N.º 42.V.ter Theil .   Le Penguin du Nord .

147

London Published Feb.º 1.º 1800 by F.P. Nodder, Newman Street.

Fig. 1.ère Le Pingouin.

Fig. 2.
Le Grand Pingouin.

(Above, left).Engraving from l'Abbé Bonnaterre's *Tableau Encyclopédique et Méthodique des trois Règnes de la Nature* (1791).

(Above, right). Engraving from Comte G.L. Buffon's *Natural History - General and Particular* (Smellie edition, 1792-3).

(Preceding page). Four Great Auks. (Top right). Hand-coloured engraving by George Edwards from his *Natural History of Uncommon Birds* (1743-51). (Top, left). Hand-coloured engraving from a German edition *Sammlung verschieden ausländischer und seltener Vögel* (1749-76). (Bottom, left). Hand-coloured engraving from an unknown source (published December 1802 and said to be drawn by Sir Charles Tinnens). (Bottom left). Hand-coloured engraving by R. Nodder and N. Street from G. Shaw's and F. Nodder's *Naturalists' Miscellany* (1793-1813). Edwards original engraving was drawn from a specimen obtained at sea, over a fishing bank, about 100 leagues from Newfoundland. The specimen has long since vanished.

(Facing preceding page). Great Auk cigarette cards. *Courtesy of David Lank.*

William Montevecchi and Leslie Tuck (1987) recount a curious anecdote concerning a man named Clark who was hauled before a certain Chief Justice Reeves. On hearing the sentence of a public flogging, Mr. Clark protested. He'd taken the eggs, he explained, to prevent his family from starving. Justice Reeves was clearly moved by these mitigating circumstances and he commuted the sentence. The flogging, he ruled, could be administered in private.

Such curious historical incidents, as well as the prehistoric traces, show that the Great Auk has always caught man's attention. Yet until the start of the nineteenth century any preoccupation with the species was fired largely by the idea of its suitability for the pot. Even its appearance on a set of particularly fine French porcelain plates made towards the end of the eighteenth century is suggestive of dinner! Only as extinction approached did other factors come to the fore and a curiosity developed showing no connection with the Great Auk's tastiness.

When Icelandic fishermen sought out the species' last refuges and finally hounded it to extinction, they did so largely under instructions from merchants keen to provide trophies for museums and collectors. Any assumption that this final onslaught - distasteful though it may have been - was the real reason for extinction is a wrong one, however. The species was brought to the brink of oblivion long before trophy collecting became an incentive and the Icelanders would have finished the process, almost as rapidly, for reasons altogether more prosaic than the procurement of museum specimens.

It is, perhaps, rather odd that the main thrust of interest in matters relating to Great Auks stemmed from the pursuit of trophies - eggs and stuffed birds. The early chroniclers of Garefowl affairs - Otto Fabricius in

Greenland, John Wolley and Alfred Newton in Britain, Michahelles, Benicken, the Reinhardts and Japetus Steenstrup on the Continent - were all dedicated collectors or curators.

By the second half of the nineteenth century the search for specimens was as much an academic pursuit as an active one and, naturally, the living bird could no longer be proceeded against. The tracing of the history of each example and the compilation of lists of owners and whereabouts of examples were preoccupations that ran hand in hand with the desire to determine a fuller impression of the species' natural history. As men found themselves unable to hunt the living bird they pursued the dead one, a symbolic chase that steadily metamorphosed into a quest for deeper knowledge and information.

Perhaps the first really serious attempt to produce a natural history of the Great Auk was that made by the Danish zoologist Japetus Steenstrup. During 1855, in a journal called *Videnskabelige Meddelelser*, he published a paper entitled 'Et Bidrag til *Geirfuglens, Alca impennis.*' This is a detailed and insightful study that laid the groundwork for all later efforts. Others - Fabricius, Faber, Michahelles and Benicken among them - had already made small stabs in this direction, but none had managed to put together anything like so comprehensive or accurate an account.

At about the same time as Steenstrup was working on his monograph, a young Englishman by the name of John Wolley was fast becoming an

(Below). Hand-coloured lithograph by J.T. Bowen (after Audubon), circa 1859.

Great Auk

(Above). The frontispiece to Darley Dale's *The Great Auk's Eggs* (1886). This illustration, by Charles Whymper, shows two schoolboys putting the finishing touches to a fake stuffed Auk that they have made.

Illustration by Harold Jones for a 1961 edition of Charles Kingsley's *The Water Babies*.

(Facing page). *The Bidwell List*. This list was circulated by Edward Bidwell during the last years of the nineteenth century. Whether he sold it to interested parties or simply gave it away is not known.

enthusiastic devotee of the Great Auk and he began planning a monograph of his own. Wolley, best remembered for his fabled egg collection, died at a comparatively early age and never got near to completing his work but shortly before his untimely death he visited Iceland in pursuit of knowledge. In the company of his friend Alfred Newton, who was to become one of the most distinguished zoologists of his day, he collected together every scrap of information he could discover and meticulously recorded it in notebooks still preserved at the Cambridge University Library. It is through his efforts that present day knowledge of the activities of those who killed the last of the Great Auks can be so complete. Alfred Newton also deserves much of the credit for it was he who ensured that Wolley's notebooks were kept and who loyally saw - after his friend's early death - that the substance of Wolley's work was put into a form suitable for publication (*see* Newton, A. Abstract of Mr. J. Wolley's Researches in Iceland respecting the Garefowl or Great Auk. *The Ibis*, 1861).

Fired by the material that Steenstrup, Wolley and their contemporaries put forward, a new generation of Auk researchers took up the challenge and attempted to fill gaps in the knowledge. Newton himself continued to amass information for the rest of his life but never produced the monograph he planned, restricting himself to publishing a few fragments in journals and magazines. It was left to Symington Grieve to write the most celebrated work in Garefowl history. In 1885 he took more established naturalists by surprise when he published *The Great Auk or Garefowl*, a book that inspired a new wave of Auk activity and one that has remained a classic to this day.

Meanwhile there had been no let up in the list making. In England, a Mr. Alfred Roberts had constructed a list and he was followed by Robert Champley and then Newton. On the Continent Wilhelm Preyer and Victor Fatio both compiled lists and they were followed by Wilhelm Blasius who made by far the most interesting one. Blasius generously gave permission for a translation of this - in somewhat reduced form - to be included as an appendix to Grieve's work.

One of the new wave of Great Auk devotees was an East Anglian gentleman by the name of Edward Bidwell. Bidwell showed a particular interest in eggs and set himself the task of photographing every example then known. After chasing this ambition for several years, he virtually achieved it and a few sets of his photographs were circulated among Garefowl enthusiasts; a set of these eventually formed the basis for a work published three quarters of a century later- *Eggs of the Great Auk* by P.M.L. and J.W. Tomkinson (1966). Bidwell also issued a printed sheet that came to be known as 'The Bidwell List' and this gives the name of every egg owner together with spaces to accommodate new discoveries or changes of ownership. Although sometimes referred to in the literature, this list - like the photographic sets - seems to be a very rare piece of Auk ephemera.

As the Great Auk figured with increasing prominence in ornithological writings, so too did it begin to appear in more mainstream literature. A book for children, *The Great Auk's Eggs* by Darley Dale (a pseudonym for Francesca Maria Steele), was published in 1886 with illustrations by Charles Whymper. It tells the story of two boys and their hunt for a Great

LIST showing present Owners of Eggs of the Great Auk.

1	BRITISH MUSEUM, NATURAL HISTORY, Cromwell Road, London.	26	Mr. ROBERT CHAMPLEY, Scarborough.	51	MUSEUM OF NATURAL HISTORY, Paris.
2	BRITISH MUSEUM, NATURAL HISTORY, Cromwell Road, London.	27	Mr. PHILIP CROWLEY, Waddon, Surrey.	52	BARON LOUIS d'HAMONVILLE, Manonville.
3	MUSEUM OF SCIENCE AND ART, Edinburgh.	28	Mr. LEOPOLD FIELD, London.	53	BARON LOUIS d'HAMONVILLE, Manonville.
4	MUSEUM OF SCIENCE AND ART, Edinburgh.	29	LORD LILFORD, Lilford Hall, Northants.	54	BARON LOUIS d'HAMONVILLE, Manonville.
5	UNIVERSITY MUSEUM, Cambridge.	30	Mr. JOHN MALCOLM, Poltallock, Argyleshire.	55	BARON LOUIS d'HAMONVILLE, Manonville.
6	UNIVERSITY MUSEUM, Cambridge.	31	Sir FREDERICK MILNER, Nunappleton, Yorks.	56	Monsieur De MEEZEMAKER, Bergues les Dunkerque.
7	UNIVERSITY MUSEUM, Cambridge.	32	Professor NEWTON, Cambridge.	57	Monsieur De MEEZEMAKER, Bergues les Dunkerque.
8	UNIVERSITY MUSEUM, Cambridge.	33	Professor NEWTON, Cambridge.	58	Monsieur M. HARDY, Perigueux.
9	UNIVERSITY MUSEUM, Cambridge.	34	Professor NEWTON, Cambridge.	59	ROYAL ZOOLOGICAL MUSEUM, Dresden.
10	UNIVERSITY MUSEUM, Oxford.	35	Mr. JOHN C. L. ROCKE, Clungunford, Salop.	60	GRAND DUCAL MUSEUM, Oldenburg.
11	ROYAL COLLEGE OF SURGEONS, London.	36	Hon. WALTER ROTHSCHILD, Tring Park, Herts.	61	HERR Th. LÖBBECKE, Dusseldorf.
12	ROYAL COLLEGE OF SURGEONS, London.	37	Mr. G. FYDELL ROWLEY, Brighton.	62	ZOOLOGICAL MUSEUM, Amsterdam.
13	ROYAL COLLEGE OF SURGEONS, London.	38	Mr. G. FYDELL ROWLEY, Brighton.	63	ZOOLOGICAL MUSEUM, Leyden.
14	DERBY MUSEUM, Liverpool.	39	Mr. G. FYDELL ROWLEY, Brighton.	64	NATIONAL MUSEUM, Lisbon.
15	NATURAL HISTORY MUSEUM, Newcastle-on-Tyne.	40	Mr. G. FYDELL ROWLEY, Brighton.	65	MUSEUM OF NATURAL HISTORY, Lausanne.
16	PHILOSOPHICAL SOCIETY'S MUSEUM, Scarborough.	41	Mr. G. FYDELL ROWLEY, Brighton.	66	ACADEMY OF NATURAL SCIENCES, Philadelphia.
17	Mr. EDWARD BIDWELL, Twickenham.	42	Mr. G. FYDELL ROWLEY, Brighton.	67	SMITHSONIAN INSTITUTE, Washington.
18	Mr. ROBERT CHAMPLEY, Scarborough.	43	Sir GREVILLE SMYTHE, Ashton Court, Somerset.	68	
19	Mr. ROBERT CHAMPLEY, Scarborough.	44	Mr. JAMES H. TUKE, Hitchin, Herts.	69	
20	Mr. ROBERT CHAMPLEY, Scarborough.	45	Mr. HENRY WALTER, Papplewick, Notts.	70	
21	Mr. ROBERT CHAMPLEY, Scarborough.	46	Mr. T. H. POTTS, (Exors. of) New Zealand.	71	
22	Mr. ROBERT CHAMPLEY, Scarborough.	47	ROYAL UNIVERSITY MUSEUM, Copenhagen.	72	
23	Mr. ROBERT CHAMPLEY, Scarborough.	48	NATURAL HISTORY MUSEUM, Angers.	73	
24	Mr. ROBERT CHAMPLEY, Scarborough.	49	MUSEUM OF NATURAL HISTORY, Paris.	74	Dr. TROUGHTON, (dec.) Model.
25	Mr. ROBERT CHAMPLEY, Scarborough.	50	MUSEUM OF NATURAL HISTORY, Paris.	75	Mr. SCALES, (dec.) Model.

THE PENGUIN, WITH THE CONC, AND OTHER SHELLS, SPONGES, &c.

The Penguin with Conch Shell. Copper engraving by E. Lacey from a drawing by J. Craig *in* R. Wilks's *The Gallery of Nature and Art. A Tour through Creation and Science* (1814). *Courtesy of Clemency Thorne Fisher.*

(Facing page). Illustrations for Charles Kingsley's *The Water Babies* by four different artists. (Top left). A.E. Jackson (1918). (Top right). Warwick Goble (1909). (Bottom left). Anne Anderson (1924). (Bottom right). Jessie Wilcox Smith (1932).

Auk's nest. Eventually, of course, they discover that the bird is extinct. Charles Kingsley, a good friend of Alfred Newton, featured the Garefowl in his famous childrens' story *The Waterbabies* (1863):

> And there [Tom] saw the last of the Gairfowl, standing up on the Allalonestone, all alone ... She had on a black velvet gown, and a white pinner and apron, and a very high bridge to her nose ... and a large pair of white spectacles on it, which made her look rather odd: but it was an ancient fashion of her house ... And the poor old Gairfowl began to cry tears of pure oil ... Soon I shall be gone, my little dear, and nobody will miss me; and then the poor stone will be left all alone.

Such an introduction into literature allowed plenty of scope for illustrators and artists who, until this time, had been largely restricted to more formal depictions of the bird.

Most of the important bird painters of the nineteenth and early twentieth centuries turned their hands to figuring the Great Auk. Audubon and J.G. Keulemans, Kuhnert, Thorburn and Fuertes were among the big names who produced pictures and some of them, most notably Keulemans,

ALCA IMPENNIS.

The re-working of Edward Lear's painting of the Great Auk. Hand-coloured lithograph by W. Hart from J. Gould's *Birds of Great Britain* (1862-73).

painted the species on several occasions. Indeed Keulemans painted an enormous - and quite uncharacteristic - oil of a Great Auk colony. This magnificent picture is one of a pair (the other shows the Great Bustard *Otis tarda*) that hang in the Natural History Museum, London. They were, apparently, donated but, with cavalier sloppiness in no way matching the generosity of the gift, the name of the donor is now forgotten at the museum. Miriam Rothschild has confirmed, however, that the pictures were bequeathed by her uncle - the celebrated Walter - who gave so much to the museum at his death in 1937.

A picture of a completely different kind is that created by Edward Lear as a plate for John Gould's *The Birds of Europe* (1832-7). As a vigorous piece of graphic design this bold image has few equals in the world of bird illustration. Although the lithographic plate clearly bears Lear's signature, Gould, with a shamelessness that is almost admirable - and certainly typical - inserted a subtext to the picture that read:

Drawn from Nature and on the stone by J & E Gould.

Hand-coloured engraving by Prideaux John Selby from his *Illustrations of British Ornithology* (1821-34).

Not content with this little conceit, he allowed the image to be copied for one of his later productions, *The Birds of Great Britain* (1862-73), making no attempt to credit the original artist.

One of the more painterly and ambitious pictures is a work called *The Great Awk's Egg* by the popular late Victorian artist Henry Stacy Marks. It was shown at the Summer Exhibition of the Royal Academy in 1892 and an engraving was subsequently made. Despite its undoubted popular success, the treatment of the subject matter is curiously flawed. The men admiring the egg so intently are dressed in the costume of a period rather earlier than the one in which obsessive interest in Garefowl eggs flourished.

There were other artistic mistakes. Great Auks were regularly shown against backdrops of icebergs and glaciers as the idea of an Arctic Garefowl stubbornly persisted. Illustrations for Anatole France's *L' île des Pingouins* (1907) show, as the title suggests, birds that are unmistakably southern hemisphere Penguins. Yet this peculiar book is concerned with Great Auks. 'Pingouin' was then still being used in France (and to a limited extent the usage continues) for *Alca impennis*. Later editions sometimes corrected the pictorial error. Anatole France, incidentally, was born in 1844, that most doom-laden of all Great Auk years, and - despite his undoubted talent and intelligence - possessed a brain of abnormally small size.

While artistic licence was cheerfully tolerated by the general public, in scholarly quarters more pedantic standards were applied. In 1912 the American ornithological artist Louis Agassiz Fuertes was asked to design a new cover for *The Auk*, journal of the American Ornithologists' Union. This

ANATOLE FRANCE

L'ÎLE DES PINGOUINS

Two steel engravings from *Cassell's Book of Birds* (1869-73).

(Page 100). *The Great Awk's Egg; the Collector's Treasure.* Oil on canvas by Henry Stacy Marks, painting no.228 at the Royal Academy Exhibition of 1892. *Courtesy of Haynes Fine Art, Broadway, Worcestershire.*

(Page 101, top). Dust wrapper illustration by Pierre Lissac for a 1941 edition of *L'île des Pingouins. Courtesy of Gaël Lagadec.*

(Page 101, below left). J.G. Keulemans (1842-1912). This renowned ornithological artist produced several important pictures of the Great Auk..

(Page 101, below right). Anatole France (1844-1924) - author of *L'île des Pingouins* (1907).

journal had shown the Garefowl on its cover since 1884 (apparently *Auk* was chosen for its brevity, being shorter than *Ibis* the title of the British journal) and Fuertes, the leading American bird painter of his day, was expected to produce a picture that was anatomically sounder. After carefully examining and measuring mounted specimens and skeletons and then applying to the work his enormous knowledge of the avian physique, Fuertes submitted a design for the approval of the AOU's ruling committee. The pomposity with which this group of individuals bore their self-appointed burden as art censors is best revealed by the message they sent back to the artist:

> The second largest bird in the background, which is waving its wings, introduces a debatable question as to the position of the wings, and we think it would be desirable to have the wings close to the body to avoid any ground for argument.

The worries and fears that beset these poor men when it came to more important issues can only be guessed at. Their fretting over the minutia of background detail brought only the gentlest of rebukes from Fuertes, who must have been a very good natured and obliging artist. After patiently accommodating the committee's wish, he simply sent back a new picture and accompanied it with a rather rueful poem:

> A wise committee in New York
> Have passed their word that this 'ere Auk
> Conforms in feature and proportion
> To the Museum's stuffed abortion.

The Great Auk in a tropical landscape. Steel engraving by S. Sargent from an edition of *Cassell's Popular Natural History* (circa 1870).

Three covers of *The Auk*- the two most recent drawn by Louis Agassiz Fuertes.

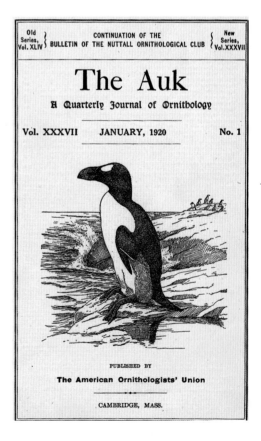

(Facing page, top left). Illustration by Sir Francis Carruthers Gould from his book *Nature Caricatures* (1929). *Photograph courtesy of Peter Blest.*

(Facing page, top right). An illustration captioned *A Relic of the Past - Great Aunt Auk* by Sir Francis Carruthers Gould for *Country Life* (April 29 1916).

(Facing page, bottom). An illustration by Linley Sambourne for an 1884 edition of Charles Kingsley's *The Water Babies.*

(Below). A cartoon drawn by Henry Stacy Marks, painter of *The Great Awk's Egg*, in the 'Garefowl Books' of John Wolley and Alfred Newton. The point of the cartoon is - apparently - that rich patrons had become as extinct as the Dodo and Great Auk. *Courtesy of the University Library, Cambridge.*

(Below, right). A postcard designed by Louis Agassiz Fuertes to commemorate the 1916 meeting of the American Ornithological Union.

As to position, wings and such
The critics don't know very much
So I just guessed my very best
Imagination did the rest.

And this we have for comfort sweet
Should doctors disagree
Nobody lives who knew the beast
And there are no more to see.

A year or two later the men of the committee were, it seems, taking themselves rather less seriously. Fuertes produced a very happy-go-lucky picture for a greetings card commemorating the AOU's annual meeting and this- as far as is known - did not meet with a howl of disapproval.

Fuertes little attempt at poetry was by no means the first celebration of the Great Auk in verse. In addition to entering the literary world of novels and childrens' books, *Alca impennis* had long since made its debut in the realm of the poem. It cannot, alas, be said that any of these rhymes were works of aesthetic beauty. In late Victorian England there was a spate of them, usually attempts at humorous lines replete with puns and rather silly jokes. One that appeared in *Punch* on August 7th 1897 began:

Oh! Talk to me not of Klondike,
Coolgardie, Peru or the Rand;
As investments they're failures alike
Compared with the latest to hand:

But give me the egg of the Auk-
The Great Auk- I ask for no more;
When it's cracked they can fill it with chalk
Till it fetches its weight in gold-ore.

Most of the poems made some play on the words 'Auk' and 'auction' and all were alike in their desperately poor quality.

Strangely the tradition of celebrating the species in verse is still alive. A recent popular song with words written by the Liverpool poet Roger McGough featured the Great Auk. Today the emphasis has shifted and the chorus strikes a darker, bleaker note:

Auk! Auk! Auk!
My name is a cry of pain,
Cull, Skull, Cudgel,
Over and over again.

The phenomenon of Great Auk collecting is, perhaps, as curious as everything else about this bird. When, in the late 1820's and the 1830's, Professor Reinhardt, of the Royal Museum, Copenhagen, began distributing Garefowl skins to collectors and museums across Europe, he fanned the flames of a craze - and began a process - that has not quite finished. Even in the last few decades there has been a considerable alteration in the whereabouts of Auk skins and eggs and such movement is only likely to end when the few remaining relics in private hands have all finally entered the great institutional collections. Right from the beginning the drift from private to public collections has been constant. Today only two stuffed birds and a handful of eggs (the whereabouts of some of which are unknown) are privately owned.

Professor Reinhardt began distributing Great Auk skins for a very good reason. He was able to considerably benefit the Royal Museum with the proceeds from his sales and exchanges. Geography and political connection gave the professor an opportunity not available to other museum curators. During the 1820's and 1830's, when specimens could still be procured from

Kakapos, Huias, Great Auks and Kiwis. Extinct and rare birds at Stevens' Rooms in November, 1934, immediately before the George Dawson Rowley Sale. *Photograph courtesy of David Wilson.*

the bird islands near Reykjavik, the Danish connection with Iceland ensured that most Icelandic goods - including Auks - entering Europe did so through the port of Copenhagen. Reinhardt was in a perfect position to obtain whatever specimens became available and the majority of preserved examples existing today are ones that passed through his hands or came because of his intervention.

By 1840 the price in England for a stuffed Auk was a pound or two. Twenty years later an interested party would have expected to pay the, then, very considerable sum of £40 - perhaps even £50 - and the curiosity of the public at large began to be aroused. From this time onwards, sales of Great Auk relics became events of general interest. The famous old auction house of Stevens' situated in King Street in London's Covent Garden became the venue for many sales of stuffed Garefowls and their eggs. Although natural history items formed only a small part of Stevens' business, this firm of auctioneers became so famous for their Auk sales that their telegraphic address was simply:

'Auks, London.'

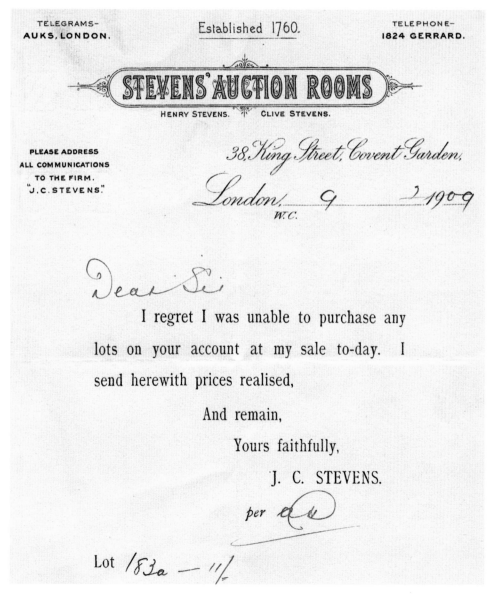

(Above, top). The auctioneer Henry Stevens in 1903 - from a photograph taken by himself.

(Above). A newspaper advertisement (*The Star*, November 10th 1916) for *Great Auk Cigarettes*.

(Left). A letter from Stevens' notifying an intending purchaser that a bid had failed.

THE DAILY GRAPHIC,
Wednesday April 24th 1895.

THE VALUE OF BEING EXTINCT.

Yesterday morning, at the auction rooms of Messrs. Stevens, in King Street, Covent Garden, a number of spectators surveyed a bird sitting or standing there with great admiration and respect. He was a big bird, perhaps of the size of a Michaelmas goose, and he had an expanse of white bosom of a spotlessness which would excite the envy of a German waiter. He had a white splash over each eye, but his head, his webbed feet and his elegantly-shaped back were black as ink. His eye was glassy, and he stood the fire of remarks and comments upon his appearance with an immovable composure, for he was stuffed. He was a great auk, and he was extinct; but neither of these facts affected his bearing, although both affected his price. When he was put up for auction Mr. Stevens made a little speech about him which was so complimentary that the casual auditor could not restrain the conviction that it was just as well that the bird was too extinct to hear him, for otherwise its head might have been turned.

What was the number of great auks when all the world was young and the plesiosaurus coquetted with the mastodon, nobody can tell. But the number of stuffed great auks now known to civilisation, said the auctioneer, was only eighty. This rarity was still more real than apparent, for, of the whole number, sixty were in museums, where they were likely to stay. Indeed, only one great auk had been put up to auction in the last twenty-six years. Twenty-four of these noble birds, pursued Mr. Stevens, were in Great Britain, but of this number ten were in museums, and the slenderness of the chance that any of them would come into the market would be quite obvious to any of his hearers. And what was the value? Well, the last great auk went at the ludicrous price of £90, but at the same sale an egg only fetched £60. Last year an auk's egg was sold in the Covent Garden auction room for 300gs.; and, therefore, concluded Mr. Stevens, triumphantly, it would be obvious that according to the rule of three the present value of a great auk should be somewhere between £500 and £600.

The audience smiled, and the auctioneer went on to declare his unhesitating belief that this was the finest specimen of a great auk in existence; it was absolutely genuine; it was not made up at all, nor had it a single false feather; and it was the property of Sir Frederick Milner, Bart. The remaining part of the sale can be conveniently cast into the form of a paragraphic monologue on the part of the auctioneer.

"It is (said Mr. Stevens) a perfect bird: as fresh and as clean as on the day it was hatched. Who'll make a bid for it? I think I am safe in putting it in at 100 guineas. 110. Thank you. 120—130—140—160."

"160: don't stop there, gentlemen: this is very slow. 180—200—220—230—250—260 guineas bid. It would be ridiculous if this bird fetched anything under 500 guineas."

"260 guineas—270—280—300 guineas. Thank you. This noble bird 300 guineas—310—320—330. Isn't it a wretched price, gentlemen?" The audience sighed sympathetically, but 330 guineas is a lot of money for a bird, and another sigh—this time of relief—went

THE GREAT AUK AND EGG WHICH WERE OFFERED FOR SALE BY AUCTION YESTERDAY.

up when another bidder said 340. This bid was followed by 350, and here the bidding stuck. In vain the auctioneer reasoned, pleaded, protested. Nobody would go a shilling better, and so Mr. Stevens had reluctantly to announce that the reserve price had not been reached, and the great auk would be withdrawn. It was a humiliating moment for everybody, but the bird came out of it unruffled. It is believed that he will be disposed of privately.

After this an egg of the great auk, not of this noble bird but of some two thousand years distant relative, was put up for sale. Mr. Stevens had exhausted much of his powers of eulogy on the bird, but he had sufficient left to declare this to be one of the finest eggs he had ever seen, and greatly superior to the phenomenon which fetched 300 guineas at these auction rooms last year. Those who were present on that celebrated occasion, perhaps, would be unable to recall anything in the auctioneer's speech which would have seemed to indicate that there was any other auk's egg superior to the one he was then putting up in general desirability, and yesterday's bidders took the same view, for the highest bid that could be obtained was only 180 guineas. At this price it was sold to the proprietor of a public-house who keeps a small private museum for the entertainment and instruction of his patrons. After this no one was surprised to find that an egg of the *Æpyornis maximus*—in size, shape, and colour very much like a new Rugby football—was sold for thirty-six guineas.

In 1924 E.G. Allingham published *A Romance of the Rostrum*, a book detailing the history of Stevens' and of all the treasures that had passed through its doors. The dust wrapper might have featured an important painting or a priceless jewel. It didn't. It carried a picture of a stuffed Great Auk.

The Auk sales that took place at Stevens' were events of considerable social importance and accounts of them - complete with descriptions of every lull in the proceedings, every nuance of expression, every bid and counter bid - were features of all the national newspapers and magazines. Even as late as 1934, photos of a Great Auk sale made the centre pages of *The Daily Sketch*, pages that were, at that time, entirely devoted to 'society' photographs and matters of corresponding interest.

Perhaps because they came onto the market more frequently (and were set, therefore, more regularly before the public eye), the preserved egg shells caught the popular imagination rather more than the mounted specimens. The slightly ludicrous spectacle of grown men, and usually very respectable ones, shelling out vast sums of money for objects that seemed - to the uninitiated - of very doubtful value, proved an irresistible entertainment. After all, the fragility of these shells was perfectly apparent and this vulnerability played a large part in the mythology that grew up around them. The mysterious cracking of an egg in the Scarborough Museum caused a great stir; so too did the dropping of an example by Lord Garvagh's footman. The most peculiar story - and one that is entirely apocryphal - is an intricate tale of a rich collector who bought an egg at auction with the deliberated intention of smashing it into a hundred pieces the minute he got his vandalous hands on it. The logic of the plan - every bit as flawed as the shell would have been - was simple. The wanton act of destruction would surely increase the rarity (and so the financial value) of an egg the gentleman already owned.

Fortunately, real life collectors saw things in a rather different light and a considerable number of them were unable to satisfy themselves with the possession of a single egg. The phenomenon of serial egg collecting became almost commonplace among the oological fraternity. Scores of two, three, four or even five - while, perhaps, remarkable - were by no means outstanding. The record in Victorian England was held by Robert Champley, Alderman of Scarborough, with a tally of nine. A certain Doctor Dick, about whom nothing is known, may have had ten but this is not verifiable and in any case the mysterious doctor probably belongs to a slightly earlier era. During the first years of the twentieth century Colonel J.E. Thayer of Lancaster, Massachusetts equalled Doctor Dick's putative score of ten but the all-time champion is the amazing Captain Vivian Hewitt who managed to gather together no less than thirteen, an achievement accomplished when the fashion for such collecting was in its death throes.

One of the best publicised collections was that belonging to Mr. T.G. Middlebrook who acquired several eggs for a small 'free museum' which formed part of his London public house *The Edinburgh Castle* in Mornington Crescent, Camden Town. The pub still survives but the eggs are long gone being among the ten obtained by Colonel Thayer; today they are in the Museum of Comparative Zoology at Harvard.

A ROMANCE OF THE ROSTRUM

BEING THE BUSINESS LIFE OF HENRY STEVENS, AND THE HISTORY OF THIRTY-EIGHT KING STREET, TOGETHER WITH SOME ACCOUNT OF FAMOUS SALES HELD THERE DURING THE LAST HUNDRED YEARS.

COMPILED BY

E. G. ALLINGHAM

WITH A PREFACE BY

THE RIGHT HON. LORD ROTHSCHILD, F.R.S.

Illustrated from Photographs.

The dust wrapper of E.G. Allingham's *A Romance of the Rostrum* (1924). The Great Auk shown is the specimen now in Edinburgh.

An account of a Stevens' sale in *The Daily Graphic* for April 24 1895.

(Facing page). *A Gallery of Auk Owners*. (Top row, left to right). John James Audubon (1785-1851), Thomas Lyttleton Powys, Lord Lilford (1833-96), Lionel Walter Rothschild, 2nd Baron Rothschild (1868-1937). (Middle row). Victor Emmanuel, King of Italy (1820-78), Rowland Ward (1848-1912), Luis I, King of Portugal (1838-89). (Bottom row). Lord Rowland Hill, inventor of the postage stamp (1800-73), John Gould (1804-81), Lucien Bonaparte (1775-1840), brother of Napoleon.

The list of men who once owned eggs reads like a roll-call of nineteenth century ornithological writers: Rowley, Potts, Hancock, Newton, Wolley, Yarrell, Lilford, Rothschild, Hardy, Temminck, Thienemann and Gould.

As far as the stuffed birds were concerned, they too had illustrious owners and many of the writers who owned eggs also owned mounted specimens. Kings and princes became Auk owners. Victor Emmanuel, King of Italy had one; so too did his son-in-law Luis I, King of Portugal. A Prince of Denmark got one and seems to have promptly given it away. The Empress Josephine may have owned an egg or two; at least Victorian researchers tried, perhaps unconvincingly, to establish that she did. Lionel Walter Rothschild, second Baron Rothschild bought two birds to match a similar number of eggs and Lord Lilford got a stuffed specimen to go with the five eggs he'd obtained. Rowland Hill, conceiver of the idea of the postage stamp, kept a mounted Garefowl at his Hawkstone estate. In

(Above). Part of T.G. Middlebrook's museum at his London public house *The Edinburgh Castle*. Although the picture shows five eggs, the historical record suggests that Middlebrook only owned four. Unfortunately, there is not sufficient detail to enable identification of the extra egg. *Photograph courtesy of John Edwards.*

(Above, right). The cover of the catalogue to Middlebrook's museum.

France the Baron d'Hamonville put together a collection of four eggs and the Baron Vilmarest was bequeathed a stuffed bird by a friend.

So much greater was demand than supply that counterfeiting became almost common. Fake Garefowls could be made from the feathers of Razorbills, although the beak, of course, needed to be entirely artificial. Although sometimes of a very high standard, these models could never deceive the discerning eye, nor, indeed were they intended to. They were made simply to satisfy the requirements of those who had neither the money nor the opportunity to buy the real thing; often they were manufactured to fill gaps in cabinets or museum collections that otherwise seemed incomplete. Similarly - and usually for similar purposes - replicas were made of the eggs, although in some cases, it seems, there was a real intention to defraud. T.H. Potts (1870) recounted such an instance:

(Facing page). Carrera Marble sculpture by Jim Sardonis. *Collection of Mr and Mrs Richard Wheeler.*

An illustration by W. Heath Robinson from an edition of Charles Kingsley's *The Water Babies*.

A fake Great Auk made by Rowland Ward Ltd. and once the property of Captain Vivian Hewitt. *Photograph courtesy of David Wilson.*

Some twenty years ago very excellent imitations of the Auk's egg were manufactured in France; they were intended to fill up the place of the real egg in the cabinets of oologists; some of the specimens crossed the English Channel, and attempts were made to pass them off as genuine. I well remember the pleasure with which a communication was received from a leading naturalist and dealer, that he was at length in possession of some eggs of the Great Auk. On examining these eggs I was at once struck by their weight, absence of pores, and the extraordinary fact that they were all alike, mark for mark; on placing one of my own specimens before my correspondent, he saw at once that he had been gulled, and admitted that he had been cheated out of £18 for half a dozen specimens in plaster- of- paris; he, however, fell back on the doubtful consolation that he was not the only sufferer, for according to a police report of a charge of obtaining money under false pretences, a brother naturalist had been similarly cajoled.

Despite the popularity of Auk relics as symbols of wealth and power, the steady movement of specimens away from private ownership and into institutional collections resulted in fewer and fewer items coming onto the market. Around the year 1900, three or four hundred pounds represented the going rate for a stuffed example and an egg would have fetched only slightly less. By the outbreak of World War I, however, the golden age of Auk collecting was over. Changing fashions, together with the fact that so few specimens remained that could ever come again onto the market, spelled the beginning of the end and public interest in Auk sales dwindled. There was a brief resurgence of curiosity during the 1930's as the last of the old Victorian collections were broken up.

At just about this time the eccentric millionaire Captain Vivian Hewitt burst onto the Great Auk scene and during the next twenty years or so this strange and enigmatic character secured nearly every egg or bird that became available, acquiring four mounted specimens to go with his thirteen eggs. In 1936 he paid the Rowland Ward Company £700 for one of his mounted birds, a sum that in real terms probably represents a financial high point. Although Auks have subsequently sold for much higher figures, £700 represented a handsome amount of money in the 1930's, enough to buy three or four modestly sized houses. The £30,000 paid recently by the Glasgow Museum for one of the last privately owned Garefowls has nothing like the same buying power. When Hewitt's collection was finally sold during the early 1970's, its dispersal represented the breaking up of the last of the great private hoardings of Auk material.

Auk relics still change hands from time to time although such occurrences are now rare events. When a stuffed bird was bought at Sotheby's, London in 1971 on behalf of the Icelandic government it realised £9,000. This price was considered a world record for a stuffed bird and it earned the transaction a place in *The Guinness Book of Records*. The bird itself was figured on the book's cover during the following year. In Iceland, children were given a half day off school and flags were flown on all government buildings to celebrate the bird's return home.

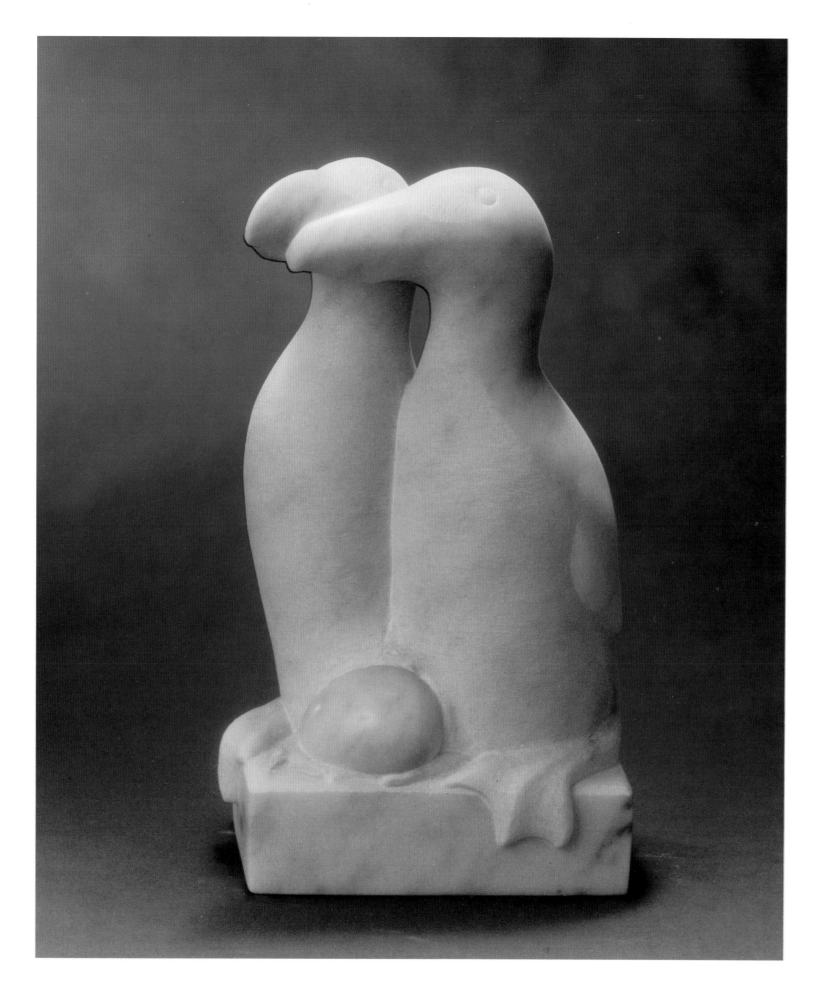

Fascination with the Great Auk is, in many respects, as strong as ever although it no longer takes the form of an obsession with the specimens. Vast numbers of papers - scientific and otherwise - have been written on the species and there is little sign of a let up in their production.

For the one hundred and fiftieth anniversary of the Great Auk's supposed date of extinction, the well known journal *New Scientist* carried an article and commissioned a picture to decorate its cover. During the previous year the name of the Great Auk was gratuitously inserted into the title of a book simply because of the appeal of its name. *Great Auk Islands* (by T. Birkhead, 1993) is certainly not an appropriate name for a book subtitled *A Field Biologist in the Arctic*, but it carries a magic that is difficult to resist and it is hard not to sympathise with the reasons for its choice. Great Auks still figure from time to time in more mainstream literature. In Patrick O'Brien's *The Surgeon's Mate* (1980) it has a typical, albeit brief, rôle. Other books have featured the Great Auk more centrally. In 1963 Allan Eckert published one of the peculiarities of Great Auk literature. Called *The Last Great Auk*, the book is written in the form of a novel that follows the fortunes of the last birds until their horrid end on the island of Eldey. Probably inspired by a rather better known novel of similar type, Fred Bosworth's *Last of the Curlews*, the action of the story hinges around the presumed migration routes that the birds may have taken.

During the early 1990's one man's obsession with the Great Auk caused him to investigate one of these supposed migration routes rather more personally. Dick Wheeler travelled from Funk Island to Cape Cod

(Facing page). Illustration by Hanife Hassan O'Keefe for the cover of *New Scientist* (May 28 1994). *Courtesy of Hanife Hassan O'Keefe and New Scientist.*

(Above). Dick Wheeler on Funk Island unfurling the flag of The International Explorers' Club before his epic adventure.

(Below, left). The mascot fastened to the prow of Dick Wheeler's canoe during his journey from Funk Island to Cape Cod.

(Below). The dust wrapper of Allan Eckert's novel *The Last Great Auk* (1963). The bird shown is Elliot's Auk.

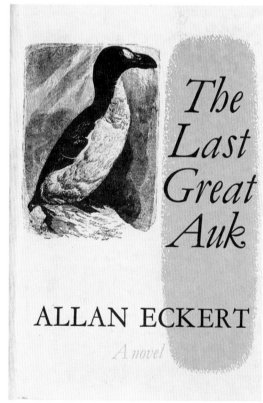

An eccentric arrangement of three Great Auk eggs, once the property of Herbert Massey.

The badge of *The Association of Women Clerks and Secretaries. Photograph courtesy of John Hammond and Peter Southon.*

A Great Auk still living in Italian territory! A tattoo created by Paolo Barbieri of Pavia for the arm of Auk expert Carlo Violani.

by canoe, an epic one man journey of hundreds of miles. Much of the voyage was filmed and the resulting film used to make a fascinating television program that was shown both in North America and in Britain.

No account of the Garefowl's peculiar relationship with the human race would be quite complete without mention of a British Trade Union- *The Association of Women Clerks and Secretaries.* AWCS for short, this union flourished between 1912 and 1941 and the badge its members wore featured a splendidly big-nosed Auk, of course.

Perhaps man's belated devotion to the cause of the Great Auk can be best summed up by an anecdote that concerns the obsessed egg collector, the Rev. F.C.R. Jourdain. Few things in life mattered more to Jourdain than the hoarding of preserved egg shells and he became such an influential figure in this field during the early years of the twentieth century that an egg collecting society was given his name, the now rather infamous *Jourdain Society.* Naturally, such a man was bound to acquire Great Auk eggs and during his lifetime he obtained two. His sister, an expert on English antique furniture, lived with the novelist Ivy Compton-Burnett and brother and sister saw little of one another; apparently, their dislike of each other was strong and mutual. On the day of the outbreak of World War II she was asked what she thought her brother might be doing at such a critical time. 'Burying his Great Auk eggs, I should imagine,' she is said to have contemptuously replied.

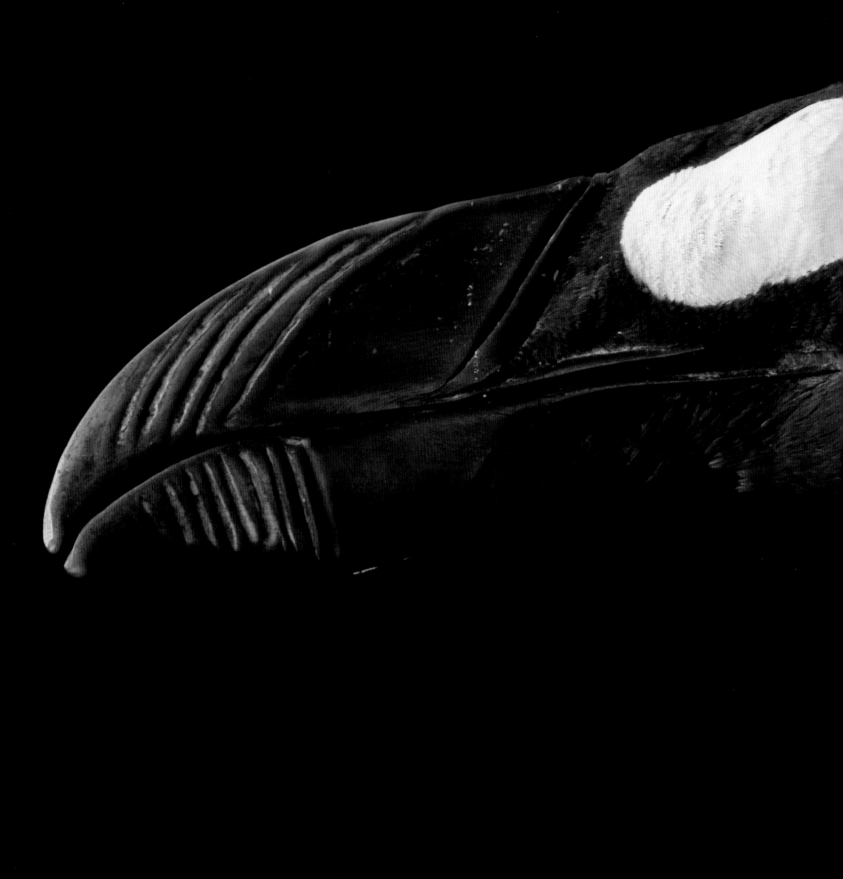

THE BIRDS

THE BIRDS

One death is a tragedy - a million deaths are a statistic.

Josef Stalin

A detailed list of 80 or so stuffed birds might seem excessive. The reason for it is simple. Each of these preserved Auks represents a little tragedy all of its own. They, along with the eggs and the bones, are our only tangible contact with the Great Auk, and each of their histories - together with the few stories we have of specific, individual birds - are all that prevent *Alca impennis* from being merely a statistic.

That stuffed Great Auks have fascinated men for more than 200 years is a surprising, but clearly demonstrable, fact. Lifetimes have been passed in gaining knowledge of them and tracing their whereabouts, fortunes have been spent in acquiring these precious relics, and enormous technical skills used to improve them or even fake them. All sorts of reasons can be advanced for this but none is quite satisfactory. After all, 80 specimens is quite a lot. Many seemingly equally intriguing birds that have become extinct in historical times have left far fewer specimens behind them. None, however, has attracted anything like the attention or the financial interest that the Great Auk has. An anonymous correspondent to a nineteenth century newspaper, writing of a specimen about to come under the auctioneer's hammer, described this attention perfectly:

> This relic of the sea is a stuffed Great Auk, one of the seventy-nine which still remain to stir spectacled old men to frenzy and inspire awe in the bosoms of those to whom the unattainable is the pinnacle of desire.

For convenience, the specimens are listed in alphabetical order according to their present whereabouts, first priority being given to the country in which they are to be found, second to the city or town. Also, they have been given numbers along with references to any numbers that previous list makers have used. To prevent, however, these particular Auks from being simply numbers on a page, they have all been named. In themselves these names may be fairly meaningless; certainly they aren't remotely objective. They are just names that it pleases me to call them, drawn variously from their histories or their locations.

Human beings have been making lists (both written ones and those simply remembered and passed down by word of mouth) of dead things since the dawn of history. It is one of our oldest, yet most profound instincts. This, then, is just another list.

Hoppa's Auk (bird no.12).

THE JOANNEUM AUK

The Joanneum Auk. *Courtesy of the Landesmuseum Joanneum.*

Bird no.1 (Grieve no. 28; Hahn no. 1) - adult in summer plumage
Origin: unknown, probably Iceland
Present location: Graz - Steiermärkisches Landesmuseum Joanneum, Raubergasse 10, A-8010 Graz, Austria.

Newton, A. 1870. On Existing Remains of the Gare-fowl. *Ibis,* p.257. Blasius, W. 1881-3. Ueber die letzen Vorkommnisse des Riesen-Alks. *Jahresberichte des Vereins für Naturwissenschaft zu Braunschweig für die Vereinsjahre 1881-2 und 1882-3,* p.114. Blasius, W. 1884. Zur Geschichte der Ueberreste von *Alca impennis. Journal für Ornithologie,* p.85. Grieve, S. 1885. *The Great Auk or Garefowl,* p.77 *and* appendix p.12. Salomonsen, F. 1944-5. Gejrfuglen et hundredaars minde. *Dyr i Natur og Museum,* p.106. Hahn, P. 1963. *Where is that Vanished Bird?* p.210.

There has been a Great Auk in the museum at Graz for a very long time, the gift - in 1839 - of a Herr Höpfner of Althofen. Like most surviving specimens this example probably originates from one of the first raids on Eldey. Although its early history is a little unclear, it is known that Höpfner obtained the bird in 1834. He got it, almost certainly, from the man who distributed so many preserved Auks to collectors and museums throughout Europe: Professor Reinhardt of the Royal Museum, Copenhagen. The close political and trading connections that existed between Denmark and Iceland made the Danish capital a great receiving house for goods coming to Europe from the large but remote Atlantic island. All but one of the many specimens that passed through Reinhardt's hands were Icelandic in origin and the Auk now in the Joanneum Museum was probably a bird he received from Iceland and then passed on almost immediately.

Its importance as a specimen has always been recognised in Graz. A letter written in December 1883 by Professor Augustus von Mojsisovics of Graz to the celebrated Auk scholar Wilhelm Blasius describes the bird as, 'the pride of the zoological collection.' Yet by September of the following year Blasius himself felt obliged to report the specimen's deterioration, detailing moth attack on the left white eye spot, on the right side, on the wing and under the jaw. Despite this and despite the ravages of two world wars and the passing of more than a century, the Joanneum Auk still exists.

A peculiarity noted by Professor Blasius (1884) was the presence of two distinct spots on the breast, but these were probably just dirt or grease marks as Dr. Adlbauer, of the Joanneum Museum, was recently (*pers. comm.* October, 1995) unable to find any trace of them.

THE VIENNESE AUK

Bird no.2 (Grieve no. 75; Hahn no. 2) - adult in summer plumage
Origin: unknown, probably Iceland
Present location: Vienna - Naturhistorisches Museum Wien, Burgring 7, A-1014, Wien, Austria (catalogue no. 1831/IV/9).

Pässler, W. 1860. Die Eier der *Alca impennis* in Deutschen Sammlungen. *Journal für Ornithologie*, p.60. Roberts, A. 1861. Skins and Eggs of the Great Auk. *Zoologist*, p.7353. Preyer, W. 1862. Der Brillenalk in Europäischen Sammlungen. *Journal für Ornithologie*, p.78. Fatio, V. 1868. Liste des divers représentants de l'*Alca impennis* en Europe. *Bullétin de la Société Ornithologique Suisse*, tome II, pt.1, p.82. Pelzeln, A. 1877. *Mittheilungen des Ornithologischen Vereins in Wien*, p.4. Blasius, W. 1884. Zur Geschichte der Ueberreste von *Alca impennis*. *Journal für Ornithologie*, pp.112-3. Grieve, S. 1885. *The Great Auk or Garefowl*, p.77 *and* appendix p.24. Sassi, M. 1947. Über einege ausgestorbene Arten in der Vogelsammlung des Naturhistorischen Museums in Wien. *Umwelt- Zeitschrift der Biologischen Station Wilhelminenberg*, heft 8, p.334. Hahn, P. 1963. *Where is that Vanished Bird?* p.210.

Throughout the nineteenth century the trade in Great Auks was a lucrative but esoteric one. It was also something of a closed shop and in the histories of the surviving specimens the names of several dealers occur over and over again. Most of these merchants are rather shadowy figures about whom little is now remembered, although they seem to have enjoyed some celebrity in their day. Two of the most familiar names are connected with the Great Auk now in Vienna.

One is Mechlenburg, usually referred to in the older records as 'The Apothecary of Flensburg.' Whether he was more of a dealer or more of a collector is uncertain but no less than eight Great Auks reputedly passed through his hands along with several eggs.

Rather more important in the history of Great Auk trading were the family Frank, but these people are also enigmatic figures about whom virtually nothing is recorded. Three members of this family made up a trading dynasty that lasted for almost the whole of the nineteenth century and spanned the continent of Europe almost from east to west. The first of the three operated out of Leipzig but his son moved to Amsterdam and continued the family business there. The youngest, the grandson of Leipzig Frank, was London based and occupied premises at the bottom of Haverstock Hill. Between them they handled many of the Great Auks that survive as specimens today but despite their undoubted importance as traders nothing is remembered concerning their personal lives.

During 1831, before he moved from Leipzig to Amsterdam, the middle Frank bought from Mechlenburg one of the apothecary's Great Auks. Very soon after acquiring it he sold it for 100 reichsthalers (or, as Frank himself later preferred to list it, 150 Dutch florins) to the Imperial Royal Zoological Court Cabinet at Vienna from which it was eventually transferred to the Naturhistorisches Museum.

The Viennese Auk. Photo by A. Schumacher. *Courtesy of the Naturhistorisches Museum, Vienna.*

How long before 1831 the bird was procured cannot be said but it is virtually certain that this was an individual from Iceland. All of Mechlenburg's specimens supposedly came from there.

THE BRUSSELS AUK

Bird no.3 (Grieve no. 15; Hahn no. 6) - adult in summer plumage
Origin: probably Eldey (perhaps June, 1844)

The Brussels Auk- perhaps one of the birds of 1844.

Present location: Brussels - Institut Royal des Sciences Naturelles de Belgique, Rue Vautier 29, B-1040 Bruxelles, Belgium.

Dubois, Ch.F. 1867. Note sur le *Plautus impennis. Archives Cosmologiques. Revue des Sciences Naturelles (Bruxelles)*, no.2, p.33. Newton, A. 1870. On Existing Remains of the Gare-fowl. *Ibis*, p.259. Selys-Longchamps, E. 1870. On various Birds observed in Italian Museums. *Ibis*, p.450. Blasius, W. 1884. Zur Geschichte der Ueberreste von *Alca impennis. Journal für Ornithologie*, p.75. Grieve, S. 1885. *The Great Auk or Garefowl*, p.77 *and* appendix p.8. Duchaussoy, H. 1897-8. Le Grand Pingouin du Musée d'Histoire naturelle d'Amiens. *Mémoires de la Société Linnéenne du Nord de la France*, p.111. Hahn, P. 1963. *Where is that Vanished Bird?* p.211.

Despite its being mentioned in more than half a dozen books and papers, the published record of this specimen adds up to just one small piece of information: the bird was obtained for the Brussels Museum during the administration of Viscount Bernard de Bus de Ghisignies, a museum Director who died in 1874.

Fortunately, among the Alfred Newton manuscripts (at the Cambridge University Library) there are several scribbled notes concerning it. These reveal that in 1845 this skin was one of two in the hands of a dealer known only as 'Israel of Copenhagen' and that Israel sold it during that year to the Hamburg dealer Lintz who disposed of it to G.A. Frank of Amsterdam. De Bus wrote himself to the English collector George Dawson Rowley explaining that he'd bought the specimen from Frank in 1847 for 100 florins.

There is little doubt then that the Brussels Auk was in Copenhagen during 1845. What makes this factor especially interesting is the implication it carries. Perhaps this Brussels bird is one of the two individuals- so often considered the last of their kind- killed on Eldey during June of the previous year. The whereabouts of the skins of these two birds has long been a mystery and nineteenth century Auk researchers were never able to satisfactorily determine what became of them.

One of the very few things known about Israel is the fact that he had good trading links with Iceland; the birds of 1844 are, of course, known to have been sent from Reykjavik to Denmark. Where else could the pair of birds that Israel had in stock have come from? There is no conclusive proof that his birds came from the last raid on Eldey; it is simply a likelihood. His second specimen is the example now in Los Angeles.

THE CLUNGUNFORD AUK

Bird no.4 (Grieve no. 18; Hahn- lost bird no.1) - adult in summer plumage
Origin: unknown but probably Iceland
Present location: Birmingham - Museum and Art Gallery, Chamberlain Square, Birmingham B3 3DH.

Hellman, A. 1860. Notizen über *Alca impennis. Journal für Ornithologie*, p.206. Preyer, W. 1862. Der Brillenalk in Europäischen Sammlungen. *Journal für*

Ornithologie, p.78. Newton, A. 1870. On Existing Remains of the Gare-fowl. *Ibis*, p.259. Blasius, W. 1884. Zur Geschichte der Ueberreste von *Alca impennis*. *Journal für Ornithologie*, p.76. Grieve, S. 1885. *The Great Auk or Garefowl*, p.87, p.91 *and* appendix p.9. Watkins, M. 1890 (May). The collection of birds at Clungunford House *and* Pilley, J. 1890 (May). The collection of birds at Clungunford House. *Transactions of the Woolhope Naturalists' Field Club*, pp.31-2 *and* p.35. Grieve, S. 1896-7. Supplementary note on the Great Auk. *Transactions of the Edinburgh Field Naturalists' and Microscopical Society*, p.247. Rothschild, W. 1928. Exhibition of a mounted Great Auk and an egg. *Bulletin of the British Ornithologists' Club*, p.9. Hahn, P. 1963. *Where is that Vanished Bird?* p.233. Spink and Son Ltd. (c1970). *Romantic Histories of Some Extinct Birds and Their Eggs*, p.11 *and* pl.3. Bourne, W. 1993. The story of the Great Auk. *Archives of Natural History*, pp.274-5.

John Rocke (1817-81).

During 1860 Mr John Rocke, owner of Clungunford House, Shropshire, England purchased a Great Auk from the celebrated ornithologist John Gould. He added this specimen to the enormous collection of stuffed birds that he kept at his home and it stayed in the possession of his family for many years. Eventually, long after Rocke's death, it was acquired by the Rowland Ward Company of Piccadilly, the taxidermy firm that handled the sale of many Great Auks and their eggs. In 1928 the company allowed Lord Rothschild to exhibit the Clungunford Auk at a meeting of *The British Ornithologists' Club* but it wasn't until 1936 that a buyer could be found.

On May 8th of that year Captain Vivian Hewitt bought it for the sum of £700 and added it to a pair of Auks he'd acquired two years previously. Twelve years later he was to get yet another specimen, bringing his

The Clungunford Auk (circa 1880), as it appeared when it was the property of John Rocke.

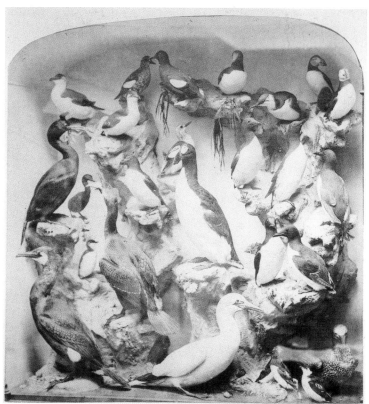

The Clungunford Auk today. *Courtesy of the Museum and Art Gallery, Birmingham.*

grand total to four. The Clungunford Auk, along with the others, stayed with Captain Hewitt for the rest of his life but after his death all were sold by the London art dealers Spink and Son Ltd. Unfortunately, the provenances of the four specimens were confused and the catalogue that Spink produced to promote sales is misleading, the illustrations not being correctly keyed to the specimen descriptions.

The Clungunford Auk, minus the correct data that might have accompanied it, was purchased from Spink for £9,000 by the Birmingham Museum and Art Gallery during February, 1971; 25 years were to pass before the identity of the museum's bird could be established. Due to the existence of a series of nineteenth century photographs of the Clungunford House collection, it has proved possible to match it with the one that John Rocke bought from Gould.

A certain amount of the specimen's early history is recorded. It is listed in the hands of G.A. Frank (the son of Leipzig Frank and father of

Frank of Haverstock Hill, London) of Amsterdam in 1835. During this year he sold it, together with another specimen, to the museum of Mainz but he got it back again 25 years later in exchange for the skin of an Indian Tapir; its companion at Mainz remained in the town museum but was destroyed during World War II. Frank sold his retrieved Great Auk almost immediately to John Gould who, in turn, quickly disposed of it to Rocke.

Although nothing is definitely known about this bird before 1835, its presence in the Amsterdam dealer's hands during that year makes it highly likely that it came from Iceland; indeed its companion in Mainz carried a label to that effect. Probably, this is one of the individuals taken during the first raids on Eldey.

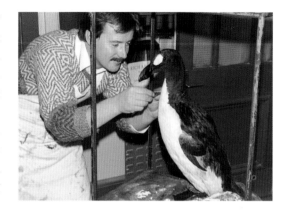

Barry Williams, taxidermist, cleaning and restoring the Clungunford Auk in 1991.

THE CAMBRIDGE AUK

Bird no.5 (Grieve no. 16; Hahn no. 8) - adult in summer plumage
Origin: unknown
Present location: Cambridge - University Museum of Zoology, Downing St., Cambridge CB2 3EJ (catalogue no. 16/Alc/9/a/1).

Jenyns, L. 1836. *Catalogue of the Collection of the Museum of the Cambridge Philosophical Society*, p.16. Champley, R. 1864. The Great Auk. *Annals and Magazine of Natural History*, p.235. Fatio, V. 1868. Liste des divers représentants de l'*Alca impennis* en Europe. *Bullétin de la Société Ornithologique Suisse*, tome II, pt.1, p.82. Blasius, W. 1884. Zur Geschichte der Ueberreste von *Alca impennis*. *Journal für Ornithologie*, p.75 and pp.127-8. Grieve, S. 1885. *The Great Auk or Garefowl*, p.77 and appendix p.8. Gadow, H. 1910. The Ornithological Collections of the University of Cambridge. *Ibis*, p.50. Hachisuka, M. 1927. *A Handbook of the Birds of Iceland*, pl.VII. Hahn, P. 1963. *Where is that Vanished Bird?* p.211. D'Errico, F. 1994. Birds of the Grotte Cosquer. *Antiquity*, 68, p.41, fig.2. D'Errico, F. 1994 (May). Birds of Cosquer Cave. *Rock Art Research (Journal of the Australian Rock Art Research Association.)*, p.46, fig.2.

The Cambridge Auk photographed by J.E. Harting in 1876. *Courtesy of the University Library, Cambridge.*

Surprisingly little is known about the large and splendid Great Auk belonging to the University Zoological Museum at Cambridge and nothing at all is known of its origin. It was first mentioned by Jenyns in 1836 in a catalogue of the collection of the Cambridge Philosophical Society; according to Newton's manuscript notes (kept at the University Library) it was acquired from the London taxidermist Leadbeater at an unknown time prior to that date. Although originally housed in the Society's Museum, it was soon transferred to the Zoological Museum of the University.

One curious incident concerning this stuffed bird occurred during the 1860's. The Newton brothers, Alfred and Edward, with their special interest in Auk remains, supervised the removal of bones from the specimen's extremities; these they needed to complete a skeleton they had. This skeleton - still at Cambridge - is illustrated in Richard Owen's paper (1865) on Great Auk bones. Before the Newtons' delicate operation, the missing bones were supplied on loan by John Hancock, the well known Newcastle taxidermist, who'd removed them from his own stuffed specimen (now in the Hancock Museum, Newcastle-upon-Tyne), and it is in this state that the skeleton was figured by E.A. Smith, Owen's lithographer. After these bones were returned to Hancock, the Newtons made good their (temporarily) incomplete skeleton by carefully withdrawing missing parts from the stuffed bird in their care. They then commissioned a Soho taxidermist, John Burton of Wardour Street, to remount the skin.

Despite this tampering, the bird is still in good condition. It shows, perhaps more clearly than any other specimen, the grey fringe on the sides of the body that Rothschild (1907) believed to be the mark of the female.

VIVIAN HEWITT'S POLTALLOCH AUK

Bird no.6 (Grieve no. 62; Hahn no. 3) - adult in summer plumage, said to be female
Origin: unknown
Present location: Cardiff - National Museum of Wales, Cathays Park, Cardiff, Wales CF1 3NP.

Grieve, S. 1885. *The Great Auk or Garefowl*, p.78, p.91, p.103, appendix p.21, *and* p.32. Anon. 1905 (Sept. 9th). The Great Auk or Garefowl. *Oban Times*. Anon. (c.1905). *Catalogue of the Ornithological Collection at Poltalloch*, pp.14-7. Hahn, P. 1963. *Where is that Vanished Bird?* p.210. Spink and Son Ltd. (c1970). *Romantic Histories of Some Extinct Birds and Their Eggs*, p.11 *and* pl.2. Bateman, J. 1971. Rare acquisitions by the Dept. of Zoology. *Amgueddfa*, 9, pp.4-5. Hywel,W. 1973. *Modest Millionaire*, pp.161-2. Bourne, W. 1993. The story of the Great Auk. *Archives of Natural History*, p.275.

Early in July of 1948, Peter Adolphe - soon to invent the table football game known as *Subbuteo* - travelled from his home in Tunbridge Wells, Kent to the wilds of Scotland on behalf of the obsessed Auk collector Captain Vivian Hewitt. Adolphe's mission was to buy the Great Auk - together with an egg - that Hewitt knew existed in a remote Highland castle. Its exact location was Poltalloch House, Poltalloch, Argyllshire and its owner was a Colonel Malcolm whose family had possessed the bird for several generations. Hewitt already owned three stuffed examples and his egg holding ran into double figures- but he intended to do even better. Adolphe (*in* Hywel, 1973) recounts his adventure and the exact nature of his brief:

> He[Vivian Hewitt] added that he must have these specimens, left it to me to argue the price and handed me a signed blank cheque! A car and driver were placed at my disposal and we made the long journey to the Highland castle ... I was ushered into the main dining room ... and from a great length away strode towards me the Laird with kilt swirling, and pointing to his right. There, in a large round case, was The Bird and, beside it, The Egg. I was informed that I could not pay the price, he subsequently adding the sum required - some 75% less than I had anticipated would be asked - so I offered a low three figure sum [in fact £500], which was accepted! We packed the egg... in a wooden box, and bundled

... the stuffed bird ... on to the back seat of the car. I shall never forget the look on the chambermaid's face as I left my hotel room ... carrying this strange and very valuable bird.

This was to be the last of Captain Hewitt's stuffed Great Auks and he kept it- along with the other three - until his death in 1965. Eventually, all four were offered for sale by Spink and Son Ltd. Unlike two of the other birds that Spink catalogued, this one was correctly listed. It was sold to the National Museum of Wales for £11,000, along with an egg and also the giant egg of an *Aepyornis*- extinct Elephant Bird of Madagascar. Although it is clearly the Poltalloch specimen that is now in Cardiff, it has been assumed that this is not the case. Bourne (1993) believed that Cardiff had acquired another of Hewitt's specimens (bird no.72) but this is incorrect.

The bird was delivered to Cardiff in a box bearing the labels of C.H. Bisshopp, taxidermist of The Esplanade, Oban. Probably, this was the box in which the specimen was originally taken to Captain Hewitt, for this

The Poltalloch Auk. *Courtesy of the National Museum of Wales.*

Great Auk was restored by Bisshopp after being damaged by fire at Poltalloch in November 1904. It is also quite likely that - in true *Citizen Kane* style - the eccentric Captain never bothered to take it from its box and that it was never properly removed until it reached Spink's premises. One of his pleasures, apparently, was to partly unpack a container, inspect its contents and then re-pack it all as carefully as he could.

There is little early history for the specimen. It belonged to the Malcolm family, who claimed it was female, for more than a century and they bought it, around 1840, from Benjamin Leadbeater, founder of a firm of natural history dealers and taxidermists that operated for many years out of Golden Square in London's Soho. Symington Grieve (1885) received direct information on the purchase from John Malcolm, the original buyer, who informed him:

> If I remember *right*, the specimen ... did not cost me more than two or three pounds, but ... I cannot be sure of this, and have no memoranda.

A written reminder of such a painful occurrence would, of course, be something that no Scotsman would wish to keep. Malcolm added to his reminiscence the opinion that the bird was brought home from, 'one of the Arctic expeditions,' but such information is·fairly meaningless.

THE EDINBURGH AUK

Bird no.7 (Grieve no. 35; Parkin no. 3; Hahn no. 10) - adult in summer plumage
Origin: unknown
Present location: Edinburgh - National Museum of Scotland, Chambers St., Edinburgh, Scotland.

The Edinburgh Auk as it appeared 100 years ago.

Newton, A. 1861. Abstract of Mr. John Wolley's Researches in Iceland. *Ibis*, p.398. Fatio, V. 1868. Liste des divers représentants de l'*Alca impennis* en Europe. *Bullétin de la Société Ornithologique Suisse*, tome II, pt.1, p.83. Blasius, W. 1884. Zur Geschichte der Ueberreste von *Alca impennis*. *Journal für Ornithologie*, p.90. Grieve, S. 1885. *The Great Auk or Garefowl*, p.78 and appendix p.14. Grieve, S. 1888. Recent information about the Great Auk. *Transactions of the Edinburgh Field Naturalists' and Microscopical Society*, pp.106-7. Bidwell, E. 1894-5. [Minutes of meeting] *Bulletin of the British Ornithologists' Club*, p.32. Traquair, R. 1895. Remains of the Great Auk in Edinburgh Museum. *Annals of Scottish Natural History*, pp.196-7. Sclater, P. 1895. [Notes on meeting of BOC] *Ibis*, p.381. Harting, J. (ed.) 1895. Sale of Great Auk and Egg. *Zoologist*, p.193. Grieve, S. 1896-8. Supplementary note on the Great Auk *and* Additional notes on the Great Auk. *Transactions of the Edinburgh Field Naturalists' and Microscopical Society*, p.248 and p.328. Parkin, T. 1911. The Great Auk. *Hastings and East Sussex Naturalist*, vol.1, pt.6 (extra paper), pp.16-7 *and* pl.V. Allingham, E. 1924. *A Romance of the Rostrum*, p.166 *and* facing plate. Hahn, P. 1963. *Where is that Vanished Bird?* p.211.

The early history of this bird, although much documented, is full of rumour, hearsay, inference and complication; certainly, the specimen changed hands a remarkable number of times in a comparatively short period of time. Its origin can be trace back only to a London afternoon in Covent Garden during September of 1844.

A man described as 'a stranger' arrived at the shop of Abraham Dee Bartlett - near to Seven Dials - carrying the preserved skin of a bird. This, he said, was an example of the Great Northern Diver (*Gavia immer*) and he wanted to sell it. Bartlett, an experienced taxidermist who rose to the position of Superintendent of the London Zoological Gardens, immediately realised that the gentleman standing in his shop was entirely mistaken and that the bird under his arm was nothing less than a Great Auk. The wily Bartlett quickly struck a deal and bought the bird at a bargain price. As the 'stranger' left the shop Bartlett, suddenly struck by the idea of obtaining some information concerning the bird's origin, called after him and rushed to the door. By the time he reached it and looked out into the street, his visitor had vanished.

This specimen's story begins, then, in a taxidermist's shop and continues like a *Who's Who* of Victorian taxidermy. From Bartlett, the bird went straight on - now, of course, correctly labelled - to another London taxidermist, the well known John Leadbeater of Golden Square who, in turn, sold it to the Shrewsbury taxidermy firm of Henry Shaw. Temporarily, it now escaped from the clutches of bird stuffers for Shaw sold it to a collector about whom little is known apart from the fact that he lived at Tickleton Court near Church Stretton, Shropshire and answered to the remarkably Dickensian name of Pinches. Mr. Pinches, sadly, had only a short time to enjoy what was, presumably, the pride of his collection. He died in 1849, at the age of 47, the bird passing to Miss Pinches, his unmarried sister. Miss Pinches was not to remain Miss Pinches for very much longer, however, and soon she became the second wife of the Reverend Joseph Buddicom, Rector of Smethcote, Shrewsbury, her Great Auk becoming her husband's property on marriage. Reverend Buddicom quickly capitalised on his new asset and returned the bird to the hands of the taxidermists by disposing of it to James Gardiner (another well known London dealer) who promptly sold it to the York stuffer David Graham, a man whose trade label - showing a Little Spotted Kiwi (*Apteryx oweni*) - clearly reveals him as something of a specialist in the rare and the curious. Regrettably, perhaps, he invented a provenance for his bird, one aimed - doubtless - at enhancing its desirability; indeed Mr. Graham developed a general reputation for being sparing with the truth. Without (as far as is known) the slightest scrap of evidence, Graham boldly announced that this particular Auk came originally from the Orkneys. Whether this British provenance actually helped him make a sale is unrecorded but he quickly found a wealthy customer. Sir William Milner of Nunappleton, Tadcaster, Yorkshire became the bird's new owner in 1856.

At some time during Sir William's ownership, the specimen was transported by train from Leeds to London and during transit its case was badly damaged. This misfortune prompted Sir William to put his bird into the hands of the highly skilled Durham taxidermist John Cullingford for a complete overhaul and re-stuffing. This instruction was quite brilliantly carried out and Cullingford, who obviously inspected every inch of the skin, declared it:

(Above, top). Abraham Dee Bartlett (1812-97).

(Above). William Pinches (1802-49).

A genuine specimen throughout, with not a single false feather in it.

On the death of Sir William the Auk passed to his son, Frederick Milner, Bart., M.P. who loaned it to the Museum of the Philosophical Society of Leeds. After several years it went back to Covent Garden where it was offered for sale at Stevens' Rooms in King Street. Here, on April 23rd 1895, the bidding reached £350, not quite enough to match a reserve price set at 360 guineas. The Great Auk was, therefore, 'bought in.' Almost immediately afterwards, however, the owner agreed to treat with the unsuccessful bidder - the Museum of Science and Art, Edinburgh - and a sale was agreed at the price reached during the auction. The specimen is now one of the great ornithological treasures of the Royal Scottish Museum.

THE GLASGOW AUK

Bird no.8 (Grieve no. 23; Hahn no. 9) - Adult in summer plumage
Origin: unknown, probably Eldey
Present location: Glasgow - Art Gallery and Museum, Kelvingrove, Glasgow, Scotland G3 8AG.

Champley, R. 1864. The Great Auk. *Annals and Magazine of Natural History*, p.235. Fatio, V. 1868. Liste des divers représentants de l'*Alca impennis* en Europe. *Bullétin de la Société Ornithologique Suisse*, tome II, pt.1, p.82. Gurney, J. 1869. Notes on the Great Auk. *Zoologist*, pp.1639-40. Blasius, W. 1884. Zur Geschichte der Ueberreste von *Alca impennis. Journal für Ornithologie*, p.79. Grieve, S. 1885. *The Great Auk or Garefowl*, p.22, p.78, p.91 *and* appendix p.10. Hahn, P. 1963. *Where is that Vanished Bird?* p.211. Bourne, W. 1993. The story of the Great Auk. *Archives of Natural History*, p.275.

This bird is first recorded in the possession of Friedrich Schultz of Dresden. Its earlier history is uncertain, except that Schultz had acquired it in Hamburg during 1835. It is likely, therefore, that it came from Eldey.

Herr Schultz sold his bird - perhaps rather surprisingly - to a resident of Doncaster in Yorkshire, a Mr. Hugh Reid of 8 Spring Gardens. Reid was something of a dealer in birds' eggs and, apparently, a considerable amount of long range business was conducted between the two men. In 1856 Reid described Schultz to the ornithological collector Canon H.B. Tristram:

I have had several hundred pounds worth of him. He is a fair dealing man.

Hugh Reid didn't keep Schultz's Great Auk for very long. According to W. Proctor, writing in 1861 from the Durham University Museum (*see* Grieve, 1885, p.22), the Reverend T. Gisborne purchased it in Doncaster for £7 or £8 around 1834 on behalf of the University. Proctor's memory was, doubtless, slightly at fault and the true date was rather later. With an especial interest in *Alca impennis* Proctor actually visited Iceland during 1837 and had the misfortune to be marooned - weather bound - on Grimsey Island for several weeks.

before stuffing.   after restuffing

An incredible transformation. Two photographs of the Glasgow Auk, one taken before the second re-stuffing and one taken soon after (circa 1900).

In accordance with his interest, and, finding the University's skin in fairly bad order, he decided to re-stuff it. For a long time afterwards it sat, in a rather forlorn and ungainly position. Procter experienced considerable trouble with the skin during the re-stuffing and only with difficulty did he prevent the feathers from slipping; he believed the bird must have been close to rotting when originally skinned. Eventually - around the turn of the nineteenth century - it was felt that a better job could be made and the Auk was taken to Rowland Ward of Piccadilly for a second re-stuffing, a reconstruction that was enormously successful.

By 1977 the authorities at the university had decided that their Great Auk was superfluous to requirements and, perhaps tempted by the well publicised price of £9,000 realised for a specimen just a few years previously, took the decision to sell. At a London auction it was knocked down to a Mr. Pilkington of the 'Dilemma' gift shop, Knightsbridge for a mere £4,200. Several years later the new owner loaned his bird to the Glasgow Museum and then, during 1993, the museum was given the opportunity of buying it. It did so for the sum of £30,000.

TUNSTALL'S IMMATURE AUK

Bird no.9 (Grieve no. 53; Hahn no.15)- immature acquiring winter plumage
Origin: unknown
Present location: Newcastle-upon-Tyne - The Hancock Museum, Barras Bridge, Claremont Rd., Newcastle-upon-Tyne NE2 4PT.

Wallis, J. 1769. *The Natural History and Antiquities of Northumberland and Durham*, vol.1, p.340. Latham, J. 1781-5. *General Synopsis of Birds*, vol.3, p.312. Donovan, E. 1794-1819. *The Natural History of British Birds*, vol.10, page following pl.243. Fox, G.T. 1827. *Synopsis of Newcastle Museum*, p.211. Fatio, V. 1868. Liste des divers représentants de l'*Alca impennis* en Europe. *Bullétin de la Société Ornithologique Suisse*, tome II, pt.1, p.82. Gurney, J. 1869. Notes on the Great Auk. *Zoologist*, pp.1839-40. Smith, J. 1879. Notice on the remains if the Great Auk...in Caithness. *Proceedings of the Scottish Society of Antiquaries*, p.94. Blasius, W. 1884. Zur Geschichte der Ueberreste von *Alca impennis*. *Journal für Ornithologie*, pp.98-100. Grieve, S. 1885. *The Great Auk or Garefowl*, p.73, p.78 *and* appendix pp.17-18. Grieve, S. 1896-7. Supplementary note on the Great Auk. *Transactions of the Edinburgh Field Naturalists' and Microscopical Society*, p.271 *and* pl.1. Howse, R. 1899. Index-Catalogue of the Birds in the Hancock Collection. *Natural History Transactions of Northumberland, Durham and Newcastle-upon-Tyne*, p.123 *and* p.138. Blasius, W. (*in* Naumann) 1903. *Naturgeschichte der Vögel Mitteleuropas*, bd.12, p.179, pl.17a *and* pl.17b. Hahn, P. 1963. *Where is that Vanished Bird?* p.216-8. Bourne, W. 1993. The story of the Great Auk. *Archives of Natural History*, p.260.

> In Mr. Tunstall's Museum is one of these, with only two or three furrows on the bill, and the oval space between the bill and the eye speckled black and white. This is probably a young bird.
>
> J. Latham (1785)

In the Hancock Museum at Newcastle-upon-Tyne are two preserved Great Auks. Only the collections of the Natural History Museum, London and the American Museum of Natural History have more. It could be argued, however, that Newcastle has the better selection, for one of its examples appears to be immature and in terms of development may be the youngest of all Garefowl specimens. In terms of historic age it may also be the oldest as its provenance takes it back well into the eighteenth century. It has other peculiarities. Alfred Newton believed that this bird and the specimen now in Strasbourg (also an example with an eighteenth century provenance) both came from Newfoundland and are the only two New World Auks in existence. This Auk's first known owner, Marmaduke Tunstall, was said by Newton to have had connections with Newfoundland but there is no real evidence pointing to an American beginning. Bourne (1993 *citing* Wallis, 1769) makes the point that Tunstall's bird may even be English in origin, coming from Farne Island about 1750. Possibly it is the individual that John Bacon kept as a pet. This bird is said to have endearingly followed its owner in the hope of receiving food. J.H. Gurney (1869) supported the idea of an English origin citing Latham (1785) who suggested

Tunstall's Auk drawn by John Hancock immediately before he re-stuffed it. *Courtesy of the Hancock Museum and the Natural History Society of Northumbria.*

(Facing page, top left). Tunstall's Auk (c.1890).

(Facing page, top right). Tunstall's Auk in pen and watercolour by Thomas Bewick (c.1800). *Courtesy of the Natural History Society of Northumbria.*

that young individuals used not infrequently to be blown ashore. This notion of an English pedigree is no better established than that of a Newfoundland one, however, for there is no evidence to suggest Tunstall acquired his bird in the north-east. He seems to have owned it during the years in which he lived in London and took it with him when he moved north in 1784.

After Tunstall's death- in 1790- his museum went to a George Allan of Darlington and in 1822 the whole collection was purchased by G.T. Fox for Newcastle's Literary and Philosophical Society. Later the Auk passed to the Hancock Museum. During March 1863 Tunstall's Auk (originally stuffed by one John Goundry) was re-mounted by John Hancock who extracted bones from the extremities while re-shaping the skin.

HANCOCK'S AUK

Bird no.10 (Grieve no. 54; Hahn no. 16) - adult in summer plumage
Origin: unknown but almost certainly Iceland
Present location: Newcastle-upon-Tyne - The Hancock Museum, Barras Bridge, Claremont Rd., Newcastle-upon-Tyne NE2 4PT.

Marmaduke Tunstall (1743-90).

Kjærbölling, N. 1851-6. *Ornithologia Danica- Danmark's Fugle*, p.415. Owen, R. 1865. Description of the skeleton of the Great Auk. *Transactions of the*

(Facing page, top left). Richard Owen with his granddaughter (circa 1870).

(Facing page, top right). John Hancock photographed by John Worsnop (circa 1880).

(Facing page, bottom). Lithograph by E.A. Smith from *Transactions of the Zoological Society of London* (1865) showing Cambridge University's Great Auk skeleton. The bones of the extremities are those removed from Hancock's stuffed bird.

Zoological Society of London, pp.317-8. Newton, A. 1861. Abstract of Mr. J. Wolley's Researches in Iceland. *Ibis*, pp.392-3. Newton, A. 1863. Remarks on the Exhibition of a Natural Mummy of *Alca impennis*. *Proceedings of the Zoological Society*, p.438. Newton, A. 1865. [Recent ornithological publications] *Ibis*, p.330. Newton, A. 1870. On Existing Remains of the Gare-fowl. *Ibis*, p.260. Blasius, W. 1884. Zur Geschichte der Ueberreste von *Alca impennis*. *Journal für Ornithologie*, pp.100-1. Grieve, S. 1885. *The Great Auk or Garefowl*, p.78 *and* appendix pp.18-9. Grieve, S. 1888 *and* 1896-7. Recent information about the Great Auk *and* Supplementary note on the Great Auk. *Transactions of the Edinburgh Field Naturalists' and Microscopical Society*, p.115 (1888) *and* p.271 *and* pl.5. Howse, R. 1899. Index-Catalogue of the Birds in the Hancock Collection. *Natural History Transactions of Northumberland, Durham and Newcastle-upon-Tyne*, p.133-4. Blasius, W. (*in* Naumann) 1903. *Naturgeschichte der Vögel Mitteleuropas*, bd.12, pl.17a. Bolam, G. 1912. *Birds of Northumberland and the Eastern Borders*, p.660. Hahn, P. 1963. *Where is that Vanished Bird?* p.218.

The apothecary Mechlenburg of Flensburg possesses a pair of birds which were killed in 1829 on the Geirfuglasker where they courageously defended their egg.

Nils Kjærbölling (1856)

It is possible, but far from certain, that this specimen- the second of those now in Newcastle- is one of the two birds referred to by Kjærbölling. The statements of the Apothecary Mechlenburg, only ever reported at second hand, cannot be relied upon, however. Perhaps he did possess two Great Auks killed at this time, perhaps he didn't. Even if he did, there is no guarantee that one of them ended up in the north of England, although several nineteenth century commentators were especially keen to suggest it. Mechlenburg himself appears not to have been one of them. He is said to have claimed - although to whom seems uncertain - that this particular bird came, together with another and two eggs, from an island off the north-east coast of Iceland just a year or two before 1844. Such a locality does not agree with any known locality for the species and the approximate date cannot be matched against the researches in Iceland of Wolley, Newton or Preyer. Why Mechlenburg should bother telling a lie (assuming he said anything at all), is a matter that it is now far too late to investigate. Perhaps he was simply mistaken or misled.

For reasons that are no longer obvious, attempting to deduce the origin of this Newcastle bird proved an irresistible temptation to Garefowl researchers and most of them expressed an opinion. Robert Champley, for instance, chose to believe that the bird was the pair to a specimen he acquired from Mechlenburg but (the origin of his own bird being itself quite uncertain) this belief has little value. Newton and Blasius both inclined to the opinion that the bird came from Eldey in 1834 but neither had solid evidence to support this idea.

Whatever the truth, it is certain that Mechlenburg sold, through the agency of a Mr. John Sewell, a stuffed Auk (one of eight said to have passed through his hands) to the Newcastle taxidermist John Hancock in April 1844. Hancock eventually presented his prize specimen, along with much

Hancock's Auk (circa 1890).

else, to the city of Newcastle but while the bird was still his property he re-stuffed it, giving it an attitude of alarm and placing an egg beneath it. According to Robert Champley, writing in 1885 to Symington Grieve:

> Mr. Hancock is the best stuffer in the world; no-one else has studied nature closer. All the birds [Great Auks] he has seen have the neck too stiff. His own specimen has the neck *pouched* so to speak. I think he is correct. He has offered to re-stuff my own bird [no.68], but great care is necessary to soften the skin.

Champley, such a dedicated collector, eventually declined Hancock's offer, doubtless fearing the risks involved in such a delicate process.

The attention Hancock paid to the throat of his bird is typical of him. He showed enormous interest in the workings of birds' throats and made a special study of this subject. Only a year or two ago a fascinating dis-covery was made among the many items he presented to the town of Newcastle. Preserved in spirits were anatomical preparations of the throats of a pair of Huias (*Heteralocha acutirostris*). This now extinct New Zealand bird was renowned for the marked difference in the shape of male and female beaks, a difference not recorded in any other bird species. Were it not for Hancock's interest in this aspect of bird anatomy, these internal organs would never have been preserved and it seems unlikely that a sim-ilar preparation exists anywhere else in the world.

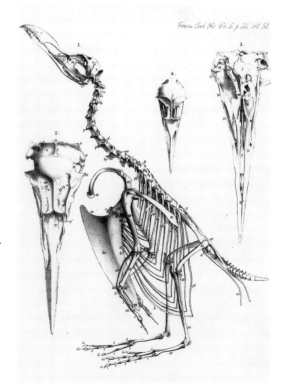

While remounting his Garefowl, Hancock removed a number of bones from the extremities. When the famous comparative anatomist Richard Owen began work on an exhaustive study of the species' skeleton, he was provided with an almost complete specimen constructed from a mummified Auk found on Funk Island. A few bones were found to be wanting, however. These, Hancock was able to supply from the pieces he'd taken from his stuffed bird. In the paper that Owen produced (1865) he remarks that Hancock's bird was female. It is not known on what grounds the usually cautious anatomist made his statement. Probably it was based on something that Apothecary Mechlenburg might or might not have said to person or persons unknown at a date that remains - similarly - enigmatic.

THE NORWICH AUK

Bird no.11 (Grieve no. 56; Hahn no. 17) - adult in summer plumage
Origin: unknown
Present location: Norwich - Castle Museum, Norwich NR1 3JU

Norwich Museum, circa 1890, with the Great Auk in its case.

Hunt, J. 1822. *British Ornithology*, vol.3, p.8 *and* facing plate. Newton, A. 1870. On Existing Remains of the Gare-fowl. *Ibis*, p.258. Blasius, W. 1884. Zur Geschichte der Ueberreste von *Alca impennis. Journal für Ornithologie*, pp.101-2. Grieve, S. 1885. *The Great Auk or Garefowl*, p.78 *and* appendix p.19.

Gurney, J. 1911. The Great Auk and its egg in Norwich Museum. *Transactions of the Norfolk and Norwich Naturalists' Society*, pp.214-5 *and* facing plate. Gladstone, H. 1917. John Hunt.1777-1842. *British Birds*, pp.135-6. Hahn, P. 1963. *Where is that Vanished Bird?* p.218. Frost, C. 1987. *A History of British Taxidermy*, p.40.

This Great Auk has a known history that goes back to 1822; for how long it was in existence before that is uncertain. The first record of it comes in the third volume of John Hunt's *British Ornithology* (published out of Norwich during 1822) in which a portrait of the bird is given. The book itself is very rare, the portrait even rarer; most surviving copies lack this particular plate.

At an unknown date before 1822, a Mr. Edward Lombe bought this Garefowl from Benjamin Leadbeater. For much of the nineteenth century members of the Leadbeater family operated as taxidermists in central London and Benjamin was probably the firm's founder. By the standards of the time he was a skilled man and, probably, he was responsible for mounting this specimen.

After buying the Auk, Lombe took it back to his family home at Melton, near Norwich and at his death it became the property of his daughter, Mrs E.P. Clarke. During 1873 she presented it to the Castle Museum where it has remained ever since.

Judging from the early date of its capture, it is likely that this bird came from the Geirfuglasker.

Engraving by J. Hunt from his *British Ornithology* (1815-22). For this very rare engraving, not present in most copies of the book, the Norwich Auk served as a model.

HOPPA'S AUK

Bird no.12 (Hahn no. 39) - adult in summer plumage
Origin: probably Eldey
Present location: Private collection

Duchaussoy, H. 1897-8. Le Grand Pingouin du Musée d'Histoire naturelle d'Amiens *and* Notes additionnelles. *Mémoires de la Société Linnéenne du Nord de la France*, p.110 *and* p.244. Grieve, S. 1897-8. Additional notes on the Great Auk. *Transactions of the Edinburgh Field Naturalists' and Microscopical Society*, p.327 *and* p.340. Smits, A. 1913 (October). Le Pingouin Brachyptère ou Grand Pingouin. *Le Chasseur Français*, pp.656-7. Scherdlin, P. 1926. A propos du Grand Pingouin du Musée zoologique de Strasbourg. *Bulletin de l'Association Philomathique d'Alsace et de Lorraine*, tome VII, fasc. prem., p.14. Hahn. P. 1963. *Where is that Vanished Bird?* p.224.

In the spring of 1830 the Geirfuglasker, island refuge of the last Great Auk colony, sank beneath the waves during violent volcanic activity. The disturbances continued into the summer and the Auks that survived the cataclysm were left homeless. Although Icelandic fishermen had regularly plundered it, the Geirfuglasker always presented major problems to them. The pounding surf, the treacherous currents and the jagged rocks made landing, in anything other than the calmest of weather, an impossibility. During the course of many summers there was no chance of launching a

Sigríður Thorláksdótter with some of the birds of 1831. A scene imagined by George Dawson Rowley and his wife Caroline some 40 years after the event. *Courtesy of Peter Rowley.*

raid and in these seasons of respite the colony of Auks could breed freely and make good the losses sustained in less happy times.

With their sanctuary gone, those that were left repaired to the nearby island of Eldey, sometimes known as Fire Island. This great block of rock jutting up from the waters of the North Atlantic held one terrible disadvantage. Although the act of landing often presented a difficult challenge for the local fisherfolk, it was nothing like the awful hazard of the Geirfuglasker. So, during the summer of 1831, the Icelanders came to Eldey. In the first raid that year, no less than 24 birds were taken. Never again would such a dreadful harvest be reaped. With each passing year the numbers of birds killed was less and less as the colony dwindled to extinction.

The birds of 1831- all 24- were taken to the mainland where a woman by the name of Sigríður Thorláksdótter was given the task of skinning 23 of them. She used an unusual method, making an incision on one of the flanks below the wing, and then, when she'd finished, she gave the bodies to her friends who cooked and ate them. The skins were sent to Copenhagen and Hamburg.

The following year, 1832, a well known French huntsman by the name of de Cossette bought from Brandt - natural history dealer of Hamburg - a Great Auk. Brandt sold the bird together with the story that it had come from Greenland. Quite why he said this is uncertain; probably he just used the term 'Greenland' carelessly. Often words like 'Greenland,' 'Iceland' or 'Lapland' were used as interchangeable descriptions that meant nothing more than 'far north' and to a limited degree this still applies today. There can be little doubt, however, that the Great Auk bought by Monsieur de Cossette is one of the haul of 24 taken on Eldey. The location of purchase, the date and the mark of an incision on the bird's side all indicate an individual from this raid. Although the specimen shows the grey fringe on its flanks that Rothschild (1907) associated with female plumage, no particular sex was ever claimed for this individual. This is entirely appropriate. Sigríður Thorláksdótter never bothered to open the dead bodies in order to determine the sex of the birds, and the value of Rothschild's opinion is uncertain.

One of de Cossette's great friends was Monsieur Le Baron de Vilmarest. The two men often went hunting together and they built up large collections of the birds they shot. At some time during their lives they made a pact: whoever died first would bequeath to the other his entire collection of stuffed birds. The winner of this particular game was Monsieur le Baron and in 1858 he got all his friend's birds including, of course, de Cossette's magnificent Great Auk.

The specimen was housed, together with all the rest of the Baron's collection, in a purpose-built museum in the grounds of his beautiful château at Nielles les Ardres near Calais. Here it stayed (with the exception of an evacuation to Nantes during World War II) for 137 years.

This specimen was unknown to Wilhelm Blasius and also to Symington Grieve when they wrote their respective monographs (1884 and 1885) and neither heard of its existence until the 1890's. It was, in fact, the very last stuffed Great Auk to come to the attention of Garefowl scholars; no discovery of an unknown specimen has since been made.

Through several generations the bird stayed in the Baron's family and eventually became the property of his great grandchild Dominique de Rosny. On September 16th 1995, Monsieur de Rosny sold his Great Auk to the author.

BULLOCK'S PAPA WESTRAY AUK

Bird no.13 (Grieve no. 42; Parkin no. 5; Hahn no. 11) - adult male in summer plumage
Origin: Papa Westray, Orkney Islands (summer, 1813)
Present location: Tring - The Natural History Museum, Sub-department of Ornithology, Tring, Hertfordshire.

Bullock's Auk.

Montagu, G. 1813. *Supplement to the Ornithological Dictionary*, p.5. Bullock, W. 1814. *Companion to the London Museum* (ed.16), pp.75-6. Latham, J. 1821-8. *A General History of Birds*, vol.10, pp.56-7. Charlton, E. 1860. On the Great Auk. *Zoologist*, p.6884. Preyer, W. 1862. Der Brillenalk in Europäischen Sammlungen. *Journal für Ornithologie*, p.77. Newton, A. 1865. The Gare-fowl and its Historians. *Natural History Review*, pp.473-4. Gurney, J. 1869. The Great Auk. *Zoologist*, p.1442. Milne, J. 1875 (April 10th). Relics of the Great Auk. *The Field*, p.370. Smith, J. Notice of the remains of the Great Auk...in Caithness. *Proceedings of the Scottish Society of Antiquaries*, p.91. Blasius, W. 1884. Zur Geschichte der Ueberreste von *Alca impennis*. *Journal für Ornithologie*, pp.93-4. Grieve, S. 1885. *The Great Auk or Garefowl*, p.10, p.78, p.91 *and* appendix p.15. Buckley, T. and Harvie-Brown, J. 1891. *A Vertebrate Fauna of Orkney*, pp.245-57. Newton, A. 1896. *A Dictionary of Birds*, pp.306-7. Newton, A. 1898. On the Orcadian Home of the Gare-fowl. *Ibis*, pp.591-2. Blasius, W. (*in* Naumann) 1903. *Naturgeschichte der Vögel Mitteleuropas*, bd.12, p.195. Parkin, T. 1911. The Great Auk. *Hastings and East Sussex Naturalist*, vol.1, pt.6 (extra paper), pp.31-2. Omond, J. 1925. *How to Know the Orkney Birds*, p.46. Gore, J. (ed.), 1938. *Mary Duchess of Bedford*, vol.1, pp.97-9. Hahn, P. 1963. *Where is that Vanished Bird?* pp.213-4. Gooders, J. (ed.), 1971. *Birds of the World*, fig. p.2888. Mearns, B. and R. 1988. *Biographies for Birdwatchers*, p.223. Birkhead, T. 1994 (May 28th). How collectors killed the great auk. *New Scientist* fig. p.27. Knox, A. and Walters, M. 1994. *Extinct and Endangered Birds in the...Natural History Museum*, p.133.

Bullock's Auk is the most celebrated of all the preserved specimens, the literature listed being by no means exhaustive and only those references of some importance or interest being included. There are several clearly definable reasons for this celebrity. First, the bird is one of the earliest of surviving examples. Second, the circumstances surrounding its capture are well known. Third, it has a British provenance and it is the British, perhaps, who have pursued the cult of the Garefowl most vigorously. Last, of all the Great Auks alleged to be male this is the only one whose sexual determination seems in any degree reliable. In this connection it might be mentioned that this is one of the few preserved specimens without the grey

East side of Papa Westray. Engraving in T. Buckley's and J. Harvie-Brown's *Vertebrate Fauna of the Ornkney Islands* (1891).

fringing below the wing that Rothschild (1907) believed to be the mark of the female.

William Bullock, the specimen's first owner, was a jeweller and silversmith who assembled an incredible collection of curiosities for his museum, situated originally in Sheffield, then at Liverpool and finally in London. In the guidebook or 'Companion' to his *London Museum and Pantherium*, the vast array of objects is described as:

EAST SIDE OF PAPA WESTRAY, LOOKING N.W.

(The arrow heads indicate the recess in which the last Great Auk lived, and from whence it was shot.)

(Facing page, top left). A drawing made in 1866 by George Dawson Rowley and sent as a gift to Alfred Newton. *Courtesy of the University Library, Cambridge.*

(Facing page, top right). The crannie in which Bullock's Auk rested shortly before it was killed. Photograph taken for T. Buckley's and J. Harvie-Brown's *Vertebrate Fauna of the Orkney Islands* (1891).

Collected during seventeen years of arduous research and at the expense of £30,000 and now open for public inspection in the Egyptian Temple, just erected for its reception in Piccadilly, London.

One of the fruits of Bullock's 'arduous research' was the stuffed skin of his Great Auk. In 1812, the pursuit of this specimen took the jeweller to the Orkneys and the island of Papa Westray. Here, by his own account, he:

Had the pleasure of examining this curious bird in its native element; it is wholly incapable of flight, but so expert a diver that every effort to shoot it was ineffectual.

Bullock pursued the Garefowl in a six-oared rowing boat for several hours but the bird's speed was so great that the chase was eventually given up. This individual had been observed, together with its mate, by Orkney residents for several years and the pair were known - just like others on lonely islands scattered across the north - as 'The King and Queen of the Auks.' The female seems to have been killed just before Bullock's visit and although the jeweller did not himself manage to capture the mate, it did not long survive his visit. It was shot the next year by one William Foulis on rocks near a location known as Fowls' Crag and it rapidly followed Bullock to London. The museum 'companion' records:

> It was taken at Papa Westray ... to the rocks of which it had resorted for several years, in the summer of 1813, and was finely preserved and sent to me by Miss Trail of that island.

J.A. Harvie-Brown (1844-1916).

Many years later, J.A. Harvie-Brown travelled to the Orkneys and was shown the actual spot that harboured the Great Auk just before it was killed. In his book (jointly authored with T.E. Buckley) *A Vertebrate Fauna of the Orkney Islands* (1891) he describes a small hole in a rock face as, 'the last resting place of the Great Auk in life' and illustrates his account with a photograph that was specially taken.

In 1819 Bullock decided to sell his entire collection. Acting as his own auctioneer, he sold some 15,000 objects (made up into 7,500 lots) during a sale that took place in his Egyptian Temple and lasted for 26 days. On May 6th, the fifth day of the sale, the Papa Westray Auk - together with an egg - was knocked down to Dr. William Leach for £16, 15 shillings and sixpence.

PLATE XXXV.

Leach was acting on behalf of the British Museum and Bullock's Auk is now the property of the Natural History Museum, London. Like three other Garefowls, it is kept at the Sub-department of Ornithology, Tring, Hertfordshire. Apart from a specimen in Dublin, this is the only Great Auk with a reliable British Isles origin.

THE VAN LIDTH DE JEUDE AUK

Bird no.14 (Grieve no. 43; Hahn no. 12) - adult in summer plumage, said to be female
Origin: unknown but probably Iceland
Present location: Tring - The Natural History Museum, Sub-department of Ornithology, Tring, Hertfordshire (Catalogue no.1858.4.27.1).

The Van Lidth de Jeude Auk (circa 1900).

Preyer, W. 1862. Der Brillenalk in Europäischen Sammlungen. *Journal für Ornithologie*, p.77. Newton, A. 1865. The Garefowl and its Historians. *Natural History Review*, p.473. Fatio, V. 1868. Liste des divers représentants de l'*Alca impennis* en Europe. *Bullétin de la Société Ornithologique Suisse*, tome II, pt.1, p.82. Blasius, W. 1881-3. Ueber die letzen Vorkommnisse des Riesen-Alks. *Jahresberichte des Vereins für Naturwissenschaft zu Braunschweig für die Vereinsjahre 1881-2 und 1882-3*, p.114. Blasius, W. 1884. Zur Geschichte der Ueberreste von *Alca impennis*. *Journal für Ornithologie*, p.93. Grieve, S. 1885. *The Great Auk or Garefowl*, p.78 *and* appendix p.15. Sharpe, R. (ed.) 1874-98. *Catalogue of the Birds in the British Museum*, vol.26, p.564. Ogilivie-Grant, W. 1910. *Guide to the Gallery of Birds in the Department of Zoology, British Museum*, p.197, pl.5. Hahn, P. 1963. *Where is that Vanished Bird?* p.214. Knox, A. and Walters, M. 1994. *Extinct and Endangered Birds in the...Natural History Museum*, p.132.

This specimen, traditionally held to be female, was acquired by the British Museum in April 1858 and was obtained - perhaps through the agency of Frank of Amsterdam, perhaps at auction - from the collection of Professor van Lidth de Jeude of Utrechte. The Professor had bought it from a certain Herr Draak who received it in 1843 (this time certainly via Frank) from the enigmatic merchant Israel of Copenhagen. Although at one time the word 'Labrador' was written on the bird's stand - and this location is claimed in the British Museum's *Guide to the Gallery of Birds* (1910) - it is altogether more likely that the specimen came from Iceland via Copenhagen at an unknown date before 1843; one of the few things known about Israel is that he had good trading links with Iceland.

Although once on display in the 'bird room' at South Kensington, the specimen was removed long ago and is now kept at the Sub-department of Ornithology at Tring. At the time of writing - and for many years now - the Natural History Museum has not seen fit to put any of its four Great Auks on display to the public.

Hand-coloured engraving, drawn by W. B. Hawkins, from J. Wilson's *Illustrations of British Zoology* (1831). Bullock's Auk served as the model for this picture.

LORD ROTHSCHILD'S AUK

Bird no.15 (Hahn no. 13) - adult in summer plumage (sometimes described as immature), said to be female
Origin: unknown
Present location: Tring - The Natural History Museum, Sub-department of Ornithology, Tring, Hertfordshire (catalogue no. 1939.12.9.537).

Grieve, S. 1896-8. Supplementary note on the Great Auk *and* Additional notes on the Great Auk. *Transactions of the Edinburgh Field Naturalists' and Microscopical Society,* pp. 248-9, p.269, pl.3 *and* pp.328-30 *and* p.340. Blasius, W. (*in* Naumann) 1903. *Naturgeschichte der Vögel Mitteleuropas,* bd. 12, p.180, pl.17a *and* pl.17b. Rothschild, W. 1907. *Extinct Birds,* p.157. Lydekker, R. 1908. *The Sportsman's British Bird Book,* fig. p.194. Ward, R. 1913. *A Naturalist's Life Study,* p.221 *and* fig. p.224. Hahn, P. 1963. *Where is that Vanished Bird?* pp.214-6. Knox, A. and Walters, M. 1994. *Extinct and Endangered Birds in the...Natural History Museum,* pp.132-3.

When Symington Grieve wrote *The Great Auk or Garefowl* (1885) this specimen was unknown to him but by 1897 - at which time he was bringing his research up to date by publishing papers in the *Transactions of the Edinburgh Field Naturalists' and Microscopical Society* - he'd discovered many things about it. Where, then, had the bird been before 1885? Even today its true history remains uncertain. What Grieve knew for sure he recounted as follows:

> This skin ... was purchased by Monsieur Boucard of Paris and London who sold it [in 1890] to Mr. Leopold Field ... who offered it for £300 to the Museum of Science and Art, Edinburgh. Before sanction could be obtained ... for its purchase, it was sold by its owner to Rowland Ward, 166 Piccadilly ... it was understood for £315. The skin is still in the possession of Mr. Ward; but the museum at Tring will eventually be its resting place as it has been purchased by the Hon. W. Rothschild. I had an opportunity of examining this skin when sent to Edinburgh on approbation. It was not in very good condition but ... has since been thoroughly overhauled and put in as good order as possible.

Leopold Field, in a letter to Grieve, gave his own account of his period of ownership:

> The skin was offered to me, I think, in the early summer [it was, in fact, January] of 1890 ... I paid £300 net cash, delivered at my rooms, and in 1893 sold it and the Potts' Egg to Mr. Rowland Ward for £630. The specimen came unmounted with little spikes in the feet. I had it handsomely displayed in a glass case with a model egg and a background of sea and sky.

Despite Leopold's 'handsome display' the case was jettisoned when the bird went on approval to Edinburgh where, according to Grieve:

> It had rather a dilapidated appearance, the plumage dirty and the webbed part of the feet somewhat worm eaten.

The poor condition that Grieve noticed may give some clue to part of the bird's history.

Tring Museum at the end of the nineteenth century. *Courtesy of The Hon. Miriam Rothschild.*

Lord Rothschild's Auk (circa 1890).

For reasons that can probably never be pin-pointed, Monsieur Boucard was reluctant to say from where he got the bird, although he'd sent Leopold Field certain details concerning its alleged early history. In a letter written from Paris on January 20th 1890, he stated that the bird was captured in Iceland in 1837 and first belonged to a Herr Eimbeck of Brunswick and afterwards to a Monsieur Bruch of Mayence (Mainz). What happened after this Monsieur Boucard was not prepared to say. Natural history merchants, like antique or art dealers, have all sorts of motives for an unwillingness to disclose this kind of information. Sometimes it is simply a fear of being bypassed in any future transaction; sometimes a dealer may be afraid that the price he paid will be discovered and the difference between buying and selling prices will prove hugely embarrassing. Occasionally, perhaps, the transaction is not quite proper or not even strictly legal. Always, the dealer can plead that his first client requires confidentiality - which usually has the effect of putting an end to all inquiries.

Unfortunately for Monsieur Boucard, when his bird finally passed to Walter Rothschild it had reached someone whose requests for data could hardly be ignored. Not only was Rothschild a rich and powerful English lord, he was a customer that any dealer would have wanted to please. His enormous wealth was used to amass a vast collection of natural history material (probably the most comprehensive ever assembled by a single individual) and his patronage could not be lightly risked. But despite requests from Rothschild and his curator Ernst Hartert, Boucard stood his ground and would only ever offer the information he'd already given- which might indicate that something shady had transpired but might equally well mean

Adolphe Boucard (1839-1905).

that he was telling the truth. With both Eimbeck of Brunswick and Bruch of Mainz long since dead, Rothschild had little choice but to accept Boucard's word. Nevertheless, he couldn't fail to wonder why the specimen had never come to the attention of Professor Wilhelm Blasius. Blasius had, of course, made exhaustive inquiries into the whereabouts of specimens and his home town happened to be Brunswick!

An Eimbeck connection is not entirely unrealistic, however. He was a dealer in natural history specimens who is known to have received two Garefowl skins from the Royal Museum, Copenhagen during 1832-3. There is no reason to suppose that he didn't acquire others.

Symington Grieve had, however, already decided on a provenance for Lord Rothschild's Auk, one that he felt was more likely. He believed it came from the Ernest Delagorgue collection. This collection did indeed once contain a Garefowl, a Garefowl that was rather quietly disposed of. On Monsieur Delagorgue's death, his stuffed birds - all in fairly bad condition - were given to the town of Abbeville, but two were held back - a Snowy Owl (*Nyctea scandiaca*) and a Great Auk. These were kept by a Monsieur Jules Barbieux. After this gentleman's own death his son-in-law found an entry in his diary for July 17th 1888, an entry that translates as:

Received from M. Maingonnat for a bird - 1,000 francs.

Whether Barbieux was genuinely empowered to sell the bird is impossible to say. Maingonnat, a natural history dealer who operated from 37 Rue Richer, Paris, died in 1893 so there was no possibility of Rothschild (or anyone else) making inquiries through him.

The time and place fit and so too does the condition of the bird. But, from where did Delagorgue get the bird. Quite possibly it came from Herr Eimbeck by way of Monsieur Bruch of Mainz!

It was not only the origin of this bird that concerned Grieve. The specimen itself particularly interested him. He believed it showed clear signs of immaturity and that it was indeed the third 'youngest' Auk known (after the immature in Newcastle and another in Prague). Rothschild took considerable exception to Grieve's opinion and in characteristic style he vigorously defended the adult status of his bird in the pages of his mighty book *Extinct Birds* (1907):

The specimen has been described as "immature" but this is a mistake. Evidently it arose from some white speckles being visible on the neck in the photograph [a picture used by Grieve to accompany one of his papers]. The specimen itself, however, shows no such white speckles, but only worn feathers, out of which the illusion arose in the photo ... The grey shade "on the body lower than the wing," mentioned by Mr. Symington Grieve is not a sign of immaturity, but appears in all adult females, though it is said to be absent in males.

Despite Rothschild's confident remark concerning the 'grey shade' on the flanks, this is no more than his opinion. Apart from Rothschild and Grieve no other writer appears to have even noticed the grey area and the meaning of its presence is unclear. Leaving aside this matter of grey flanks, there is no other known pointer to this individual's sex. As far as the speckles on the neck are concerned, an inspection of the specimen shows quite clearly that here Rothschild was correct. They are a photographic illusion.

Although Rothschild chose to ignore it, Grieve's main reason for supposing the bird immature was its small size. It was, apparently, the smallest Garefowl he'd seen but this may or may not have significance.

In 1932 Walter Rothschild sold his huge bird collection to the American Museum of Natural History but this bird- probably a favourite-he kept back; its appearance had been considerably improved since the early days. Another stuffed Great Auk that Rothschild owned did, however, cross the Atlantic. On the English lord's death all that was left of his vast natural history collection, including his museum building at Tring, his magnificent library and his remaining Great Auk, passed to the British people. The Auk is still at Tring in the Natural History Museum's Sub-department of Ornithology.

Iceland is, of course, the best guess for the specimen's origin, simply because it is the locality from which almost all surviving mounted Garefowls come. Any Eimbeck connection would be added confirmation of this.

LORD LILFORD'S AUK

Bird no. 16 (Grieve no. 44; Hahn no. 14) - adult in summer plumage
Origin: unknown but certainly Iceland
Present location: Tring - The Natural History Museum, Sub-department of Ornithology, Tring, Hertforshire (catalogue no.1947.17.1).

Newton, A. 1870. On Existing Remains of the Gare-fowl. *Ibis*, p.258. Blasius, W. 1884. Zur Geschichte der Ueberreste von *Alca impennis. Journal für Ornithologie*, p.94. Grieve, S. 1885. *The Great Auk or Garefowl*, p.78 *and* appendix p.15. Grieve, S. 1896-7. Supplementary note on the Great Auk. *Transactions of the Edinburgh Field Naturalists' and Microscopical Society*, p.259. Lilford, Lord. 1885-98. *Coloured Figures of the Birds of the British Isles*, vol.6, p.137. Hahn, P. 1963. *Where is that Vanished Bird?* p.216. Knox, A. and Walters, M. 1994. *Extinct and Endangered Birds in the...Natural History Museum*, p.133.

> I need hardly say that I never had any personal acquaintance with this extinct and much-lamented fowl in life.
>
> Lord Lilford (1898)

With these words Thomas Lyttleton Powys, 4th Baron Lilford, began his account of the Great Auk for his seven volume work *Coloured Figures of the Birds of the British Isles*. Although Lilford claimed no experience of the living bird, he certainly knew about stuffed ones. A little further on he wrote:

> I am the fortunate possessor of a fine stuffed specimen ... which from the date of a ship's bill of lading written in Danish and Icelandic, found inside the skin by the late H. Ward, of Vere St., to whom it was sent to be mounted, was probably obtained on the coast of Iceland about the year 1833.

This, then, is probably a bird from Eldey taken in the years following the destruction of the Geirfuglasker. According to the calculations of Alfred

Lord Lilford's Auk (circa 1870). *Courtesy of the University Library, Cambridge.*

Copy of the bill of lading found inside Lilford's Auk. *Courtesy of the University Library, Cambridge.*

Newton (1861) approximately thirteen individuals were killed in raids on Eldey during 1833; perhaps this is one of them.

When the specimen was placed in the hands of Henry Ward (father of Rowland, who was later to handle the sale of many Great Auks) for re-stuffing, it belonged to Lilford's brother-in-law Arthur W. Crichton who lived at Broadwater Hall, Shropshire. He'd received it as a gift from a neighbour, Sir Harford Brydges of Bouttisbrook, Presteign but how Sir Harford came by a Garefowl is not known.

On Crichton's death Lord Lilford bought it for his collection at Lilford Hall, Oundle, Northamptonshire. Before displaying it at his country seat, he loaned it for a while to the offices of the British Ornithological Union, then at Tenterden St., London. In 1947 it passed to the Natural History Museum from Lilford's estate.

THE FOLJAMBE AUK

Bird no.17 (Grieve no. 58; Hahn no. 18) - adult in summer plumage
Origin: unknown, but probably the Geirfuglasker
Present location: Worksop - Osberton Hall, Worksop, Nottinghamshire.

Newton, A. 1870. On Existing Remains of the Gare-fowl. *Ibis*, p.258. Blasius, W. 1881-3. Ueber die letzen Vorkommnisse des Riesen-Alks. *Jahresberichte des Vereins für Naturwissenchaft zu Braunschweig für die Vereinsjahre 1881-2 und 1882-3*, p.110. Blasius, W. 1884. Zur Geschichte der Ueberreste von *Alca impennis*. *Journal für Ornithologie*, pp.103-4. Grieve, S. 1885. *The Great Auk or Garefowl*, p.78 *and* appendix p.20. Grieve, S. 1896-7. Supplementary note on the Great Auk. *Transactions of the Edinburgh Field Naturalists' and Microscopical Society*, p.247. Hahn, P. 1963. *Where is that Vanished Bird?* p.219.

The most celebrated action of the conflict between England and Denmark during the Napoleonic Wars was the Battle of Copenhagen (1801). Although, by any standard, a mighty victory, it owes its real celebrity to one particular moment during the struggle. The signal to retreat was raised by Horatio Nelson's commanding officer, Admiral Sir Hyde Parker, and the fact was brought to the attention of the great British hero. Determined to pursue his offensive through to final victory (yet not wishing to disobey the order of a higher ranking officer), Nelson slowly raised his telescope to his blind eye. "Signal," he is supposed to have called, "I see no signal!" Then he proceeded to destroy the Danish Fleet.

If this is the best known incident of the war with Denmark, one of the least known had a consequence that was entirely extraordinary. It hastened the end of the Great Auk in its final decline to extinction.

In 1808, a privateer, *The Salamine*, sailed to the Faroes to subdue Danish interests and then it voyaged on to Iceland where the crew promptly sacked the town. As the vessel left Icelandic waters it passed the Geirfuglasker. This passage coincided with exceptionally calm weather; it also happened at the season when the seabirds were nesting. The crew spent a day slaughtering birds before heading off for the Faroes where they re-patriated a man they had pressed into acting as their pilot.

Five years later the Faroese were reduced to a most pitiful condition. Prevented by the Danes from trading with the English enemy, they were enduring a terrible famine. A ship was sent to Iceland for urgently needed supplies and leading the expeditionary force was none other than the man the English had press-ganged into being their guide. He knew all about the Geirfuglasker and its plentiful supply of birds - and, of course, his people were hungry! Landing on the reef in the month of July (once again the weather was calm), the Faroese killed all the Great Auks they could find, carrying off most of these but leaving many dead ones behind as the wind changed and they felt the need to abandon the island rapidly. Some of the birds were salted down but it is said that there were at least 24 additional carcasses on board, one of which the Faroese gave to Vidalín, Bishop of Reykjavik, on their arrival in Iceland. Later in the year (1813) the Bishop passed it on to a friend in England.

The bird now at Osberton Hall, Worksop has been owned by the Foljambe family since early in the nineteenth century and it is known that it was purchased from a Mr. Walker in Liverpool during October of 1813 and stuffed by B. Corbet. Alfred Newton, in his manuscript notes kept at the Cambridge University Library, speculated that this Foljambe bird corresponds with the one the Faroese gave to Bishop Vidalín. Even if it is not the Bishop's bird, it seems likely that it came from the Faroese haul.

The Foljambe Auk as it appears today. *Courtesy of Mr. and Mrs. R. Budge.*

For many, many years it has occupied one of a series of glass-fronted cabinets set into the fabric of the building at Osberton Hall. These house an extensive collection of British birds. Recently, the house passed out of the possession of the Foljambe family. As these stunning and unique cabinets could not sensibly be removed, Mr. Foljambe presented them and their contents to the new owner, a Mr. Budge, but retained ownership - if not actual possession - of the Great Auk; for the time being, therefore, the Great Auk remains at Osberton Hall.

FREDDY BELL'S YORK AUK

Bird no.18 (Grieve no. 76; Hahn no. 19) - adult in summer plumage
Origin: unknown
Present location: York - The Yorkshire Museum, Museum Gardens, York YO1 2DR.

Yorkshire Philosophical Society Annual Report, 1853, p.9. Gurney, J. 1869. Notes on the Great Auk. *Zoologist*, p.1640. Fatio, V. 1868. Liste des divers représentants de l'*Alca impennis* en Europe. *Bullétin de la Société Ornithologique Suisse*, tome II, pt. 1, p.82. Blasius, W. 1884. Zur Geschichte der Ueberreste von *Alca impennis*. *Journal für Ornithologie*, p.113 *and* p.128. Grieve, S. 1885. *The Great Auk or Garefowl*, p.78 *and* appendix p.24. Hahn, P. 1963. *Where is that Vanished Bird?* p.219. Denton, M. 1995. *Birds in the Yorkshire Museum*, p.6 *and* pp.121-2.

The Yorkshire Museum at York is particularly fortunate in owning two Great Auks. The first, donated in 1853 and described by Gurney (1869) - and later by Grieve (1885) - as being in excellent condition, is sometimes recorded as coming from the William Rudston Read collection - despite the fact that this gentleman never actually owned it. The confusion arose simply because Rudston Read presented a large number of ornithological specimens to the Yorkshire Philosophical Society and when - later - a Great Auk was donated it was added to Rudston Read's collection.

The specimen was, in fact, given by a Mr. Frederick Bell of Thirsk who had inherited it from his uncle, J. Bell, M.P. Rudston Read was deeply involved in the Philosophical Society's acquisition of it, however. Soon after Freddy came into his uncle's property (so pleased was he by this apparent alteration in circumstance that he changed his name from McBean to Bell), Rudston Read chanced on him while both were out hunting. Knowing that he'd inherited the Auk and that he was unlikely to be interested in it, Read asked whether the Museum of the Philosophical Society might have it.

"You shall have it if I've got it," was the spirited reply that the new Mr. Bell is alleged to have given. At the earliest possible date the York taxidermist David Graham was sent over to Thirsk to claim the bird. With him went the instruction that he was not to come away without it! Alfred Newton scribbled down (among his notes at the Cambridge University Library) what he was told by Mr. Graham of this outing:

The case was found in a room at the top of the house, with its face turned towards the wall. Mr. Bell began to consider about it. "Now tell me Mr. Graham, is this bird worth £10?" "Yes sir, more." He seemed not inclined to let it go till I told him that gentlemen always keep their promises ... Afterwards [he] much regretted having parted with it.

(Above, left). Arthur Strickland's Auk (circa 1870). *Courtesy of the University Library, Cambridge.*

(Above, right). Freddy Bell's Auk (circa 1870). *Courtesy of the University Library, Cambridge.*

Had poor Freddy really been a 'gentleman' he would, of course, have known that the business about promises and gentlemen is very far from the truth.

Whether Frederick McBean/Bell regretted his rash promise or not, the fact is he went through with it and it is to his generosity - and, perhaps, to Rudston Read's determination - that the Yorkshire Museum owes one of its greatest treasures. Strangely - but quite typically for a British Museum - the Curators have decided to hide it away in a storehouse.

Unfortunately the history of this bird cannot be traced back to an origin. Before Frederick inherited it, his Uncle John had owned it for several years. He'd bought it for the sum of £7, 12 shillings and sixpence, probably in the year 1845, at the sale of the effects of a Mr. Thomas Allis. According to Allis's son the specimen was obtained in London, 'ready stuffed,' at some time prior to 1840 from a Mr. Warwick, Keeper of the Surrey Zoological Gardens. Here the trail of the bird, as it moves backwards in time, is lost.

153

ARTHUR STRICKLAND'S AUK

Bird no.19 (Grieve no. 77; Hahn no. 20) - adult in summer plumage
Origin: unknown
Present location: York - The Yorkshire Museum, Museum Gardens, York
YO1 2DR.

Yorkshire Philosophical Society Annual Report, 1866, p.9. Newton, A. (ed.) 1867.
[Notices], *Ibis,* p.384. Gurney, J. 1869. Notes on the Great Auk. *Zoologist,*
pp.1641-2. Fatio, V. 1868. Liste des divers représentants de l'*Alca impennis*
en Europe. *Bullétin de la Société Ornithologique Suisse,* tome II, pt.1, p.82.
Blasius, W. 1884. Zur Geschichte der Ueberreste von *Alca impennis. Journal
für Ornithologie,* p.113 *and* p.128. Grieve, S. 1885. *The Great Auk or Garefowl,*
p.78 *and* appendix p.24. Hahn, P.1963. *Where is that Vanished Bird?* p.219.
Denton, M. 1995. *Birds in the Yorkshire Museum,* p.6 *and* pp.121-2.

The *Report* of the *Yorkshire Philosophical Society* for 1866 records:
> The Curator ... has to report the presentation of a very perfect and
> beautiful collection of British Birds, by Mrs Trevenen, the sister of the
> late Arthur Strickland ... of Bridlington Quay, one of the most zealous
> and scientific ornithologists of his day... Mrs Trevenen's attachment to
> her brother induced her to purchase it from his widow for a large sum,
> that she might present it to the Society in memory of this naturalist.

The collection contained many interesting and unusual pieces, but one
great treasure - a stuffed Garefowl. Unfortunately, nothing is known of the
bird's origin. There is a rumour, noted by G. Dawson Rowley, that this was
a St Kilda bird but the only definite record of it before it came to
Strickland indicates that it belonged to a Mr. Grenville. Strickland is supposed to have bought it about 1835 for £9 or £10. One curious fact that
emerged when the specimen was cleaned, soon after its acquisition by the
Society, regards the manner of its preparation. It was found to be hollow
with no stuffing inside, the skin being held in position by wires. This hollow state - as well, perhaps, as the bird's presence in Yorkshire - might point
to the handiwork of the celebrated Charles Waterton, nineteenth century
eccentric, creator of the Nondescript and author of *Wanderings in South
America* (1828). The connection is, of course, extremely tenuous.

J.H. Gurney (1869) reported that this 'un-stuffed' bird (it seems to have
since been 'filled') was in an excellent state of preservation but just a few
years later Grieve (1885) deplored the fact that its feathers were ruffled and
broken and that it looked as though it had been badly exposed to dust.

Like its companion at the Yorkshire Museum, this specimen is stored -
at the time of writing - in a box and is not available for public viewing.

AUDUBON'S AUK

Bird no.20 (Grieve no. 63; Hahn no. 67) - adult in summer plumage
Origin: unknown
Present location: Toronto - Royal Ontario Museum, 100 Queen's Park,
Toronto, Canada M5S 2C6.

Audubon, J. 1827-38. *The Birds of America*, vol.4, pl.341. MacGillivray, W. 1852. *A History of British Birds*, vol.5, p.359. Baird, S. 1860. *The Birds of North America*, p.902. Newton, A. 1861. Abstract of Mr. J. Wolley's Researches in Iceland. *Ibis*, p.398. Coues, E. 1868. A Monograph of the Alcidae. *Proceedings of the Academy of Natural Sciences, Philadelphia*, p.16. Newton, A. 1870. On Existing Remains of the Gare-fowl. *Ibis*, p.259. Orton, J. 1871. Notes on some birds in the Vassar College. *American Naturalist*, p.717. Blasius, W. 1884. Zur Geschichte der Ueberreste von *Alca impennis*. *Journal für Ornithologie*, p.105. Grieve, S. 1885. *The Great Auk or Garefowl*, p.81 *and* appendix p.21. Belknap, J. 1960 (May). *The Kingbird*, vol.10, no.1, p.12. Hahn, P. 1963. *Where is that Vanished Bird?* pp.230-2. Baillie, J. 1965 (Christmas). Two Vanished Birds. *Meeting Place. Journal of the Royal Ontario Museum*, pp.92-5. Audubon. J. 1966. *The Original Watercolour Paintings...for The Birds of America*, pl.169. Baillie, J.1969 (Sept.). Audubon- and his Great Auk. *Ontario Naturalist*, p.12-5. Bourne, W. 1993. The story of the Great Auk. *Archives of Natural History*, p.275. Ralph, R. 1993. *William MacGillivray*, pl.V.

John James Audubon, probably the most flamboyant of all bird painters, and certainly the most celebrated, once owned a stuffed Great Auk. This larger than life character escaped from a life of grinding poverty to produce the most spectacular and romantic of all bird books - the

Watercolour by J.J. Audubon on which the illustration in his *Birds of America* (1827-38) is based. Audubon used his stuffed bird as a model for this painting.

four volume, double elephant folio *Birds of America* (1827-38). To create this stunningly beautiful work, Audubon tramped and vagabonded his way backwards and forwards across North America, confronting one setback after another, for a period of more than 40 years. Nothing he encountered could deflect him from his one great purpose - to depict in its natural size every species of North American bird. In the course of these years of wandering, he encountered nearly 400 different species - but he never saw a living Great Auk! Just once, it seems, he may have come close. During 1833, while visiting the coast of Labrador, he was told that Great Auks still bred upon a low, rocky islet south-east of Newfoundland where, apparently, large numbers of their young were regularly killed for bait. The story is doubtful and the Garefowls were probably long gone but, in any case, the information came to Audubon too late in the season and he was unable to check its accuracy. He never got another chance.

Instead he had to content himself with his stuffed Auk - a specimen he is supposed to have bought in London during 1836 - and, using this bird as reference, he produced his famous plate for *The Birds of America*. To say that Audubon produced this plate is, actually, slightly misleading.

Two views of Audubon's Auk as it appeared before its 'rescue.'

He painted only the figures of the two birds, leaving (as he often did) his engraver - Robert Havell - to create the memorable and atmospheric background.

Audubon's Auk (leading) and Bonaparte's Auk photographed (c. 1925) immediately after their restoration.

The chequered history of Audubon's Auk is almost as romantic as the picture itself. The artist gave his bird to a friend, Jacob Post Giraud Jr., author of *The Birds of Long Island* (1844). In 1867, three years before his death, Giraud presented his entire bird collection, including the Auk, to Vassar College, Poughkeepsie, New York. Despite all attempts to prise it away, the bird remained the property of this institution for almost a century.

Audubon's Auk after restoration.

Leonard Sanford (seated) at the opening of the Sanford Bird Hall. *Courtesy of Mary LeCroy and The American Museum of Natural History.*

One of these attempts is described by Hahn (1963) who relates a story told him by Professor Joseph Hickey of the University of Wisconsin. The tale outlines Dr. L.C. Sanford's quest to obtain the bird:

With a purchase price in mind, Sanford ... journeyed to Poughkeepsie to see the specimen and separate it from its present owners. At Vassar, the head of the biology department casually recalled that his dept. had such a bird, and he led Dr. Sanford to a classroom where the mounted auk reposed - dirty and neglected - on the floor under a laboratory sink! Although much embarrassed by this circumstance, the Vassar biologist told Dr. Sanford that the college could not sell him the ... Auk because the bird was a gift to the college and because such a sale might offend the family of the donor. Dr. Sanford was not, however, to be put off. He pointed out the rarity of the specimen, its value to science, and the obligation that Vassar had to give the specimen the highest degree of protection. Having made his host still more uncomfortable, he finally proposed the following solution to Vassar's difficulties: Vassar could give the auk to him on indefinite loan (an ancient device of museum people who cannot afford to purchase great rarities) so that it could be housed under insect-proof and fire-proof conditions. Dr. Sanford would hire the finest bird artist alive [in the event Louis Agassiz Fuertes was chosen] to paint a picture of the great auk that would then be turned over to Vassar (thus enabling the college to discharge its teaching responsibilities in the absence of the specimen). And finally, Dr. Sanford would hire the finest taxidermist in the land to dismount the skin, clean it and make it into a study specimen (thus relieving the college of its responsibility to science). This masterly proposition had no loopholes ... and so to Dr. Sanford went Audubon's Great Auk.

Sanford first sent the bird to the Museum of Comparative Zoology at Cambridge, Mass., where Mr. George Nelson set about renovating it. He did not, in fact, turn it into a study skin but instead completely re-mounted it. During this process a significant discovery was made - the Auk was stuffed with old German newspapers. Several American authorities had shown a reluctance to accept that Audubon's Auk was a bird of European origin, preferring the (almost) patriotic - and fanciful - belief that it had come instead from Newfoundland. Here now was proof that such an idea was wrong; a German provenance is indicative of an Icelandic origin.

After the renovation, which took place in October of 1921, the Auk was sent to the American Museum of Natural History where it remained for many years housed, according to Hahn (1963):

In a bronze and glass case which is kept inside a locked and dust proof case.

It was still, of course, technically the property of Vassar College. As far as the promised painting by Fuertes is concerned, Vassar never got it. The school chose instead to receive a fake Great Auk specially made by the

The fragment of German newspaper found inside Audubon's Auk.

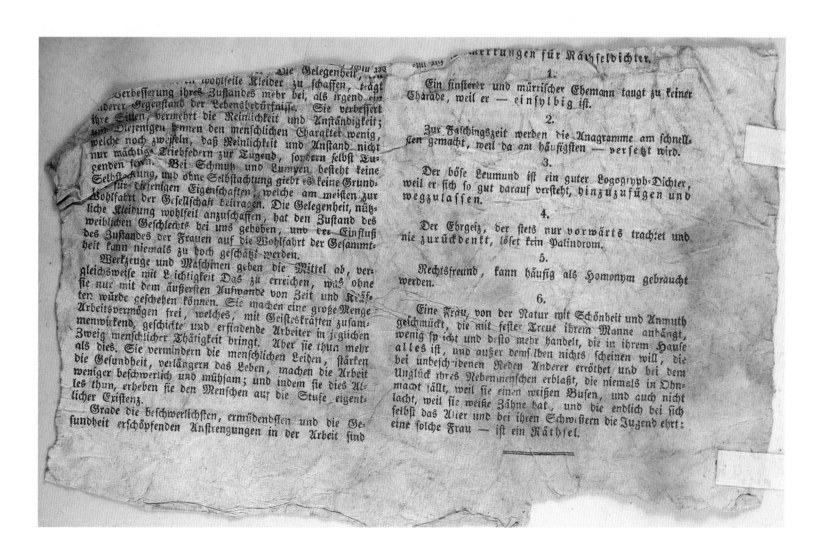

same George Nelson who re-mounted the original bird and this still exists at Poughkeepsie.

During the early 1960's, stirred into action by the activities of Paul Hahn, the Royal Ontario Museum began to search for a Great Auk that could be purchased for its collection. Hahn died before this could be achieved but James L. Baillie took up the hunt and in 1964 a chance meeting with Dr. Ralph Palmer acted as the catalyst that brought the search to an end.

Dr. Palmer had taught at Vassar and was well known to influential members of the school board. Although the College had hitherto resisted all propositions to relinquish its ownership (actual possession had, of course, passed from it long before), Dr. Palmer was of the opinion that were enough money to be proffered, any resolve to keep the bird might crumble. According to Baillie (1969):

> Dr. Palmer ... thought ... a fresh approach might be made to Vassar, pointing out that their bird hadn't even been in the college for 43 years, that the American Museum had three great auks of their own and might relish the opportunity of terminating an agreement whereby they were forever held responsible for the well-being of some other institution's property, that the Royal Ontario Museum was a responsible organisation that could be expected to give the auk constant curatorial attention and that, most of all, there was no great auk in Canada, but nine in the United States.

By the end of 1964, with the plan receiving the blessing of the American Museum of Natural History, Vassar succumbed to the financial inducement of $12,500 and Audubon's Auk got a new owner for the first time since 1867. A specimen of the extinct Labrador Duck (*Camptorhynchus labradorius*) was included in the sale.

THE PRINCE OF DENMARK'S AUK

Bird no.21 (Grieve no. 64; Hahn no. 21) - adult in summer plumage
Origin: Unknown, probably Eldey
Present location: Prague - Narodni Museum, Praha, Czech Republic.

Fric (Fritsch), A. 1853-70. *Naturgeschichte der Vögel Europa's*, taf.59, fig.9. Newton, A. 1861. Abstract of Mr. J. Wolley's Researches in Iceland. *Ibis*, p.390. Blasius, W. 1884. Zur Geschichte der Ueberreste von *Alca impennis*. *Journal für Ornithologie*, p.106. Grieve, S. 1885. *The Great Auk or Garefowl*, p.77 and appendix p.21. Grieve, S. 1896-7. Supplementary note on the Great Auk. *Transactions of the Edinburgh Field Naturalists' and Microscopical Society*, p.248. Salomonsen, F. 1944-5. Gejrfuglen et hundredaars minde. *Dyr i Natur og Museum*, p.106. Hahn, P.1963. *Where is that Vanished Bird?* p.219.

Purchased by Dr. Anton Fric (sometimes spelled Fritsch) for the Bohemian National Museum in 1854 at an auction in Carlsbad for the equivalent of £5, this specimen has been in Prague ever since. Fric supposed it to be a bird presented to Baron Feldegg by the Prince of Denmark and he cata-

Chromolithograph from A. Fric's *Naturgeschichte der Vögel Europa's* (1853-70) showing the two specimens in Prague.

Fig.1.Mergulus alle, Bp. Fig.2.Grylle columba, Bp. Fig.3.Grylle columba, Bp. Fig.4.Grylle Mandtii, Fig.5.Uria bringvia Fig.6.Uria arra, Pall. Fig.7.Alca torda, Fig.8.Alca impennis, L.(hiem) Fig.9.Alca impennis, L.(ad).

logued it thus. If he was correct then this is probably - though not necessarily - a bird killed on Eldey during 1834 and given to the Prince on the occasion of his visit to Iceland. This provenance has been claimed for other examples, however, so it cannot be regarded as certain.

THE PRAGUE IMMATURE AUK

Bird no.22 (Grieve no. 65; Hahn no. 22) - immature, perhaps in early autumn plumage
Origin: unknown
Present location: Prague - Narodni Museum, Praha, Czech Republic.

Fric (Fritsch), A. 1853-70. *Naturgeschichte der Vögel Europa's*, taf.59, fig.8. Fric (Fritsch), A. 1863. Notiz über *Alca impennis* in Böhmen. *Journal für Ornithologie*, p.296. Blasius, W. 1884. Zur Geschichte der Ueberreste von *Alca impennis*. *Journal für Ornithologie*, p.106. Grieve, S. 1885. *The Great Auk or Garefowl*, pp.73-5, p.77, appendix p.17 *and* p.21. Grieve, S, 1896-7. Supplementary note on the Great Auk. *Transactions of the Edinburgh Field Naturalists' and Microscopical Society*, p.268 *and* pl.2. Blasius, W. (*in* Naumann) 1903. *Naturgeschichte der Vögel Mitteleuropas*, bd.12, p.179, pl.17a *and* pl.17b. Salomonsen, F. 1944-5. Gejrfuglen et hundredaars minde. *Dyr i Natur og Museum*, p.106. Hahn, P. 1963. *Where is that Vanished Bird?* p.219. Bourne, W. 1993. The story of the Great Auk. *Archives of Natural History*, p.260. Lyngs, P. 1994. Gejrfuglen- Et 150 års minde. *Dansk Ornitologisk Forenings Tidsskrift*, fig. p.61.

The Prague Immature Auk (circa 1890).

This curious example has been in the Narodni Museum for well over a century. The plumage is not typical and it seems to be a young individual, although Bourne (1993) believes it to be a bird in full adult winter plumage. Usually, it is considered the second most juvenile (after the Garefowl in Newcastle) of all existing specimens but it has at least some claim to be regarded as the 'youngest.' It lacks any white spot before the eye, the white plumage of the breast reaches half way up the throat then speckles a darker surface as it continues upwards; similarly, there is a speckled appearance where the white feathers meet the dark ones on the side of the neck, and an extension below the wing of the dark plumage onto the white.

Unfortunately there is absolutely no indication as to where the bird originally came from. It was unknown until 1863 when Professor Anton Fric, Director of the Bohemian National Museum, drew attention to it in the Woboril collection. Some time later, this collection was purchased by a Prague sugar refiner named Anton Richter who generously presented the Auk to Fric's museum.

COPENHAGEN'S SUMMER AUK

Bird no.23 (Grieve no. 33; Hahn no. 24) - adult in summer plumage
Origin: Iceland (possibly summer 1823, but more likely 1832)

Present location: Copenhagen - Universitetets Zoologiske Museum, Universitetetsparken 15, DK-2100 København Ø, Denmark.

Preyer, W. 1862. Der Brillenalk in Europäischen Sammlungen. *Journal für Ornithologie*, p.78. Fatio, V. 1868. Liste des divers représentants de l'*Alca impennis* en Europe. *Bullétin de la Société Ornithologique Suisse*, tome II, pt.1, p.82. Blasius, W. 1884. Zur Geschichte der Ueberreste von *Alca impennis*. *Journal für Ornithologie*, pp.87-90. Grieve, S. 1885. *The Great Auk or Garefowl*, p.76 *and* appendix p.14. Salomonson, F. 1944-5. Gejrfuglen et hundredaars minde. *Dyr i Natur og Museum*, p.105 *and* pl.1. Hahn, P. 1963. *Where is that Vanished Bird?* p.220. Lyngs, P. 1994. Gejrfuglen- Et 150 års minde. *Dansk Ornitologisk Forening Tidsskrift*, fig. p.57 *and* fig. p.60.

Many of the Great Auk specimens that survive today passed through the hands of Professor J.H. Reinhardt of the Royal Museum, Copenhagen. Knowing he could strike attractive deals with collectors and museums all over Europe he acquired, whenever opportunity arose, freshly killed birds as they arrived at the Danish port from Iceland. In 1831, for instance, he received 20 of the 24 birds taken that year on Eldey. Due to a shortage of official funds, he was often obliged to purchase birds himself and hope for re-imbursement later. With the proceeds of sales he was then able to make, he built up the museum's collection in areas that seemed weak.

Of all the Great Auks that passed through the Royal Museum, Reinhardt and his successors chose to keep just two. One is in winter plumage and the other a bird in typical summer dress. It may be assumed that the retained summer bird was - and is - a particularly fine example .

Copenhagen's Summer Auk photographed for J.T. Reinhardt during 1878.

COPENHAGEN'S WINTER AUK

Bird no.24 (Grieve no. 34; Hahn no. 23) - adult in winter plumage
Origin: unknown, said to be Qeqertarsuatsiaat (Fiskenæsset), Greenland
Present location: Copenhagen - Universitetets Zoologiske Museum, Universitetetsparken 15, DK-2100 København Ø, Denmark.

Donovan, E. 1794-1819. *The Natural History of British Birds*, vol.10, pl.243. Boie, F. 1822. Ornithologische Beiträge. *[Oken's] Isis*, p.871. Benicken, F. 1824. Beiträge zur nordischen Ornithologie. *[Oken's] Isis*, pp.886-7. Holbøll, C. 1842-3. Ornithologiske Bidrag til den Gronlandske Fauna. *Kroyer's Naturhistorisk Tidsschrift*, IV, p.457. Reinhardt, J. 1861. List of the Birds hitherto observed in Greenland. *Ibis*, p.16. Newton, A. 1861. Abstract of Mr. J. Wolley's Researches in Iceland. *Ibis*, p.376. Preyer, W. 1862. Der Brillenalk in Europäischen Sammlungen. *Journal für Ornithologie*, p.78. Fatio, V. 1868. Liste des divers représentants de l'*Alca impennis* en Europe. *Bullétin de la Société Ornithologique Suisse*, tome II, pt.1, p.82. Blasius, W. 1881-3. Ueber die letzten Vorkommnisse des Riesen-Alks. *Jahresberichte des Vereins für Naturwissenschaft zu Braunschweig für die Vereinsjahre 1881-2 und 1882-3*, p.92, p.109 *and* p.114. Blasius, W. 1884. Zur Geschichte der Ueberreste von *Alca impennis*. *Journal für Ornithologie*, pp.87-90. Grieve, S. 1885. *The Great Auk or Garefowl*, p.75, p.79 *and* appendix p.14. Newton, A. 1896. *A Dictionary of*

The head of Copenhagen's Winter Auk. Photograph by Geert Brovad. *Courtesy of Morten Meldgaard and the Zoologisk Museum, Copenhagen.*

Hand-coloured engraving from E. Donovan's *Natural History of British Birds* (1794-1819).

Birds, p.305. Parkin, T. 1911. The Great Auk. *Hastings and East Sussex Naturalist*, vol.1, pt.6 (extra paper), pp.3-5 *and* pp.30-1. Salomonsen, F. 1944-5. Gejrfuglens et hundredaars minde. *Dyr i Natur og Museum*, p.101, p.105 *and* pl.12. Greenway, J. 1958. *Extinct and Vanishing Birds of the World*, p.272 *and* pp.290-1. Hahn, P. 1963. *Where is that Vanished Bird?* p.220. Meldgaard, M. 1988. The Great Auk...in Greenland. *Historical Biology*, p.145, p.159, fig.9 *and* fig.13. Lyngs, P. 1994. Gejrfuglen- Et 150 års minde. *Dansk Ornitologisk Forening Tidsskrift*, fig. p.53, fig. p.56 *and* fig. p.60.

For more than 150 years speculation about the origin of this bird has been rife. Did it, or did it not, come from Greenland? If it did, then it is the only Greenland specimen in existence. Yet despite the popularity that this idea has now achieved, a Greenland provenance is by no means certain.

There seems little reason to doubt that a Great Auk in winter dress was killed around the year 1815 by an Inuit hunter operating from a kayak. This happened near to Qeqertarsuatsiaat (Fiskenæsset) and the bird was delivered to the Director of the Royal Greenland Trading Company, a man by the name of Heilman. No-one knows what Heilman did with his specimen and its subsequent history is a matter of conjecture although it has been identified - sometimes tentatively, sometimes more dogmatically - with the winter bird now in Copenhagen.

Some six years after the capture a stuffed bird in winter plumage turned up in the collection of the ornithologist F. Benicken of Schleswig. It was suggested at the time that this was an individual caught at Disko, Greenland in 1821 but this record of a bird wandering so far to the north is usually discounted (Disko being well within the Arctic Circle and no other Great Auk being reliably recorded from anywhere so northerly).

The extreme rarity of winter plumaged birds and the discounting of the Disko data might, ordinarily, clinch the issue; naturally, it makes a Qeqertarsuatsiaat origin for Benicken's specimen seem likely. There is one factor that has been overlooked by all Garefowl researchers, however. During the six year gap in the assumed history of Benicken's Auk another winter bird became available, one that looks - in a surviving portrait - suspiciously like the example now in Copenhagen.

This is the celebrated Leverian Auk, a stuffed bird with a provenance taking it well back into the eighteenth century and an individual allegedly killed in Britain. It was auctioned in London on May 6th 1818, after which it mysteriously disappeared. There is no record of who bought it, where it went to or how much was paid for it. All that appears to remain is a pic-

Copenhagen's Winter Auk. Photo by Geert Brovad. *Courtesy of Morten Meldgaard and the Zoologisk Museum, Copenhagen.*

ture in volume ten of Edward Donovan's *Natural History of British Birds* (1794-1819). Although the portrait is drawn in a romantic style, certain parallels between the individual depicted and the Copenhagen bird are immediately striking; even details of the interior of the slightly parted beak are similar. All this may be no more than co-incidence, of course; one winter Auk might look very much like another. What is certain is that a specimen of unknown provenance came into the possession of Benicken soon after the Leverian Sale and that this has often been correlated with the bird taken in Greenland in 1815.

At an unknown date between 1830 and 1842 a Danish ornithologist, Emil Hage, bought Benicken's bird and on December 20th 1842 he sold it to Professor J. Reinhardt of the Royal Museum.

Japetus Steenstrup, writing to Symington Grieve in 1885, certainly believed in the Greenland tradition but despite his bold tone his words show that the Copenhagen museum curators had no real knowledge of their bird's origin. His opinion is clearly based on what he believed was probable rather than what he really knew:

> Our individual in winter dress, I have no doubt is identical with Benicken's, and with Heilman's got in 1815, and is the same individual seen by Faber in Copenhagen before his voyage to Iceland ... Reinhardt, senior, bought this not very good specimen at a price three times higher than the price paid for the excellent specimens from Iceland offered to the museum in the same year. [There were, in fact, no birds from Iceland in 1842].

The implication that Reinhardt paid heavily for this specimen simply because of its alleged Greenland provenance is certainly misleading. Reinhardt probably saw more specimens than any other ornithologist and well understood the importance and rarity of a bird in winter feathering. Presumably, he paid through the nose for the state of the bird's plumage as much as for any alleged locality data. That later curators failed to see the significance of the bird's dress is shown by the fact that they considered this an 'inferior' example and allowed it to be routinely handled by students during their teaching programs.

THE ABBEVILLE AUK

Bird no.25 (Grieve no. 3; Hahn no. 31) - adult in summer plumage
Origin: unknown, but almost certainly Iceland
Present location: Abbeville - Musée Boucher des Perthes, Rue de Beffroi, Abbeville, France.

Canivet, E. 1843. *Catalogue des Oiseaux du Départment de la Manche*, p.31. Degland, C. 1849. *Ornithologie Européene*, vol.2, p.529. Degland, C. 1855. Première liste des ornithologists de France. *Naumannia*, p.423. Newton, A. 1865. The Gare-fowl and its Historians. *Natural History Review*, p.475. Newton, A. 1870. On Existing Remains of the Gare-fowl. *Ibis*, p.258. Blasius, W. 1884. Zur Geschichte der Ueberreste von *Alca impennis. Journal für Ornithologie*, p.71. Grieve, S. 1885. *The Great Auk or Garefowl*, p.79 *and* appendix pp.5-6. Gadeau de Kerville, H. 1890-2. *Faune de la Normandie*, fasc.3, pp.499-502 *and* pl.1. Dresser, H. 1871-96. *A History of the Birds of Europe*, vol.8, p.565. Duchaussoy, H. 1897-8. Le Grand Pingouin du Musée d'Histoire naturelle d'Amiens *and* Notes additionnelles. *Mémoires de la Société Linnéenne du Nord de la France*, pp.115-8, *and* pp.245-6 *and* pl.2. Salomonsen, F. 1944-5. Gejrfuglen et hundredaars minde. *Dyr i Natur og Museum*, p.106. Hahn, P. 1963. *Where is that Vanished Bird?* p.222. Voisin, C. & J. 1991. Sur quelques spécimens remarquables de la Musée Boucher de Perthes. *Bulletin de la Société d'Émulation Historique et Littéraire d'Abbeville*, tome 27, fasc. 1, p.108.

The Abbeville Auk (circa 1890). Photograph by H. Gadeau de Kerville.

This bird came to the Musée Boucher des Perthes during the nineteenth century from the collection of a Monsieur de la Motte. There are two conflicting tales regarding its previous history. The first - and by far most likely - is told by Alfred Newton in a letter (1883) to his fellow Garefowl enthusiast Wilhelm Blasius of Brunswick:

> The specimen ... was got in Iceland, and sent in the year 1831 to M. de la Motte from the Royal Museum in Copenhagen as I have learned from my good old friend Reinhardt.

Newton's 'old friend,' the son of the Professor Reinhardt who sent preserved Garefowls all over Europe, was clearly in a position to know. That he was correct in believing a Great Auk was sent to de la Motte was recently confirmed by Gerd Malling (*pers. comm.* 1996) who combed the Copenhagen archives and found a record of the transaction. On 15th May 1831 a skin was sent to Abbeville for de la Motte and Monsieur de Cossette (*see* bird no.12) who were evidently partners. The specimen had been received by the Royal Museum on 21st October of the previous year and was shot, together with another individual by one Marskal V. Moltke on:

> Det længst udliggende Gyrskjaer [the most distant Garefowl Island].

Although there is no reason to doubt this record, it is just possible that de la Motte disposed of this bird elsewhere, there being evidence to show that he regularly traded in natural history specimens.

The second story was promoted by Côme-Damien Degland (one-time owner of the Garefowl now in Lille) who believed that at an unknown date during the first decade of the nineteenth century three Great Auks were killed at Cherbourg and that one of these was subsequently acquired by de la Motte. This is not the only tale of Garefowls dying on the French coast that Degland retailed. He maintained that two birds turned up near Dieppe (one of which he correlated with the specimen once in the town's museum) during the same decade as the Cherbourg birds. These stories of individuals washing ashore in northern France are not necessarily untrue but there is little likelihood that their remains still exist.

THE AMIENS AUK

Bird no.26 (Grieve no. 4; Hahn no. 32) - adult in summer plumage
Origin: unknown, probably Iceland
Present location: Amiens - Parc Zoologique de la Petite Hotoie, 139 Rue du Faubourg de Hem, 80000 Amiens, France.

The Amiens Auk (circa 1890). Photograph by Nocq.

Newton, A. 1870. On Existing Remains of the Gare-fowl. *Ibis*, p.258. Blasius, W. 1884. Zur Geschichte der Ueberreste von *Alca impennis*. *Journal für Ornithologie*, p.72. Grieve, S. 1885.*The Great Auk or Garefowl*, p.79 *and* appendix p.6. Grieve, S. 1896-7. Supplementary note on the Great Auk. *Transactions of the Edinburgh Field Naturalists' and Microscopical Society*, p.245. Duchaussoy, H. 1897-8. Le Grand Pingouin du Musée d'Histoire naturelle d'Amiens *and* Notes additionnelles. *Mémoires de la Société Linnéenne du Nord de la France*, p.88, pp.118-5, pl.1, *and* p.244, *and* p.247. Salomonsen, F. 1944-5. Gejrfuglen et hundredaars minde. *Dyr i Natur og Museum*, p.106. Hahn, P. *Where is that Vanished Bird?* p.222.

Herr Mechlenburg, the Apothecary of Flensburg, is rumoured to have bought and sold no less than eight Great Auks during his lifetime. All of his birds supposedly came from Iceland and one of them he sold to a Monsieur A. Delahaye, one-time Director of the St. Omer Natural History Museum.

It is assumed that this specimen is identical to the one now in Amiens and that it was donated to the town's museum by Delahaye's son Antoine. Alfred Newton saw it in the museum in 1862 and it probably arrived there at some time during the previous five years.

THE AUTUN AUK

Bird no.27 (Grieve no. 17; Hahn no. 33) - adult in summer plumage, said to be male
Origin: unknown
Present location: Autun - Muséum d'Histoire Naturelle, 14 Rue Saint-Antoine, 71400 Autun, France.

Blasius, W. 1884. Zur Geschichte der Ueberreste von *Alca impennis*. *Journal für Ornithologie*, p.75. Olphe-Galliard, L. 1884. *Contributions à la Faune Ornithologique de l'Europe Occidentale*, fasc.1, p.27. Grieve, S. 1885. *The Great Auk or Garefowl*, p.79 *and* appendix pp.8-9. Duchaussoy, H. 1897-8. Le Grand Pingouin du Musée d'Histoire naturelle d'Amiens *and* Notes additionnelles. *Mémoires de la Société Linnéenne du Nord de la France*, p.114 and p.243. Grieve, S. 1897-8. Additional notes on the Great Auk. *Transactions of the Edinburgh Field Naturalists' and Microscopical Society*, p.328. Hahn, P. 1963. *Where is that Vanished Bird?* p.223.

This Great Auk was presented to the town of Autun in 1895. Before that it was in the collection of Dr. B.F. de Montessus of Chalon-sur-Sâone. Montessus acquired his bird - a badly stuffed one reduced to a cabinet skin-from Dr. L.W. Schaufuss of Dresden by responding to a circular, issued in January 1873, in which the Dresden doctor offered it for sale. The Frenchman paid 3,000 francs despite the specimen's pitiful state; head and feet were detached from the body. The two men had done business together for a number of years and, with the Auk being so in need of improvement, the skins of two Razorbills were included in the bargain; doubtless these were intended for use as patching.

The earlier history of the specimen is unknown although it is probable that it came from Iceland. There is a suspicion that it corresponds with a damaged bird that the Apothecary Mechlenburg left among his effects when he died in 1861 but it could just as easily have come from another source.

Despite the descriptions of the specimen's poor condition, Monsieur G. Pacaud, Conservateur of the Autun Museum, was able to report (*pers. comm.*, 1995) that it appears in a reasonably good state and is on show to the public; clearly, the patching worked. Paul Hahn (1963) recorded that a resident of Chicago (probably on behalf of the Field Museum) offered $1,000 for this bird in 1958 but was turned down.

MALHERBE'S AUK

Bird no.28 (Grieve no. 48; Hahn no. 36) - adult in summer plumage
Origin: Eldey (1831)
Present location: Le Havre - Museum d'Histoire Naturelle, Place du Vieux-Marché, 76600 Le Havre, France.

Newton, A. 1870. On Existing Remains of the Gare-fowl. *Ibis*, p.258. Blasius, W. 1881-3. Ueber die letzen Vorkommnisse des Riesen-Alks. *Jahresberichte des Vereins für Naturwissenschaft zu Braunschweig für die Vereinsjahre 1881-2 und 1882-3*, p.106 *and* p.114. Blasius, W. 1884. Zur Geschichte der Ueberreste von *Alca impennis. Journal für Ornithologie*, p.95. Grieve, S. 1885. *The Great Auk or Garefowl*, p.79 *and* appendix p.16. Grieve, S. 1888. Recent information about the Great Auk. *Transactions of the Edinburgh Field Naturalists' and Microscopical Society*, p.108. Duchaussoy, H. 1897-8. Le Grand Pingouin du Musée d'Histoire naturelle d'Amiens. *Mémoires de la Société Linnéenne du Nord de la France*, pp.110-1. Scherdlin, P.1925. A propos du Grand Pingouin du Musée Zoologique de Strasbourg. *Bulletin de l'Association Philomathique d'Alsace et de Lorraine*, tome VII, fasc. prem., pp.16-7. Salomonsen, F. Gejrfuglen et hundredaars minde. *Dyr i Natur og Museum*, p.106. Hahn, P. 1963. *Where is that Vanished Bird?* p.223.

Malherbe's Auk. Photograph by T. Vincent. *Courtesy of the Museum d'Histoire Naturelle, Le Havre.*

"His wife liked money better than birds."
The wife was that of the recently deceased French ornithological writer Alfred Malherbe and the disparaging words, scribbled on January 21st 1871, belonged to a disgruntled Englishman, J.H. Gurney. What had fired Gurney with such self-righteous disapproval was the manner of the disposal of Malherbe's collection of stuffed birds. Not only did the grieving widow sell off the birds almost as soon as her poor husband was cold, it seemed she made an even greater transgression against Gurney's delicate sensibilities. Many of the specimens hadn't even lasted that long. She'd started the sell-off while her husband was still breathing!

For all anyone knows, the dying man was perfectly content for his wife to secure her future, but Gurney's dissatisfaction is expressed in a series of letters (kept today at the Zoological Museum, Cambridge) written during a lengthy stay in Malherbe's home town of Metz. The letters were to Alfred Newton and in them Gurney reported on all matters concerning Great

Auks. Monsieur Malherbe, formerly a judge of the Civil Tribunal at Metz, had owned a stuffed one, so the fate of his collection was of enormous interest to Newton and to Gurney - both, of course, avid Auk fans.

What had become of the Auk? That, naturally, was the chief concern of the two Englishmen. Was it the example that both gentlemen knew existed in the town museum? The Curator didn't seem sure! But then, he had little interest in his museum's Great Auk. According to an earlier letter of Gurney's:

It is not in a case by itself and the Curator did not seem in the least to know that it was a rare bird.

It soon became apparent that the specimen was indeed the bird once owned by Malherbe. The Metz Museum had purchased many of his birds and Gurney found an inscription on the Auk's stand indubitably written in the French judge's hand. It has since been established that it was purchased in 1857 for the sum of 65 francs.

The origin of Malherbe's Auk is clear. He received it from Professor Reinhardt of Copenhagen - perhaps during 1833, more probably in 1842 - together with the information that it came from Iceland in 1831. Almost certainly, therefore, it is one of the 24 individuals taken that year on Eldey.

Forty years or so after this event, Gurney - during his visit to Metz - made an interesting observation concerning the white slashes on the bird's beak. In his opinion this colour was natural; they had not been painted. Gurney couldn't help noticing something else:

The neck is dreadfully stretched and appears to have been cut open ... It is miserably stuffed and very mouldy.

A little later (1888) Symington Grieve added to the literature another description of the bird's sad state, quoting a letter he'd just received from the youngest of the Frank family, G.A. Frank of Haverstock Hill, London:

I have seen this summer the specimen in Metz ... but I fear that this bird will not last many more years, as the moths have got in it.

Despite these reported ravages this Garefowl has survived, although it is no longer in Metz having been transferred to the museum at Le Havre.

Alfred Malherbe is best remembered not for his Great Auk but for his sumptuous and exceedingly rare work on woodpeckers, *Monographie des Picidees*, (1859-62), generally regarded as one of the jewels of the golden age of bird books. Ironically, his own collection of woodpeckers was one of the first of his possessions to vanish during his wife's pruning exercise and they went to Count Ercole Turati of Milan, a man who, incidentally, was also once the owner of a stuffed Great Auk.

J.H. Gurney Jnr, author of several papers on the Great Auk.

THE LILLE AUK

Bird no.29 (Grieve no. 39; Hahn no. 35) - adult in summer plumage
Origin: unknown
Present location: Lille - Musée d'Histoire Naturelle, 19 Rue de Bruxelles, 59000 Lille, France (catalogue no. 535a).

Macquet-Degland. 1857. *Catalogue Raisonné de la Collection D'Oiseaux D'Europe de Côme-Damien Degland, Acquise par la Ville de Lille*, p.273. Olphe-

Galliard, L. 1862. [Letter to the Editor] *Ibis*, p.302. Newton, A. 1870. On Existing Remains of the Gare-fowl. *Ibis*, p.258. Blasius, W. 1884. Zur Geschichte der Ueberreste von *Alca impennis*. *Journal für Ornithologie*, p.92. Grieve, S. 1885. *The Great Auk or Garefowl*, p.79 *and* appendix p.15. Duchaussoy, H. 1897-8. Le Grand Pingouin du Musée d'Histoire naturelle d'Amiens *and* Notes additionnelles. *Mémoires de la Société Linnéenne du Nord de la France*, p. 113 *and* p.244. Hahn, P. 1963. *Where is that Vanished Bird?* p.223.

During the first half of the nineteenth century, and up until 1855, Dr. Côme-Damien Degland formed an impressive museum collection of European birds. After his death these were acquired by the town of Lille. By far the most important specimen was a Great Auk. According to a letter (preserved in the Newton archive at Cambridge) from the middle Frank (G.A. Frank of Amsterdam), this bird was a gift he made to Degland in return for many favours he'd received. Frank makes it clear in his letter that the specimen was faulty and, among other defects, had only one leg.

Confirmation of this tale is given in one of H. Duchaussoy's papers (1897-8). He quotes from a letter written on July 19th 1897 by Monsieur de Méezemaker (once the owner of two Garefowl eggs). Méezemaker saw the specimen soon after it was presented to Degland and recorded that - although the head and beak were good - many of the feathers on the body were ruffled and there were large areas of bald or torn skin; also, a foot was lacking. Frank had believed that the damage might be too bad to effect a successful repair but Degland, ever hopeful, took the bird to Monsieur Semet, the taxidermist at the Lille Museum. With great skill, this gentleman entirely transformed it. De Méezemaker claimed to be the only man then living (1897) who knew this and claimed also that he was present during much of the restoration process, providing Razorbill skins to facilitate the repair. The missing foot was reconstructed from cardboard or *papier mâché*.

These stories fit together very neatly, except for on thing. During 1994, the specimen was X-rayed and the radiography revealed not only (as might be expected) the presence of a skull and wing bones but also the presence of the bones of the feet. Probably, de Méezemaker was wrong about the *papier mâché* foot (he was a very old man at the time he recalled the restoration) and the taxidermist had, perhaps, substituted a leg of similar shape and size (that of a Diver, for instance). Such a substitution might not prove easy to detect on an X-ray.

It is likely that, like most surviving specimens, this one is Icelandic and that it came from Eldey. A detailed catalogue of the Degland collection was published in 1857 - just after Lille acquired it - but this is not very helpful concerning the Great Auk's provenance. The entry reads:

Male - Mai 1835- plumage de noces et d'adulte.

The date of 1835 probably relates to the bird's death rather than to Degland's acquisition of it, though this is not completely clear. The Lille Musuem holds a note that the specimen came from Greenland but this is unlikely to be correct. It may result from a superficial reading of the rest of Degland's catalogue entry - which consists of a few general words relating to the species' range, habits and rarity:

The Lille Auk. *Courtesy of Bertrand Radigois and the Musée d'Histoire Naturelle, Lille.*

Le Groenland, la baie de Baffin, le nord-ouest de l'Island. Il niche dans les grandes crevasses des rochers. C'est un oiseaux excessivement rare, et qui fend à le devenir encore davantage; aussi, bien peu de collections possèdent.

The entry concludes with a few words on the spectacular behaviour of Englishmen:

Ses œufs sont vendus excessivement cher; ainsi, on cite un Anglais qui en a acheté un pour le prix de 600 francs.

THE VIAN AUK

Bird no. 30 (Hahn no. 37) - adult in summer plumage
Origin: Iceland
Present location: Nantes - Muséum d'Histoire Naturelle, 12 Rue Voltaire, 44000 Nantes, France.

Olphe-Galliard, L. 1884. *Contributions à la fauna ornithologique de l'Europe occidentale*, fasc.1, p.26. Grieve, S. 1885. *The Great Auk or Garefowl*, p.79, p.95 *and* appendix p. 20. Grieve, S. 1896-7. Supplementary note on the Great Auk. *Transactions of the Edinburgh Field Naturalists' and Microscopical Society*, pp.250-2, p.265, p.270 *and* pl.IV. Duchaussoy, H. 1897-8. Le Grand Pingouin du Musée d'Histoire naturelle d'Amiens- notes additionnelles. *Mémoires de la Société Linnéenne du Nord de la France*, pp.247-51. Bureau, L. 1933. Le Pingouin Brachyptère...de la Collection Jules Vian. *Bullétin de la Société des Sciences Naturelles de l'Ouest de la France*, pp.165-72, pl.2 *and* pl.3. Hahn, P. 1963. *Where is that Vanished Bird?* pp.223-4.

When 24 birds were taken on Eldey during the summer of 1831, one individual was kept alive - more or less out of curiosity - and carried, by the fishermen who caught it, back to the town of Keflavik wrapped in a cloth. Here it passes out of historical record although a note made by John Wolley suggests that it was, perhaps, killed by a man named Thomsen. It is possible, however, that this is not quite what happened and that this captured bird can be correlated with the mounted specimen now in Nantes.

The origin of this Nantes specimen is documented in an intriguing and comparatively thorough manner, despite the fact that its existence became known to researchers only at a fairly late date. Symington Grieve, for instance, found out about it just before publication (1885) of his famous monograph. The owner, Jules Vian of 42 Rue de Petits Champs, Paris, Honorary President of the Société Zoologique de la France, soon became one of Symington's correspondents and the Scotsman speedily discovered a great deal about the bird. Monsieur Vian (1815-1904) owned an almost complete museum collection of European birds and was only too pleased to send Grieve a long and full account of his Garefowl. Grieve's translation (1896-7) from Vian's French outlines the tale:

The following information was given me 50 years ago by the stuffer of the bird. He said ... it was one of a colony which laid their eggs on an inaccessible islet [evidently the Geirfuglasker]; that in 1830 a volcanic eruption covered this islet with debris; that several specimens, dead or dying, were driven by the waves on the shores of Iceland, and were eaten; that he (my friend the stuffer) himself secured the subject of this letter while it was still alive. He kept it for several days on the ship which was returning to France ... it died on the way, and ... he stuffed it during the voyage. In fact, its feathers, and the exact retention of its outlines and undulations, indicate a bird mounted while the skin was newly removed from the flesh ... It was in 1847 that I saw the *Alca* ... of which I am at present the owner. It formed part of the collection of Monsieur Oursel (father) of Havre. That collection was not very numerous, but it was composed of birds perfectly mounted, and indicated study of the forms... especially of the heads... Monsieur Oursel, who had picked it up in Iceland and mounted it on the return voyage, furnished me at that time with circumstantial details, but I did not take any notes as I did not then imagine that the bird would become part of my collection. I have nothing but recollections of fifty years ago, and my memory, which has been at work now for eighty-one years, has no longer the strength of my early days. Monsieur Oursel (father) died a few years afterwards. I acquired the bird from Monsieur Oursel (son) who I believe was not born in 1830, and who had kept his father's collection without having personally any taste for natural history, and was not able to give me any information. I do not know whether Monsieur Oursel (son) is still alive, but he sold all his collection in 1881.

This, then, is one of the individuals disturbed and made homeless by the destruction of the Geirfuglasker. Although Monsieur Vian makes it clear that his bird died during 1830 and was an individual driven ashore by the volcanic activity, this may not be the case. As Vian was relying on his memory of a conversation that took place long before, he'd quite possibly muddled the tale slightly; his bird was, perhaps, one of the 24 disorientated individuals taken on Eldey during 1831. The Garefowl that escaped the massacre on the island and was taken to Keflavik was kept alive with a deliberate intention. The fisherfolk wanted to see whether the merchants they dealt with would send the bird - still living - to foreign parts. Perhaps they did, or perhaps the fishermen themselves sold it to Monsieur Oursel. It is, of course, just as likely that Vian's story is entirely accurate and has no bearing on the birds of 1831 whatsoever.

In 1932, the descendants of Jules Vian gave their splendid specimen to the museum at Nantes where it is regarded as a great treasure.

The Vian Auk. *Courtesy of the Muséum d'Histoire Naturelle, Nantes.*

THE PARIS AUK

Bird no.31 (Grieve no. 59; Hahn no. 38) - adult in summer plumage
Origin: unknown but said to be from the coasts of Scotland
Present location: Paris - Muséum National d'Histoire Naturelle, 55 Rue de Buffon, 75005 Paris, France.

Preyer, W. 1862. Der Brillenalk in Europäischen Sammlungen. *Journal für Ornithologie*, p.77. Champley, R. 1864. The Great Auk. *Annals and Magazine of Natural History*, p.235. Fatio, V. 1868. Liste des divers représentants de

l'*Alca impennis* en Europe. *Bullétin de la Société Ornithologique Suisse*, tome II, pt.1, p.82. Blasius, W. 1884. Zur Geschichte der Ueberreste von *Alca impennis*. *Journal für Ornithologie*, p.104. Grieve, S. 1885. *The Great Auk or Garefowl*, p.79, appendix p.20 *and* p.41. Milne-Edwards, M. & Oustalet, M. 1893. Notice sur Quelques Espèces d'Oiseaux Actuellement Éteintes...dans les collections du Muséum. *Volume Commémoratif du Centenaire de la Fondation du Muséum d'Histoire Naturelle*, pp.57-8 *and* pp.61-2. Duchaussoy, H. 1897-8. Le Grand Pingouin du Musée d'Histoire naturelle d'Amiens *and* Notes additionnelles. *Mémoires de la Société Linnéenne du Nord de la France*, p.113 *and* p.243. Didier, R. 1934 (Jan.). Le Grand Pingouin. *La Terre et La Vie*, fig.1 *and* fig.2 (p.14). Berlioz, J. 1935. Notice sur les spécimens naturalisés d'oiseaux éteintes existant dans les collections du Muséum. *Archives de Muséum d'Histoire Naturelle*, p.487. Jouanin, C. 1962. Inventaire des Oiseaux Éteintes ou en voie d'Extinction Conserves au Museum de Paris. *La Terre et la Vie*, p.272. Hahn, P. 1963. *Where is that Vanished Bird?* p.224.

The Great Auk acquired by the Paris Museum during 1832 allegedly came from Scottish coasts. An old inscription on its stand read 'Des Côtes de l'Ecosse' and this origin has been defended by successive curators at the museum. Milne-Edwards and Oustalet (1893) pointed out that the inter-

esting yet seemingly unlikely data is justified by the known capture of a bird in Irish waters (1834) and the record of a bird at St Kilda in 1821. Other late records of birds in Scottish waters do exist. An individual was perhaps washed ashore at Gourock and the bird captured at St Kilda in 1821 escaped into the Firth of Clyde. The Paris curators rejected Blasius's belief (1884) that the specimen might be the one figured by Brisson (1760) belonging to the Réamur Collection, a specimen that passed into the *Cabinet du Roi* and was soon lost sight of. That they were right to reject this idea is clear. Brisson figured a swimming bird rather than one standing in the manner of the Paris example. Just as telling is the condition of the bird. It is far too good for a specimen with a history reaching back to the middle of the eighteenth century. The confidence of the Paris curators in a Scottish origin is difficult to share, however. The date of 1832 is highly suggestive and points to the likelihood of this being one of the 24 individuals taken on Eldey during the previous year. But, bearing in mind the ancient historical association between Scotland and France, one can never be sure.

(Facing page, top). Two views of the Paris Auk (circa 1930).

(Facing page, bottom). Hand-coloured engraving by F. Martinet from M. Brisson's *Ornithologie* (1760).

JOSSE HARDY'S AUK

Bird no.32 (Grieve no. 20; Hahn no. 34) - adult in summer plumage
Origin: unknown, said to be Dieppe, thought to be Iceland
Present location: Rouen- Museum d'Histoire Naturelle, 198 Rue Beauvoisine, 76000 Rouen, France.

Hardy, J. 1841. Catalogue of Birds observed in the Department of Seine-Inférieure. *Annuaire de l'Association de la Basse Normandie*, p.298. Degland, C. 1849. *Ornithologie Européene*, tome II, p.529. Champley, R. 1864. The Great Auk. *Annals and Magazine of Natural History*, p.235. Fatio, V. 1868. Liste des divers représentants de l'*Alca impennis* en Europe. *Bullétin de la Société Ornithologique Suisse*, tome II, pt.1, p.82. Blasius, W. 1884. Zur Geschichte der Ueberreste von *Alca impennis*. *Journal für Ornithologie*, p.78. Grieve, S. 1885. *The Great Auk or Garefowl*, p.79 *and* appendix pp.9-10. Dresser, H. 1871-96. *A History of the Birds of Europe*, vol.8, p.565. Duchaussoy, H. 1897-8. Le Grand Pingouin du Musée d'Historie naturelle d'Amiens *and* Notes additionnelles. *Mémoires de la Société Linnéenne du Nord de la France*, p.114 *and* p.245. Grieve, S. 1897-8. Additional notes on the Great Auk. *Transactions of the Edinburgh Field Naturalists' and Microscopical Society*, p.338. Hahn, P. 1963. *Where is that Vanished Bird?* p.223.

Several Great Auks were reported on the coasts of the English Channel during the early years of the nineteenth century. Traditionally, these reports have met with scepticism, but they may be true. The Reverend W. Whitear in his *Calendar* (posthumously published in *Transactions of the Norwich Natural History Society*, 1880-1) suggests that a Great Auk was captured near Hastings between 1814 and 1816. Côme-Damien Degland (1849) recorded that three individuals were killed at Cherbourg, probably in the first years of the century. The same writer maintained that in the first decade of the nineteenth century, during the month of April (but in two different years), two Great Auks were discovered on the shore near Dieppe; one was found dead, the other was killed. Of this pair nothing else is known, except that one of them is rumoured to have passed into the collection of a local resident - the ornithologist Josse Hardy.

Both Alfred Newton and H.E. Dresser took exception to this locality data (as they did over Cherbourg data for the Abbeville Auk) and believed - on a balance of likelihood, for they had no evidence - that Hardy's bird came from Iceland. As there is no mention of a Great Auk in an 1841 catalogue of the Hardy collection, it is assumed that he acquired his bird after this date and this is said to argue against the possibility of a Dieppe locality and point strongly to an Icelandic origin. In reality it does neither. There is every reason to suppose that a bird taken in Dieppe might stay for years in the town and be acquired by its most avid stuffed bird collector only after the passing of several decades. On the other hand, Iceland is a fairly safe guess for the origin of any Great Auk with doubtful locality data attached to it.

Hardy's own daughter, Madame Le Bœuf, regarded the Dieppe locality as false; she stated, after her father's death, that the specimen was the gift of Temminck, first Director of the Leiden Museum. She made a similar claim for an egg that Hardy once owned, but whether she really knew the truth concerning these specimens, whether she was confusing the two or whether she was completely mistaken, cannot now be said. It is certainly possible that Temminck had a Great Auk to give away. Long before his association with Leiden, he was a keen collector of stuffed birds with a home completely full of them. If the bird was really his, there is no way of knowing from where he might have obtained it; it might even have come from Dieppe!

Whatever the truth concerning its origin, the bird's more recent history is well enough known. Before his death - which occurred towards the end of 1863 - Hardy donated most of his ornithological material (including the Great Auk) to the town of Dieppe and the collection became the property of the Dieppe Museum. For many years the specimen stayed in the town but it has now passed to the Museum d'Histoire Naturelle at Rouen.

THE STRASBOURG AUK

Bird no.33 (Grieve no. 69; Hahn no. 40) - adult in summer plumage
Origin: unknown, rumoured to be Newfoundland
Present location: Strasbourg - Musée Zoologique de l'Université Louis Pasteur et de la Ville de Strasbourg, 29 Place de la Victoire, F67000 Strasbourg, France.

Preyer, W. 1862. Der Brillenalk in Europäischen Sammlungen. *Journal für Ornithologie*, p.78. Fatio, V. 1868. Liste des divers représentants de l'*Alca impennis* en Europe. *Bullétin de la Société Ornithologique Suisse*, tome II, pt.1, p.82. Newton, A. 1870. Recent Ornithological Publications. *Ibis*, p.518. Selys-Longchamps, E. 1876. Note sur un voyage scientifique fait en Allemagne. *Comptes-rendues des séances de la Société entomologique de Belgique*, p.LXVII. Blasius, W. 1884. Zur Geschichte der Ueberreste von *Alca impennis*. *Journal für Ornithologie*, pp.108-10. Grieve, S. 1885. *The Great Auk or*

Garefowl, p.80 *and* appendix pp.22-3. Blasius, W. (*in* Naumann) 1903. *Naturgeschichte der Vögel Mitteleuropas*, bd.12, p.206. Scherdlin, P. 1925. A propos du Grand Pingouin du Musée Zoologique de Strasbourg. *Bulletin de l'Association Philomathique d'Alsace et de Lorraine*, tome VII, fasc. prem., pp.10-17. Hahn, P. 1963. *Where is that Vanished Bird?* pp.224-5. Bourne, W. 1993. The story of the Great Auk. *Archives of Natural History*, p.260.

The Strasbourg Auk is an ancient and very battered one. Writing in 1862, W. Preyer felt obliged to describe it as the worst example he knew of. Eight

The Strasbourg Auk (circa 1920). *Courtesy of Marie-Dominique Wandhammer and the Musée Zoologique, Strasbourg.*

P.S. Pallas (1741-1811).

The Berlin Auk photographed circa 1900.

years later this already badly damaged stuffed bird survived the Franco-Prussian War and the bombardment of Strasbourg in circumstances that went largely unrecorded but are described as 'almost miraculous' (Newton, 1870, *Ibis*, p.518 *and* Grieve, 1885).

After another thirteen years Dr. Döderlein, one time Director of the Strasbourg Museum, reported further on the specimen's deteriorating state:

> Our specimen is in a truly pitiable condition. Its head, wings and posteriors reveal suspiciously bare places; on the lower jaw the whole hornsheath of the bill is wanting; the rest is tolerable; the upper jaw is genuine; the feet are very well preserved.

Although Dr. Döderlein knew of no information concerning the specimen's origin, and in any case distrusted the records kept at Strasbourg, he expressed the hope that:

> Perhaps there are, hidden in the archives of the Museum, some facts that may at some future time cast light upon its history.

Unbeknown to Döderlein, his English colleague Alfred Newton had already unearthed just such information: a catalogue entry that almost certainly relates to this bird. According to Newton, the Strasbourg Auk was entered into a catalogue in 1776 and was given by the naturalist and explorer P.S. Pallas - after whom a number of bird species are named - to Dr. Jean Hermann, the founder of the Strasbourg Museum, in 1760. Its place of origin was given as 'Mers du Nord' ('northern seas'). From this information Newton - always so quick to find fault with the speculations of others - jumped to the conclusion that Strasbourg's Auk came from Newfoundland. This it may well have done, but it could easily have come from elsewhere. Perhaps the balance of probability lies a little in favour of Newfoundland.

This historic old Auk still survives, on display to the public, at the Musée Zoologique.

THE BERLIN AUK

Bird no.34 (Grieve no. 6; Hahn no.26) - adult in summer plumage
Origin: Iceland, either the Reykjanes coast or Eldey
Present location: Berlin - Museum für Naturkunde der Humboldt-Universität, Invalidenstraße 43, 10115 Berlin, Germany (catalogue no. 14466).

Lichtenstein, H. 1854. *Nomenclator Avium*, p.105. Preyer, W. 1862. Der Brillenalk in Europäischen Sammlungen. *Journal für Ornithologie*, p.87. Fatio, V. 1868. Liste des divers représentants de l'*Alca impennis* en Europe. *Bullétin de la Société Ornithologique Suisse*, tome II, pt.1, p.81. Blasius, W. 1884. Zur Geschichte der Ueberreste von *Alca impennis*. *Journal für Ornithologie*, pp72-3. Grieve, S. 1885. *The Great Auk or Garefowl*, p.79 *and* appendix p.6. Schillings, C. 1907. *In Wildest Africa*, plate on p.521. Salomonsen, F. 1944-5. Gejrfuglen et hundredaars minde. *Dyr i Natur og Museum*, p.106. Stresemann, E. 1954. Ausgestorbene und Aussterbende Vogelarten, Vertreten im Zoologischen Museum zu Berlin. *Mittheilungen aus dem Zoologischen Museum in Berlin*, p.40. Hahn, P. 1963. *Where is that Vanished Bird?* p.221. Luther, D. 1986. *Die Ausgestorbenen Vögel der Welt*, p.82.

During the early summer of 1830 the Great Auks of south-west Iceland returned from their season at sea to find that their breeding station - the Geirfuglasker - had vanished. Due to volcanic activity it had simply disappeared beneath the waves. Henceforth, the remaining birds chose to resort to the nearby, but much more accessible, island of Eldey where their last encounter with man was so soon to occur. The volcanic activity was not entirely over at the time of the birds' return and it continued for many months causing violent disturbances in the sea. A number of disorientated birds were reported washed ashore where they were promptly dispatched by the Icelanders who found them. The Great Auk now in Berlin is one of these poor individuals, or else it may be a bird killed on Eldey when the island was first raided.

On December 28th 1830 Professor Reinhardt of Copenhagen wrote to H. Lichtenstein at the Berlin Museum advising him of the year's crop of birds. Due to the tragedy that occurred earlier in the year, Reinhardt had specimens surplus to requirements. Roughly translated, his letter runs:

> Several years ago you ordered a Great Auk for the Berlin Museum. If you still need a specimen I can now oblige you. I cannot dispose of it for under 24 reichthalers... The skin is good, the feathers tight, the bare parts complete and there is only a little dirt and grease. It is obvious that the skin was stretched when the bird was skinned by a fisherman from Iceland, but this may be corrected when the bird is stuffed.

Lichtenstein clearly agreed the deal for he acknowledged safe receipt of the skin in a letter dated March 15th 1831. The specimen has remained in Berlin ever since.

M.H.C. Lichtenstein (1780-1857).

THE FLOORS CASTLE AUK

Bird no.35 (Grieve no. 24; Hahn no. 69 *and* lost bird no. 2) - adult in summer plumage
Origin: unknown
Present location: Bonn - Zoologisches Forschungsinstitut und Museum Alexander Koenig, Adenauerallee 160, 53113 Bonn, Germany.

Gibson, J. 1882-3. On a hitherto Unrecorded Specimen of the Great Auk in the Collection of the Duke of Roxburghe. *Proceedings of the Royal Physical Society, Edinburgh*, pp.335-8 *and* fig. p.336. Blasius, W. 1884. Zur Geschichte der Ueberreste von *Alca impennis*. *Journal für Ornithologie*, pp.83-4. Grieve, S. 1885. *The Great Auk or Garefowl*, p.78, pp.92-4 *and* appendix p.11. Bolam, G. 1912. *Birds of Northumberland and the Eastern Borders*, p.660. Niethammer, G. 1954. Bälge, Skelette und Eier ausgestorbener (oder sehr seltener) Vögel im Museum Koenig in Bonn. *Bonner Zoologische Beiträge*, p.192. Hahn, P. 1963. *Where is that Vanished Bird?* p.232 *and* pp.233-4. Luther, D. 1986. *Die Ausgestorbenen Vögel der Welt*, p.83.

The Great Auk belonging to the Alexander Koenig Museum is, without doubt, the specimen once held at Floors Castle, Kelso, Scotland and reported missing by Paul Hahn (1963). Although aware of the existence of a

Two views of the Floors Castle Auk as it appears today. *Courtesy of Dr. G. Rheinwald and the Alexander Koenig Museum, Bonn.*

Great Auk in Bonn, Hahn was unable to correlate it with the example once belonging to the Duke of Roxburghe at Floors. Only through the help of Dr. Goetz Rheinwald, who combed the Bonn Museum archives and provided photographs, has it been possible to now make this identification.

Dr. Rheinwald unearthed correspondence between Ernst Hartert (Walter Rothschild's German-born curator at the Tring Museum) and Alexander Koenig concerning the possibility of acquiring (and the subsequent acquisition of) a Great Auk for Bonn. Unfortunately, only the Hartert half of the correspondence is available, the Koenig responses being missing. Hartert's letter of February 16th 1905 can be translated as follows:

> In spring a mounted Auk will be sold but it is suspected that this will go to America since in England at the moment nobody is willing to purchase Auks. If you are interested...

A subsequent letter, probably written in May of the same year, says:

> I write in a hurry. The Auk is entrusted to Rowland Ward for selling. He has had an offer of £400 but for this price would prefer to keep it in Europe.

Later, Hartert wrote:

> After your telegram I ordered immediately and now have it in front of me.

On May 30th he sent Koenig more information:

> I do not know the history of the Auk but Ward has promised to give it me immediately it is paid for. I will take care he sticks to his promise.

In a fashion that seems quite typical of him, Rowland Ward never did keep the promise he made. Although he assured Hartert that he possessed the full history of the bird, the only information he communicated was that the former owner was a Marquis.

In Britain this term often refers, quite specifically, to the son of a Duke and Ward, with a well known tendency to snobbery, would - almost certainly - have used the term accurately. From this information alone it is fairly safe to assume that the bird was sold on behalf of the son of the Duke of Roxburghe. No other British Duke owned a stuffed Garefowl at this time and, in any case, there is no other known specimen from this period that is unaccounted for.

An old picture of the Floors Castle specimen, closely drawn from a photograph (*see* Gibson, 1882-3), provides additional proof of the identity of the bird now in Bonn. Although it has been skilfully re-stuffed and re-shaped during its time in Germany, the set of the head remains the same. A major difference - and one that might seem hard to account for - is a downward curve to the gape. This difference can only be explained in terms of an artistic invention designed to give emphasis to the mouth.

Engraving of the Floors Castle Auk from *The Proceedings of the Royal Physical Society, Edinburgh* (1882-3).

The photograph on which the engraving is based. *Courtesy of Peter Meadows and the University Library, Cambridge.*

Fortunately, a print of the original photograph on which the drawing is based has recently been found - by Peter Meadows - in the Alfred Newton archive at Cambridge. Not only does it prove how skilfully (in general!) the artist copied but also that he did indeed invent part of the structure of the mouth. Magnification of this photograph conclusively confirms that the configuration of the ridges on the beak conforms to the pattern shown by the Bonn specimen.

It is not known how a Great Auk came to be at Floors Castle, a location used, incidentally, during the filming of *Greystoke - The Legend of Tarzan*. Although the specimen was apparently purchased for Floors at some time during the 1830's, its existence there was known only to a very few people prior to 1883 when John Gibson publicised it by featuring it in a paper for the *Proceedings of the Royal Physical Society*. Gibson became aware of the bird through his acquaintance with Mr. Andrew Brotherston, naturalist of Kelso and curator of the Duke's museum.

In 1882 the then Duke of Roxburghe expressed the belief that his father had bought the Great Auk in Edinburgh but there is no certainty of this and it is even possible that the Duke obtained the bird in Iceland, which country he is known to have visited. Whether or not this is so, it is very likely, of course, that it has an Icelandic origin. The word 'male' was once inscribed on the bird's stand but this description is of very uncertain value.

By the time Paul Hahn compiled his catalogue of Great Auks (1963), no-one at Floors remembered the specimen. Hahn quoted from a letter he received from the Estate Office at Kelso dated May 10th 1960:

> His Grace thinks it quite likely that his specimen may have been sold during his father's lifetime, that is prior to 1932. I am afraid His Grace has no idea to whom this specimen might have been sold.

THE BREMEN AUK

(Below and facing page). Two views of the Bremen Auk - perhaps one of the birds of 1844. *Courtesy of Jochen Lempert.*

Bird no. 36 (Grieve no. 10; Hahn no. 71) - adult in summer plumage
Origin: unknown, probably Eldey
Present location: Bremen - Übersee Museum Bremen, Bahnhossplatz 13, 2800 Bremen 1, Germany.

Bolle, C. 1862. Notiz, *Alca impennis* betreffend. *Journal für Ornithologie*, p.208. Sclater, P. 1864. Notes on the Great Auk. *Annals and Magazine of Natural History*, p.320. Fatio, V. 1868. Liste des divers représentants de l'*Alca impennis* en Europe. *Bullétin de la Société Ornithologique Suisse*, tome II, pt.1, p.81. Blasius, W. 1884. Zur Geschichte der Ueberreste von *Alca impennis*, *Journal für Ornithologie*, pp.73-4. Grieve, S. 1885. *The Great Auk or Garefowl*, p.79, p.91 *and* appendix p.7. Hartlaub, G. 1895. Ein Beitrag zur Geschichte der ausgestorbenen Vögel der Neuzeit. *Abhandlungen von Naturwissenschaftlichen Vereine zu Bremen*, p.21. Salomonsen, F. 1944-5. Gejrfuglen et hundredaars minde. *Dyr i Natur og Museum*, p.106. Hahn, P. 1963. *Where is that Vanished Bird?* p.232. Luther, D. 1986. *Die Ausgestorbenen Vögel der Welt*, p.83.

In a footnote to a brief account of this specimen, Symington Grieve (1885) mentions a belief concerning it held by the Danish museum curator Japetus

Steenstrup. Arguing against a suggested - and slightly earlier - origin for the bird, Steenstrup stated that he:

> Prefers a certain tradition met with here [Copenhagen] that the skin sold to Bremen in 1844 belonged to one of ... [the] last individuals ... got in 1844.

Possibly, therefore, Bremen holds the skin of one of the Great Auks strangled on Eldey in June 1844 during man's last definite encounter with the species. Although the internal organs of these two unfortunate individuals went to the Royal Museum, Copenhagen (where they remain preserved in spirits), the skins themselves went missing and have never been properly accounted for. Steenstrup and his colleagues in Copenhagen were well placed to form an opinion concerning the fate of these skins but their knowledge of such matters was not always as comprehensive as might be imagined. There are at least two other specimens (those now in Brussels and Los Angeles) with an equal claim to the dubious honour of being the last of the Great Auks.

What is known for certain about the Bremen Auk is that it was purchased by Dr. Hartlaub for the equivalent of £6 at the time of the Bremen Congress of German Naturalists (Autumn, 1844). Hartlaub made the purchase from Salmin, a Hamburg dealer who occasionally obtained Garefowl skins from Iceland and who handled three birds killed on Eldey in the summer of 1840 or 1841. It is quite possible that Salmin still had one of these birds and perhaps it was this specimen that he sold to Hartlaub.

THE BRUNSWICK AUK

Bird no. 37 (Grieve no. 8; Hahn no. 70) - adult in summer plumage
Origin: unknown
Present location: Brunswick - Staatliches Naturhistorisches Museum, Pockelsstraße 10a, D-38106 Braunschweig, Germany.

Newton, A. 1870. On Existing Remains of the Gare-fowl. *Ibis*, p.257. Blasius, W. 1881-3. Ueber die letzen Vorkommnisse des Riesen-Alks. *Jahresberichte des Vereins für Naturwissenschaft zu Braunschweig für die Vereinsjahre 1881-2 und 1882-3*, pp.98-9 and p.113. Blasius, W. 1884. Zur Geschichte der Ueberreste von *Alca impennis*. *Journal für Ornithologie*, p.73. Grieve, S. 1884. *The Great Auk or Garefowl*, p.79 and appendix p.7. Hahn, P. 1963. *Where is that Vanished Bird?* p.232. Luther, D. 1986. *Die Ausgestorbenen Vögel der Welt*, p.83.

The town of Brunswick, home of the Garefowl scholar Wilhelm Blasius, is connected with the history of several Great Auks, one of them Blasius's own property. His bird is long gone (it is now in Moscow) but another still belongs to the town museum. Its origin is unknown but from the mode of preparation - and from what is known of its history - it may be judged Icelandic. Probably between 1830 and 1840 (but certainly before 1842) it was purchased from one of the Franks. Newton, in his manuscript notes, suggests it came from Copenhagen around 1832 through the agency of Herr Eimbeck, a man who was clearly something of a dealer and whose name crops up enigmatically in connection with several other Great Auks.

During World War II the Brunswick Auk was removed from the town museum, perhaps on loan or perhaps to store it more securely. Hahn (1963) recorded that it was still unreturned in May 1959, but today the bird is safely back in the Staatliches Naturhistorisches Museum.

THE DRESDEN AUK

(Facing page). The Dresden Auk. Photograph by H. Höhler. *Courtesy of Siegfried Eck and the Staatliches Museum für Tierkunde, Dresden.*

Bird no.38 (Grieve no. 21; Hahn - no number, assumed destroyed) - adult in summer plumage, said to be female
Origin: unknown
Present location: Dresden - Staatliches Museum für Tierkunde, Augustusstraße 2, 01067 Dresden, Germany.

Reichenbach, H. 1836. *Das Königliche Naturhistorische Museum*, p.22. Reichenbach, H. 1848. *Vollständigste Naturgeschichte des in und Auslands*, pt.1, sect.2, *Vogel*, bd.1, taf.3. Preyer, W. 1862. Der Brillenalk in Europäischen Sammlungen. *Journal für Ornithologie*, p.78. Fatio, V. 1868. Liste des divers représentants de l'*Alca impennis* en Europe. *Bullétin de la Société Ornithologique Suisse*, tome II, pt.1, p.81. Blasius, W. 1884. Zur Geschichte der Ueberreste von *Alca impennis*. *Journal für Ornithologie*, p.78. Grieve, S. 1885. *The Great Auk or Garefowl*, p.79 *and* appendix p.10. Hahn, P. 1963. *Where is that Vanished Bird?* p.234. Luther, D. 1986. *Die Ausgestorbenen Vögel der Welt*, p.82.

For many years - particularly in the English-speaking world - it was believed that the Great Auk belonging to the Dresden Museum was lost. It was supposed that it was either destroyed outright or else looted at the end

Hand-coloured engraving from H.G.L. Reichenbach's *Die Vollständigste Naturgeschichte der Vögel des in und Auslands* (1845-63). It is said that the Dresden Auk served as a model for this picture - but this seems doubtful.

Alca impennis *L.*
Brillenalk (ausgestorben)

Früher gemein an den Küsten Islands und Neu Fundlands,
diente den Seefahrern als Nahrung. Unfähig zum Flie-
gen, wurde er am Land und an seinen Brutplätzen leicht
gefangen und ungefähr im Jahr 1840 ganz ausgerottet.

12632   *Nördliches Eismeer*

of World War II. According to Hahn (1963), it was destroyed by air attacks during the terrible fire-bombing of the city; according to Tomkinson and Tomkinson (1966), writing of the Dresden Egg, the museum's most precious objects were stored in twelve large cases in the historic fortress of Königheim and these cases vanished during the Russian occupation.

The truth is that both the bird and the egg have survived and both are safely back in the Staatliches Museum für Tierkunde. At the end of World War II they were taken to St. Petersburg (then Leningrad) where they remained until 1982. In that year Siegfried Eck, Curator of Birds at the Dresden Museum, was allowed to travel to the USSR and reclaim the Great Auk together with the egg and many other important specimens (including the missing example of New Zealand's *Notornis*, purchased during the nineteenth century and only the third collected) for his famous old museum.

The Great Auk has belonged to Dresden for a very long time and its origin is uncertain. It is ticketed 'Eismeer' (Polar Seas) and is said to be female. Its first mention in ornithological literature came in 1836 when H.G.L. Reichenbach listed it as a possession of the Königliche Museum at Dresden. Probably, the specimen corresponds to one sent from the Royal Museum, Copenhagen to Dr. Thienemann in Dresden during 1832. The Dresden Egg once belonged to Thienemann so it is likely that the bird did too. Assuming this is the case, the Dresden Auk is probably one of the 24 birds killed during the raid on Eldey that took place in the summer of 1831.

THE FRANKFURT AUK

Bird no.39 (Grieve no. 26; Hahn no. 72) - adult in summer plumage
Origin: unknown
Present location: Frankfurt am Main - Forschungsinstitut und Naturmuseum Senckenberg, 60325 Frankfurt 1, Germany.

Preyer, W. 1862. Der Brillenalk in Europäischen Sammlungen. *Journal für Ornithologie*, p.78. Homeyer, A. 1862. Notiz zu *Alca impennis*. *Journal für Ornithologie*, p.461. Fatio, V. 1868. Liste des divers représentants de l'*Alca impennis* en Europe. *Bullétin de la Société Ornithologique Suisse*, tome II, pt.1, p.81. Blasius, W. 1884. Zur Geschichte der Ueberreste von *Alca impennis*. *Journal für Ornithologie*, p.84. Grieve, S. 1885. *The Great Auk or Garefowl*, p.79 *and* appendix p.12. Hartert, E. 1891. *Katalog der Vogelsammlung im Museum der Senckenbergischen*, p.246. Parrot, C. 1895. [Letter to the Editor] *Ibis*, p.165. Strassen, O. 1910. Der Riesenalk. *Berichte Senckenbergischen Naturforschenden Geseltschaf im Frankfurt am Main*, 41, pp.184-90 *and* pl.2. Martens, R. and Steinbacher, J. 1955. Die im Senckenberg-Museum vorhandenen Arten ausgestorbener, aussterbender oder seltener Vögel. *Senckenbergiana Biologica*, bd.36, no.3 &4, p.248. Hahn, P. 1963. *Where is that Vanished Bird?* pp.232-3. Luther, D. 1986. *Die Ausgestorbenen Vögel der Welt*, p.83.

The origin of this specimen is something of a mystery. When Wilhelm Blasius (1884) was painstakingly compiling his monograph he was unable to discover anything significant about it. It was labelled 'Northern Europe'

The Frankfurt Auk (circa 1910).

and the curator at Frankfurt could only tell him that nothing was known, 'as to whence, when or from what hands it came into the possession of the Museum.' Subsequent research by Ernst Hartert (1891) established that the museum taxidermists Erckel held definite beliefs about the bird. According to them it was received in Frankfurt during November 1837 as part of an exchange with Professor Fries of Stockholm. The story is certainly curious. Professor Fries is known to have been involved in the exchange of a Great Auk at about this time but the exchange he made was with Florence not with Frankfurt. As there is no record of Fries having parted with two specimens, it is probable that the Erckel's memories were at fault.

On the authority of Professor Reinhardt, Alfred Newton was in no doubt that this Great Auk came from the Royal Museum, Copenhagen in 1834 and a scribbled note to this effect exists in his archive at Cambridge. Such a provenance would make the bird's origin Icelandic and probably it came from Eldey.

THE GOTHA AUK

Bird no.40 (Grieve no. 27; Hahn no. 27) - adult in summer plumage
Origin: unknown, said to be Greenland but probably Iceland
Present location: Gotha - Museum der Natur Gotha, Parkallee 15, PSF 217, 99853 Gotha, Germany.

The Gotha Auk. *Courtesy of the Museum der Natur, Gotha.*

Hellman, A. 1860. Notizen über *Alca impennis. Journal für Ornithologie*, p.206. Preyer, W. 1862. Der Brillenalk in Europäischen Sammlungen. *Journal für Ornithologie*, p.78. Fatio, V. 1868. Liste des divers représentants de l'*Alca impennis* en Europe. *Bullétin de la Société Ornithologique Suisse*, tome II, pt.1, p.82. Blasius, W. 1884. Zur Geschichte der Ueberreste von *Alca impennis. Journal für Ornithologie*, p.84. Grieve, S. 1885. *The Great Auk or Garefowl*, p.79, p.91 *and* appendix p.12. Hahn, P. 1963. *Where is that Vanished Bird?* p.221. Zimmerman, W. 1965. Zur Herkunft eines Präparates von *Pinguinus impennis* im Naturkundemuseum Gotha. *Journal für Ornithologie*, pp.106-8. Luther, D. 1986. *Die Ausgestorbenen Vögel der Welt*, p.82. Samietz, R. 1990. Der Riesenalk des Museums der Natur Gotha. *Neue Museums Kunde*, pp.170-1

Several writers, including Wilhelm Blasius (1884), have reported that this specimen was purchased in 1835 by the museum at Gotha from the Leipzig dealer Frank (oldest of the three Franks who traded in natural history items) but this seems not to be the case. Blasius went to some trouble to trace the history of this bird and, after writing his paper, came to believe it derived instead from a Gotha bank clerk, F.B. Knapp (1819-92).

He discovered that on December 7th 1843, the Gotha Museum paid this gentleman 65 thalers for thirteen bird skins, one of which was the Great Auk. According to Zimmerman (1965) this price is incorrect and resulted from a misreading of a catalogue entry; apparently, 20 thalers was paid for the Great Auk alone and the other twelve birds were bought separately. At the time of sale Knapp claimed that all his birds came from a missionary in Greenland and were accumulated between 1839 and 1842, being purchased through a friend, Herr Ramann, who acted as an intermediary. Many years later, in 1890, Knapp confirmed - in writing - that Greenland, unlikely though it seemed, was indeed the actual source of the Auk. Probably it wasn't and the Gotha Auk, like almost all the other preserved examples, came from Iceland. Possibly Knapp was misled over the bird's origin, the Greenland data being invented to give the specimen a more interesting provenance; perhaps the term 'Greenland' was used casually and Knapp, understandably, took it literally. It may even be that this bird came, as originally supposed, from Frank and then passed through several hands before reaching Herr Knapp. The truth is now unlikely to be determined.

During 1979 this Great Auk, along with those from Leipzig and Köthen, was shown in a travelling exhibition that toured several towns in the old DDR. Each bird was specially cleaned for this purpose by the Leipzig museum preparator Horst Spicale and, during the cleaning process, he discovered that at an unspecified time in the past, goose feathers were used to patch small areas of damage.

THE HANOVER AUK

Bird no.41 (Grieve no. 29; Hahn no. 73) - adult in summer plumage
Origin: unknown but said to be Iceland
Present location: Hanover - Niedersächsisches Landesmuseum Hannover, Am Maschpark 5, 30169 Hannover, Germany.

Fatio, V. 1868. Liste des divers représentants de l'*Alca impennis* en Europe. *Bullétin de la Société Ornithologique Suisse*, tome II, pt.1, p.82. Blasius, W. 1884. Zur Geschichte der Ueberreste von *Alca impennis. Journal für Ornithologie*, p.86. Grieve, S. 1885. *The Great Auk or Garefowl*, p.79 *and* appendix p.13. Hahn, P. 1963. *Where is that Vanished Bird?* p.233. Luther, D. 1986. *Die Ausgestorbenen Vögel der Welt*, p.83.

Very little is known about the Hanover Auk. Dieter Luther (1986) states that it came from Iceland in 1830, although Blasius (1884) recorded that it was not prepared in the usual Icelandic manner. Either via Frank of Leipzig or else through the agency of a Herr Kaufmann (possibly the bird was handled by both gentlemen), it reached the Clausthal Museum and was later transferred to Hanover where, according to Dr. R. Schumacher (*pers. comm.*, 1995) it is 'still a special attraction for our visitors.'

THE SCHLESWIG-HOLSTEIN AUK

Bird no.42 (Grieve no. 31; Hahn no. 74) - adult in summer plumage
Origin: unknown
Present location: Kiel - Zoologisches Museum Christian-Albrechts-Universität zu Kiel, Hegewischstraße 3, D-24105 Kiel, Germany (cat. no. A 0585).

Blasius, W. 1881-3. Ueber die letzen Vorkommnisse des Riesen-Alks. *Jahresberichte des Vereins für Naturwissenschaft zu Braunschweig für die Vereinsjahre 1881-2 und 1882-3*, p.94, p.105 *and* p.114. Blasius, W. 1884. Zur Geschichte der Ueberreste von *Alca impennis. Journal für Ornithologie*, pp.86-7. Grieve, S. 1885. *The Great Auk or Garefowl*, p.79 *and* appendix p.13. Hartert, E. 1894. [Remarks made at meeting of the British Ornithologists' Club] *Ibis*, p.549. Hahn, P. 1963. *Where is that Vanished Bird?* p.233. Luther, D. 1986. *Die Ausgestorbenen Vögel der Welt*, p.84.

> Only three men... ever... understood the Schleswig-Holstein Question: the Prince Consort - who is dead, a German professor - who has gone mad, and I - who have forgotten all about it.
>
> Lord Palmerston (1863)

The Schleswig-Holstein Question was a complex political problem that dogged European diplomacy for a number of years during the nineteenth century and came to a head in the 1860's. Since then generations of unwilling students have been forced to ponder and attempt to unravel this rather dry piece of history. What few of them will ever have known, while wrestling with the tedious complexities, is that Schleswig-Holstein had its own Great Auk whose story is as difficult to pin down as the 'Question' that their teachers took such grim delight in bringing to their attention.

Even the records of the specimen's acquisition by the University of Kiel are contradictory. Paul Hahn (1963) states that it came into the possession of an important official of Schleswig-Holstein during 1844 and that he presented it to the Zoological Museum of the University. Wilhem Blasius (1884) believed that it was purchased in 1844 by a Professor Behn out of a large grant that the 'Prelates and Proprietors' of Schleswig-Holstein made available to the museum.

The date of 1844, being an emotive one, creates the suspicion that this might be one of the individuals killed in June of that year. Other specimens, notably those in Brussels, Los Angeles and Bremen, have claims that are as good, however, and the date doesn't necessarily signify anything; this could just be an 'older' example purchased in 1844 rather than a bird freshly killed. Japetus Steenstrup, the Danish Garefowl scholar, knew of nothing to connect the Schleswig-Holstein Auk to the slaughter of the last known pair of birds. On March 15th 1885, he wrote to Symington Grieve:

> If really purchased in 1844, it might perhaps be the second of those two Garefowls got in 1844, but traditionally I never heard that mentioned.

On the other hand, Professor Reinhardt believed - perhaps rightly, perhaps wrongly - that the skins of the last two Great Auks were sold at the Congress of German Naturalists held in Bremen during late 1844. Geographically, this congress was very handily placed for the convenience of the dignitaries of Kiel.

The Schleswig-Holstein Auk is as much a puzzle as the Schleswig-Holstein Question. Like Steenstrup, Reinhardt, Grieve, Hahn, Blasius - as well, of course, as Palmerston's audience - we can only wonder.

NAUMANN'S AUK

The logo of the Köthen Museum.

(Below, left). Naumann's Auk. *Courtesy of the Naumann-Museum, Köthen.*

(Below, right). Hand-coloured engraving from J. F. Naumann's *Naturgeschichte der Vögel Deutschlands* (1822-60).

Bird no.43 (Grieve no. 32; Hahn no. 28) - adult in summer plumage
Origin: Iceland
Present location: Köthen - Naumann-Museum, Schloßplatz 4, PF 1454, 06366 Köthen/Anhalt, Germany.

Naumann, J. F. 1822-60. *Naturgeschichte der Vögel Deutschlands*, vol.XII, pl.337. Olphe-Galliard, L. 1862. [Letter to the Editor] *Ibis*, p.302. Preyer, W. 1862. Der Brillenalk in Europäischen Sammlungen. *Journal für Ornithologie*, p.82. Fatio, V. 1868. Liste des divers représentants de l'*Alca impennis* en Europe. *Bullétin de la Société Ornithologique Suisse*, tome II, pt.1, p.82. Blasius, W. 1884. Zur Geschichte der Ueberreste von *Alca impennis*. *Journal für Ornithologie*, p.87. Grieve, S. 1885. *The Great Auk or Garefowl*, p.79 *and* appendix p.13. Salomonsen, F. Gejrfuglen et hundredaars minde. *Dyr i Natur og Museum*, p.106. Hahn, P. 1963. *Where is that Vanished Bird?* p.221. Luther, D. 1986. *Die Ausgestorbenen Vögel der Welt*, p.82. Samietz,

R. 1990. Der Riesenalk des Museums der Natur Gotha. *Neue Museums Kunde*, p.171.

The Naumann's, father and son, were responsible for one of the great land-marks of ornithological literature, a work that became a saga taking more than a century to reach its final form. Between the years 1795 and 1817 J.A. Naumann (1744-1826) published *Naturgeschichte der Land- und Wasser-Vögel des Nördlichen Deutschlands und Angränzender Länder, Nach Eigenen Erfahrungen Entworfen*. His son, J.F. Naumann (1780-1857), published a new edition between 1822 and 1860 under the title *Naturgeschichte der Vögel Deutschlands, Nach Eigenen Erfahrungen Entworfen*. Finally, long after the deaths of both men, the work was re-written and re-issued as *Naturgeschichte der Vögel Mitteleuropas* (1896-1905), a book which Sitwell, Buchanan and Fisher (1953) describe in *Fine Bird Books* as:

> The most scholarly and complete ornithological text book of its time (perhaps of any time).

So important were the Naumann's to the development of European bird study that the German Ornithological Society named its journal *Naumannia* for the first few years of issue, although subsequently it merged with the *Journal für Ornithologie*. As might be expected, the Naumanns put together a large collection on the basis of which the Naumann Musuem at Köthen was founded. Perhaps the most important single item is the Great Auk that J.F. Naumann received from Iceland - via Copenhagen - in 1830. Nothing more is known of it. Probably, it is one of the disorientated birds taken after the destruction of the Geirfuglasker.

J.F. Naumann (1780-1857).

THE LEIPZIG AUK

Bird no.44 (Grieve no. 37; Hahn no. 29) - adult in summer plumage
Origin: Iceland
Present location: Leipzig - Naturkundemuseum Leipzig, Lortzingstraße 3, 04105 Leipzig, Germany.

Naumann, J. F. 1822-60. *Naturgeschichte der Vögel Deutschlands*, vol XII, p.646. Blasius, W. 1884. Zur Geschichte der Ueberreste von *Alca impennis*. *Journal für Ornithologie*, pp.91-2. Grieve, S. 1885. *The Great Auk or Garefowl*, p.79, p.96 *and* appendix pp.14-15. Hahn, P. 1963. *Where is that Vanished Bird?* p.221. Luther, D. 1986. *Die Ausgestorbenen Vögel der Welt*, pp.82-3. Samietz, R. 1990. Der Riesenalk des Museums der Natur Gotha. *Neue Museums Kunde*, p.171.

There has been a Great Auk in Leipzig for more than 150 years. A bird from Iceland is entered into a museum catalogue under the date of February 24th 1841 but it was possibly acquired a year or two earlier. The property of the Universität Leipzig (although kept at the city's museum), it was presented by a Herr D. Uckermann.

In the great ornithological work of Naumann (1822-60) it is stated that a skin from Iceland came to Leipzig via England around 1820, being sold for an enormous price. Possibly, this specimen is identical with the one still in the city, but the connection of the older Frank with Leipzig means that a number of Great Auks passed through. Indeed, Alfred Newton had a record of a specimen being sold to Leipzig by Frank, but Blasius (1884) felt unable to confirm this. It is quite likely that Frank sold a Garefowl to Uckermann who purchased it with the intention of eventually presenting it to the museum of his home city; he is known to have been a considerable contributor to the museum collection.

THE MUNICH AUK

Bird no.45 (Grieve no. 50; Hahn no. 76) - adult in summer plumage, said to be female
Origin: unknown, but probably Eldey
Present location: assumed to be Munich - Zoologische Staatssammlung, Münchhausenstraße 21, D-81247 München 60, Germany.

Preyer, W. 1862. Der Brillenalk in Europäischen Sammlungen *and* Ueber *Plautus impennis. Journal für Ornithologie,* p.78 *and* p.119. Fatio, V. 1868. Liste des divers représentants de l'*Alca impennis* en Europe. *Bullétin de la Société Ornithologique Suisse,* tome II, pt.1, p.82. Blasius, W. 1884. Zur Geschichte der Ueberreste von *Alca impennis. Journal für Ornithologie,* pp.95-8. Grieve, S. 1885. *The Great Auk or Garefowl,* p.80 *and* appendix p.16. Parrot, C. 1895. [Letter to the Editor] *Ibis,* p.165. Hahn, P. 1963. *Where is that Vanished Bird?* p.233. Luther, D. 1986. *Die Ausgestorbenen Vögel der Welt,* p.84.

Through much of the nineteenth and twentieth centuries there were two Great Auks in the collection of Munich Zoological Museum. Just over a decade ago both were included in a list of specimens in German museums (*see* Luther, 1986), but today there seems to be only one bird left. Dr. Reichholf, Curator of Birds at Munich, reported (*pers. comm.,* Jan. 1995) that he could find only a single specimen. Subsequent inquiries have failed to produce a fuller response. Whether one of the Auks has vanished in dubious circumstances or whether it is lying overlooked in the museum's store remains a mystery. So too does the precise identity of the missing bird.

One of the Auks was acquired by the museum during the 1830's. It was labelled 'Eismeer' (Polar Seas), 1836 and probably came from Eldey. The bird was thought to be female but the grounds for this belief are unknown.

THE DUKE OF LEUCHTENBERG'S MUNICH AUK

Bird no.46 (Grieve no. 49; Hahn no. 75) - adult in summer plumage, said to be male
Origin: Iceland, probably Eldey, 1833
Present location: assumed to be Munich - Zoologische Staatssammlung, Münchhausenstraße 21, D-81247 München 60, Germany.
All references are identical to those of The Munich Auk.

The first of the Munich Auks was mounted in a standing position, the other- according to Blasius (1884)- was posed 'sitting.' This sitting bird is alleged to be male but, as with its 'female' companion, the grounds for the designation of sex remain unclear. It once belonged to the Duke of Leuchtenberg and passed at an early date into the Zoological Museum of the Royal Bavarian Academy of Science. In 1895 C. Parrot noted that a label then present on the specimen indicated it came from Iceland in 1833. This Icelandic origin was previously noted by Blasius who was told by Dr. Pauly of Munich that both Munich birds had passed through the hands of the ornithological writer Karl Michahelles and that he'd got them from Reinhardt of Copenhagen.

Whether it is this specimen or the other that has vanished remains uncertain.

THE OLDENBURG AUK

Bird no.47 (Grieve no. 57; Hahn no. 77) - adult in summer plumage
Origin: Iceland, probably Eldey
Present location: Oldenburg - Staatliches Museum für Naturkunde und Vorgeschichte, Damm 40-44, D-26135 Oldenburg, Germany.

Preyer, W. 1862. Der Brillenalk in Europäischen Sammlungen. *Journal für Ornithologie*, p.78. Fatio, V. 1868. Liste des divers représentants de l'*Alca impennis* en Europe. *Bullétin de la Société Ornithologique Suisse*, tome II, pt.1, p.82. Blasius, W. 1884. Zur Geschichte der Ueberreste von *Alca impennis*. *Journal für Ornithologie*, p.102. Grieve, S. 1885. *The Great Auk or Garefowl*, p.80 *and* appendix p.19. Salomonsen, F. 1944-5. Gejrfuglen et hundredaars minde. *Dyr i Natur og Museum*, p.106. Hahn, P. 1963. *Where is that Vanished Bird?* p.233. Luther, D. 1986. *Die Ausgestorbenen Vögel der Welt*, p.84.

According to Herr C.F. Wiepkin, once Director of the Oldenburg Museum, this specimen was purchased in 1840 or 1841 from the Hamburg natural history dealer Salmin. The likelihood of this story is confirmed by the fact that Salmin did indeed handle three freshly killed birds, taken from Eldey, at around this time. Although it is probable that one of these individuals can be correlated with the Oldenburg specimen it is by no means certain. Salmin had obtained other Icelandic Great Auks in previous years and Oldenburg perhaps acquired a bird he'd had in stock for some time.

THE STUTTGART AUK

Bird no.48 (Grieve no. 70; Hahn no. 78) - adult in summer plumage
Origin: Iceland
Present location: Stuttgart - Naturwissenschaftliche Sammlungen (Rosenstein Museum), Stuttgart, Germany (catalogue no.5934).

Newton, A. 1870. On Existing Remains of the Gare-fowl. *Ibis*, p.258. Selys Longchamps, E. 1876. Note sur un voyage scientifique fait en Allemagne.

The Oldenburg Auk. *Courtesy of the Staatliches Museum für Naturkunde, Oldenburg.*

Comptes-rendues des séances de la Société entomologique de Belgique, p.lxx. Blasius, W. 1884. Zur Geschichte der Ueberreste von *Alca impennis*. *Journal für Ornithologie*, pp.110-1. Grieve, S. 1885. *The Great Auk or Garefowl*, p.80 *and* appendix pp.23-4. Parrot, C. 1895. [Letter to the Editor] *Ibis*, p.165. Salomonsen, F. 1944-5. Gejrfuglen et hundredaars minde. *Dyr i Natur og Museum*, p.106. Hahn, P. 1963. *Where is that Vanished Bird?* p.233. Luther, D. 1986. *Die Ausgestorbenen Vögel der Welt*, p.84.

The Royal Cabinet of Natural History Stuttgart obtained this specimen in May 1867 from Baron John Wilhelm von Müller of Württemberg. Baron Müller was a particularly acquisitive traveller who brought back interesting collections from his journeys and put together a large assemblage of African birds. In the hope of securing some of these, Japetus Steenstrup and Professor Forchhammer presented the Baron - in either 1849 or 1850 - with the last of the duplicate Great Auk skins belonging to the Royal Museum, Copenhagen. Whether or not they got anything in return is unclear but they got none of the African material they'd hoped for and complained rather bitterly in the years that followed.

During the 1880's the Auk was re-stuffed and this process revealed that it was originally skinned using a cut along the side, a characteristic Icelandic method of preparation.

COUNT RABEN'S ICELANDIC AUK

Bird no.49 (Grieve no. 1; Hahn no. 25) - adult in summer plumage
Origin: Iceland
Present location: Reykjavik - Náttúrufraedistofnun Íslands (Icelandic Museum of Natural History), IS-125 Reykjavík, Iceland.

Faber, F. 1825. *Ueber das Leben der hochnordischen Vögel*, p.48. Faber, F. 1827. Beitrage zur Artischen Zoologie. *Isis*, pp.682-3. Charlton, E. 1860. On the Great Auk. *Zoologist*, pp.6885-6. Newton, A. 1861. Abstract of Mr. J. Wolley's Researches in Iceland. *Ibis*, pp.386-7. Blasius, W. 1884. Zur Geschichte der Ueberreste von *Alca impennis*. *Journal für Ornithologie*, pp.67-80. Grieve, S. 1885. *The Great Auk or Garefowl*, p.78, pp.94-5 *and* appendix pp.4-5. Helms, O. 1923. Nogle ornithologiske oplevelser fra de senere aar. *Dansk Ornitologisk Forening Tidsskrift*, pl. p.84. Salomonsen, F. Gejrfuglen et hundredaars minde. *Dyr i Natur og Museum*, pp.104-5. Hahn, P. 1963. *Where is that Vanished Bird?* p.221. Tinker, J. 1971 (11th March). Extinct mounted Birds. *New Scientist*, p.578. *Guinness Book of Records* (1972-nineteenth ed.), p.155. Bárðarson, H. 1986. *Birds of Iceland*, p.20, pp.23-4 *and* pl. p.21. Fuller, E. 1987. *Extinct Birds*, p.98. Bourne, W. 1993. The story of the Great Auk. *Archives of Natural History*, p.275. Petersen, A. 1995. Brot úr sögu Geirfuglsins. *Náttúrufræðingurinn*, 65 (1-2), p.59, p.66, pl.1, pl.3 *and* pl.4.

During the early summer of the year 1821, Count F.C. Raben left his home at Aalholm Castle and headed for the shores of Iceland. His express intention was to obtain a Great Auk to stuff and add to the collection of mounted birds kept in the halls of his ancestral home. During June, in the company of the Danish ornithologist Friedrich Faber and a botanist named Mörck, he attempted to reach the Geirfuglasker to get at the colony of Auks that - at this time of the year - were sure to be there. So treacherous did the native Icelanders judge the seas that the Danes were at first unable to find a single seaman prepared to take them; only in the calmest weather did the locals think it worthwhile to attempt a landing. Eventually, at considerable cost, the Danish party managed to hire a fishing smack named *The Villinger* at Keflavík. On June 29th they reached the Geirfuglasker (one of the last ever visits before the skerry vanished during the volcanic activity of 1830) but, to their surprise, found no Great Auks. Not easily to be put off, they sailed on to the Geirfugladrángr where, despite heaving seas and a furious surf, the Count insisted on getting ashore. This he did, coming very close to drowning when he fell between boat and rocks. His daring effort was wasted. There were no Garefowls and all he managed to collect were a few strands of seaweed.

The Count did not go back to Denmark empty-handed, however. At some point he parted from Faber (who later maintained that the trip to Iceland had proved a failure as far as Great Auks were concerned) and after

Count Raben's Auk. Photograph from Hjálmar Bárðarson's *Birds of Iceland* (1986). *Reproduced by kind permission of Hjálmar Bárðarson.*

this separation somehow managed to obtain the creature he was so desperately in search of. It is sometimes recorded that Raben killed his bird himself by hitting it with a pole or an oar, but this story seems to rest on a misunderstanding. During this particular Icelandic summer two or three individuals were - according to Newton (1861) - shot and one was killed with a sail pole or gaffe by a man named Jón Jónsson and his son Sigurðr. Probably, it was this last bird that was sold to the Count, although Newton (*see* Grieve, 1885) believed this not to be the case. Whatever the details of his acquisition of the specimen, there is little doubt that he did actually return home with it, a circumstance confirmed by his descendant Count Raben-Levetzau (*see* Bárðarson, 1986).

The preserved bird remained at the Count's home at Aalholm, Nysted, Laaland, Denmark for many, many years. In the early part of 1971 it was moved and taken to Sotheby's in London where it was put up for auction on March 4th. The bidding reached £9,000, at that time considered a record price for a stuffed bird and enough to ensure an appearance on the cover of the *Guinness Book of Records* during the following year. (The Birmingham Museum had, in fact, paid Spink and Son Ltd. a similar amount for another Great Auk just a few weeks previously). The successful bidders were representatives of Iceland's Natural History Museum and the money to make the purchase was, apparently, raised by public subscription in just three days. The Icelandic Airline donated two first class seats for the homecoming, one for Finnur Guðmundsson, the gentleman who'd handled the purchase, and the other for the bird itself. At the airport in Reykjavik a red carpet was put down and a band played. Children were given a half day off school and flags were flown on public buildings. Just three months short of an exile lasting 150 years, the bird had come home.

Cover illustration for the 19th edition (1972) of *The Guinness Book of Records*.

Finnur Guðmundsson bringing the Great Auk home. Photograph by Guðjón Einarsson.

THE IRISH AUK

Bird no.50 (Grieve no. 22; Hahn no. 30) - young female moulting into winter plumage
Origin: captured close to Waterford Harbour, May 1834 (died the following September)
Present location: Dublin - University of Dublin, Trinity College, Dublin 2, Ireland.

Thompson, W. 1849-56. *Natural History of Ireland*, vol.3, p.238. Newton, A. 1865. The Gare-fowl and its Historians. *Natural History Review*, pp.474-5. Fatio, V. 1868. Liste des divers représentants de l'*Alca impennis* en Europe. *Bullétin de la Société Ornithologique Suisse*, tome II, pt.1, p.82. Gurney, J. 1868. The Great Auk. *Zoologist*, pp.1449-53. Milne, J. 1875 (April 10th). Relics of the Great Auk. *The Field*, p.370. Selys Longchamps, E. 1876. Note sur un voyage scientifique fait en Allemagne. *Comptes-rendues des séances de la Société entomologique de Belgique* (Oct. 7th) p.lxx. Blasius, W. 1884. Zur Geschichte der Ueberreste von *Alca impennis*. *Journal für Ornithologie*, pp.78-9. Grieve, S. 1885. *The Great Auk or Garefowl*, p.23, pp.70-1, p.75, pp.77-8 *and* appendix p.10. Ball, R. 1893 (Aug. 5th). *The Daily Graphic*. Dresser, H. 1871-96. *A History of the Birds of Europe*, vol.8, p.564. Barrett-Hamilton, G. 1896. *Irish*

Two photographs of the Irish Auk taken for Alfred Newton (circa 1870). *Courtesy of the University Library, Cambridge.*

Naturalist, p.122. Grieve, S. 1896-7. Supplementary note on the Great Auk. *Transactions of the Edinburgh Field Naturalists' and Microscopical Society*, pp.245-6. Ussher, R. & Warren, R. 1900. *The Birds of Ireland*, pp.358-61. Blasius, W. (*in* Naumann) 1903. *Naturgeschichte der Vögel Mitteleuropas*, bd.12, p.180 *and* pl.17b. Armstrong, E. 1944. *Birds of the Grey Wind*, p.163. Greenway, J. 1958. *Extinct and Vanishing Birds of the World*, p.272 *and* p.284. Bannerman, D. & Lodge, G. 1953-63. *The Birds of the British Isles*, vol.12, p.84. Hahn, P. 1963. *Where is that Vanished Bird?* pp.221-2. Bourne, W. 1993. The story of the Great Auk. *Archives of Natural History*, p.260.

This is among the most interesting of all the stuffed Garefowls. The living bird was kept in captivity for some time and its history is very thoroughly recorded. Perhaps the story is too well recorded; the existing accounts conflict over details. What seems to be the truth is that a pair of Great Auks wandered close to the entrance of Waterford Harbour during the month of May in the year 1834. Many years later Ussher and Warren (1900) ascertained from a man named David Hardy that the actual spot was several miles west of the entrance and was close to the cliffs between Ballymacaw and Brownstown Head. One of these birds was a young female and, seemingly, it was starving for it came to the side of a fishing boat looking for food. A man named Kirby threw a few sprats to it and managed to haul it on board. The other individual was caught too (some authors doubt that this bird ever existed) but soon it either died or was killed and was not preserved. The young female survived, however.

Kirby kept it for ten days or so then sold it to a Mr. Francis Davis who put it into the care of one Jacob Gough of Horetown, County Wexford. At first the Auk was reluctant to feed but then, in very Irish fashion, it was given potatoes mashed in milk. Surprisingly, it gobbled this mixture rather greedily and, sufficiently encouraged to eat further, it turned its attention to trout and other freshwater fish, seeming to prefer these to fish caught at sea. Some accounts say that when presented with food it liked, the bird would stroke its head with its foot but this story probably rests on a misunderstanding. The carefully compiled notes of Dr. Burkitt, who never saw the bird in life but was instrumental in the preservation of its remains, leave little doubt on this point:

> This Auk stood very erect, was a very stately-looking bird, and
> had a habit of frequently shaking its head in a peculiar manner,
> more especially when any particularly favourite food was presented
> to it ... Fish were swallowed entire. It was rather fierce.

It is surprising to find that the Auk showed something of an aversion to water and this is difficult to explain; perhaps only fresh water was made available to it.

One of the people who saw the bird at Jacob Gough's house was a Captain John Spence who, according to J.H. Gurney (1868):
> Bespoke it for Dr. Burkitt's collection should it die.

Die, of course, the poor thing did - a little over four months after its capture and while in the process of moulting. During September 1834, its voracious appetite suddenly subsided and a day or two later it was dead. The body was sent, as previously arranged, to Dr. Burkitt who believed that without the intervention of Captain Spence it would surely have been lost. Gurney (1868) remarked:

> No notice whatever would have been taken of the Great Auk at Horetown, more than as an odd-looking bird "of the penguin tribe," and it would in all probability have been thrown away when dead.

This sad little tale of - perhaps - the last Great Auk to be seen in Irish waters, and one of the very few preserved specimens showing traces of winter plumage, had - at least from the point of view of Dr. Burkitt - a rather unsatisfactory ending. Having stuffed his bird and then kept it for several years, the doctor presented it - in 1844 - to Trinity College, Dublin. A number of writers record that he sold it for £50 but this seems not to be the case. In a long letter to J.H. Gurney dated April 27th 1868 (and preserved for posterity by Alfred Newton who copied it - complete with underlinings- into one of his 'Garefowl Books' now kept at Cambridge University Library), Burkitt expressively poured out his grievances and regrets:

> This Great Auk has been a <u>very sore subject</u> with me for some years past because I cannot but consider myself very <u>badly treated indeed</u> in this matter, and have often regretted this piece of credulity, and <u>extreme</u> stupidity on my part, in placing <u>implicit confidence</u> in whatever was told me... by interested individuals...These... <u>credibly</u> informed [me] that, "the Auk was only worth a few shillings!" Without making any further inquiry whatever, and <u>while under</u> this <u>delusion</u> I <u>presented</u> this valuable specimen, along with some other rarities ... to the Dublin University Museum, to the <u>extreme deprivation</u> of my own private collection, and to a most ... <u>thankless</u> body for I have been informed by friends who have visited the ... Museum, that the authorities there have not had the <u>common courtesy even</u> to note my name as the donor on the tickets attached to this and the other objects, so very foolishly on my part, and it now appears so very <u>thanklessly bestowed</u> on them.

Perhaps this injustice was eventually attended to. There is a story that, 40 years after the donation, in recognition of his efforts to save some tangible piece of the last Irish Auk, the University conferred upon the (by now) poor and aged doctor nothing more nor less than ... a Great Auk Pension!

A photograph of the Irish Auk taken for George Dawson Rowley (circa 1870). *Courtesy of Peter Rowley.*

THE FLORENCE AUK

Bird no.51 (Grieve no. 25; Hahn no. 41) - adult in summer plumage
Origin: unknown, but probably Iceland
Present location: Florence - Museo Zoologico de "La Specola," Sezione del Museo di Storia Naturale, Via Romana 17, Firenzi 50125, Italy (catalogue no. *Ornith. Coll.* 274).

Champley, R. 1864. The Great Auk. *Annals and Magazine of Natural History*, p.235. Fatio, V. 1868. Liste des divers représentants de l'*Alca impennis* en Europe. *Bullétin de la Société Ornithologique Suisse*, tome II, pt.1, p.82.

Saunders, H. 1869. On the Ornithology of Italy and Spain. *Ibis*, p.393. Selys Longchamps, E. 1870. Notes on various Birds observed in Italian Museums in 1866. *Ibis*, pp.450-1. Blasius, W. 1884. Zur Geschichte der Ueberreste von *Alca impennis*. *Journal für Ornithologie*, p.84. Grieve, S. 1885. *The Great Auk or Garefowl*, pp.80-1, pp.97-9 *and* appendix p.11. Lönnberg, E. 1926. The Ornithological Collection of the Natural History Museum in Stockholm. *Auk*, p.438. Germiny, Conte G. de. 1938. Catolago delle Collezione Ornitologica Generale del R. Museo di Firenzi. *Rassegna Faunistica*, 5, no.3/4, p.99 *and* pl. p.101. Hahn, P. 1963. *Where is that Vanished Bird?* p.225. Violani, C. 1975. L'Alca impenne...nelle Collezioni Italiane. *Natura- Revista di Scienze Naturali*, p.16, p.18 *and* pl.2.

The story of the Florence Auk holds many doubtful elements. According to Professor Enrico Giglioli, writing in 1884 to Symington Grieve from the Museo Zoologico, it was obtained by exchange with Professor Sundevall of Stockholm - the swap being effected sometime between 1830 and 1837, during which period the Stockholm Museum had acquired a duplicate Garefowl (from Iceland via the Royal Museum, Copenhagen) to add to one already there. Sundevall disclaimed any knowledge of the transaction, which is hardly surprising. He didn't take charge at Stockholm until 1839 when he replaced a Professor Fries. Giglioli's implication of Sundevall was probably the result of a faulty memory; after all, he was writing almost 50 years behind the event. Records at Stockholm suggest (*see* Lönnberg, 1926) that the swap was arranged during 1835 or 1836, at which time Fries sent away a Great Auk and received in exchange ... a stuffed Platypus!

Time has not been kind to the deal that poor Professor Fries organised. Today such an exchange seems almost ludicrously one-sided but during the 1830's it was, doubtless, perfectly fair. Stockholm would have had every reason to suppose that more Auks would be forthcoming whereas the Platypus (*Ornithorhynchus anatinus*) - a great curiosity by any standard - was a completely unknown quantity.

The exact date of the exchange is in some doubt. Although the Swedish records mention 1835 or 1836, Carlo Violani (1975) found the Great Auk listed in an accessions catalogue at Florence dated 1833. It is entirely possible that the somewhat vague Swedish record was registered a while after the actual event, but it is equally likely that the date of 1833 on the Florence catalogue refers to the year when the record book was started; later acquisitions may have been added to the list in subsequent years. The date of 1833 - being the year when Stockholm actually obtained a duplicate Great Auk from Copenhagen - seems fairly probable, however.

Although the Stockholm provenance provides by far the most likely history for the Florence Auk, other sources are possible.

There was once an old inscription on the bird's stand that read 'Schulz Schaufuss.' This probably related to two Dresden residents who were both keen naturalists. The presence of their names may be misleading, however. Schulz (usually spelled Schultz) was a well known dealer in natural history wares and the pedestal may simply be one that he supplied. Originally, perhaps, some other stuffed creature sat on it (museum specimens being often removed from their original stands and put on others). 'Schaufuss' presumably refers to Dr. L.W. Schaufuss. He and Schulz knew one another and both shared a lively interest in Great Auks, although none of their specimens can be correlated with the bird now in Florence.

Alfred Newton, in his manuscript notes and also in a footnote to a paper by Howard Saunders (*see Ibis*, 1869, p.393), expressed the belief that Florence's Auk passed through the hands of Apothecary Mechlenburg of Flensburg, G.A. Frank of Amsterdam and the ornithological writer Karl Michahelles of Nuremburg before its eventual arrival in Italy. Newton jumbled the whole question of Auks in Stockholm, Florence and Pisa, however, and there is little doubt that he confused the identity of the Florence Auk with that of the one in Pisa. Perhaps it hardly matters. The bird from Stockholm had an Icelandic origin and all the Great Auks belonging to Dr. Michahelles were believed to be Icelandic also.

A curious footnote to the story of the Florence Auk concerns a scrap of paper that the Scarborough collector Robert Champley discovered among his notes. Champley visited Florence in 1861 and maintained, more than twenty years afterwards, that while there he'd seen part of an Auk preserved in spirits. By the time he made his claim no-one at the museum had any recollection of such a specimen - which might be expected to end the

(Facing page, above). *Dodo, Auk and Platypus.* Watercolour by Angel Dominguez for a 1996 edition of Lewis Carroll's *Alice's Adventures in Wonderland. Reproduced by permission of the artist.*

(Facing page, below). Hand-coloured lithograph from C.J. Sundevall's *Svenska Foglarna* (1856-69).

The Florence Auk. *Courtesy of Annamaria Nistri and the Museo Zoologico, Florencee.*

matter. Champley was insistent, however. He recalled that during his visit someone in Florence had scribbled a note for him concerning this item and, eventually, he found it. Written in French on May 21st 1861, it is reproduced by Grieve (1885):

> Le seul individu de l'*Alca impennis* dans la Collection ornithologique du Rl. Musée de Physique et d'Histoire Naturelle de Florence, fut acheté dans l'an 1837 du marchand naturaliste Étienne Moricaud de Genève. [The sole individual of *Alca impennis* in the collection of the Florence Museum was purchased in 1837 from the natural history dealer Étienne Moricaud of Geneva]

This note was signed by two museum officials, Federigo Bruscoli and Ferdinand Piccioli. What it all means is unclear. Grieve thought perhaps the note related to the stuffed bird but Champley remained convinced he'd seen a specimen in pickle. Who now can say what the truth is?

COUNT TURATI'S AUK

Bird no.52 (Grieve no. 46; Hahn no. 42) - adult in summer plumage, said to be male
Origin: unknown but probably Iceland
Present location: Milan - Museo Civico di Storia Naturale di Milano, Corso Venezia 55, 20121 Milano, Italy.

Champley, R. 1864. The Great Auk. *Annals and Magazine of Natural History*, p.235. Fatio, V. 1868. Liste des divers représentants de l'*Alca impennis* en Europe. *Bullétin de la Société Ornithologique Suisse*, tome II, pt.1, p.82. Selys Longchamps, E. 1870. Notes on various Birds observed in Italian Museums in 1866. *Ibis*, p.450. Salvadori, T. 1881. [Obituary of Count Turati] *Ibis*, p.609. Blasius, W. 1884. Zur Geschichte der Ueberreste von *Alca impennis*. *Journal für Ornithologie*, p.95. Grieve, S. 1885. *The Great Auk or Garefowl*, p.81 *and* appendix p.16. Sclater, P. (ed.) 1888. The Turati Collection. *Ibis*, p.150. Hahn, P. 1963. *Where is that Vanished Bird?* p.225. Violani, C. 1975. L'Alca impenne...nelle Collezioni Italiane. *Natura- Revista di Scienze Naturali*, p.14, p.16 *and* pl.1. Violani, C. et al. 1984. Uccelli Estinti e Rari nei Musei Naturalistici. *Revista Italiana di Ornitologie*, p.144.

The Auk now in Milan was once a duplicate specimen in the great ornithological collection at Leiden. On April 30th 1860, feeling that one of his two examples was surplus to requirements, Hermann Schlegel, Director of Leiden Museum, took the decision to dispose of it. It was sold to G. A. Frank of Amsterdam, the natural history dealer who, along with his father in Leipzig and his son in London, negotiated the sale of so many Great Auks. The price was paid with bird skins to an equivalent of the value of 220 Dutch guilders.

Frank sold the bird later in the same year to Count Ercole Turati of Milan, the owner of a magnificent collection of mounted birds that contained - according to P.L. Sclater (1888) - exactly 20,618 specimens. From the Count's collection the Great Auk eventually passed to Milan's Natural History Museum where it is still exhibited.

Count Ercole Turati.

Count Turati's Auk. *Courtesy of Carlo Violani.*

THE PISA AUK

Bird no. 53 (Grieve no. 60; Hahn no. 43) - adult in summer plumage
Origin: unknown but probably Iceland
Present location: Pisa - Università di Pisa, Museo di Storia Naturale e del Territorio, Certosa di Calci, Via Roma, 103, Pisa, Italy.

The Pisa Auk. *Courtesy of Marco Zuffi and the Museo di Storia Naturale, Pisa.*

Selys Longchamps, E. 1870. Notes on various Birds observed in Italian Museums in 1866. *Ibis*, p.450. Newton, A. 1870. On Existing Remains of the Gare-fowl. *Ibis*, p.258. Blasius, W. 1884. Zur Geschichte der Ueberreste von *Alca impennis. Journal für Ornithologie*, pp.104-5. Grieve, S. 1885. *The Great Auk or Garefowl*, p.81 *and* appendix p.20. Hahn, P. 1963. *Where is that Vanished Bird?* p.225. Violani, C. 1975. L'Alca impenne...nelle Collezioni Italiane. *Natura- Revista di Scienze Naturali*, p.13, p.18, p.20 *and* pl.2.

Dr Karl Michahelles (1807-1834) was a naturalist from Nuremburg who, during his short lifetime, travelled widely in south-east Europe. In 1834 he left Germany for Greece as doctor to a Bavarian regiment sent to subdue an uprising but while at Nauplia he contracted dysentery and died aged just 27. Among a number of ornithological interests, he'd developed a passion for Great Auks and owned at least three stuffed ones. Auks rightly or wrongly associated with the Michahelles name exist in several museums and uncovering the true histories of the doctor's birds baffled nineteenth century Garefowl specialists. Even as careful a researcher as Alfred Newton became entirely confused by the history of one of them. The identity of the specimen now in Florence (which probably has nothing to do with Michahelles) he mixed up with Pisa's bird, an example which certainly did pass through the German doctor's hands.

Dr. Marco Zuffi (*pers. comm.*, 1995) has confirmed that the Pisa Auk was purchased from Michahelles in 1833 for 200 florins and that it was re-mounted in 1870 by Paolo Savi (after whom Savi's Warbler *Locustella lusciniodes* is named). It probably belonged to Apothecary Mechlenburg of Flensburg and then G. A. Frank of Amsterdam before reaching Michahelles. Almost certainly, it has an Icelandic origin; all Michahelles birds are thought to have come from there. A Baffin Bay locality - mentioned by Hahn (1963) - can be discounted. The suggestion of Baffin Bay seems to derive from a few general remarks in the Pisa Museum's records that concern the species' range. Greenland and Iceland are also mentioned.

PASTOR BREHM'S AUK

Bird no 54 (Grieve no. 72; Hahn no. 44) - adult in summer plumage
Origin: Iceland, probably Eldey
Present location: Rome - Comune di Roma Servizio Giardino Zoologico e Museo di Zoologia, Roma, Italy (catalogue no.5310 Z).

Olphe-Galliard, L. 1862. [Letter to the Editor] *Ibis*, p.302. Fatio, V. 1868. Liste des divers représentants de l'*Alca impennis* en Europe. *Bullétin de la*

Société Ornithologique Suisse, tome II, pt. 1, p.81. Saunders, H. 1869. On the Ornithology of Italy and Spain. *Ibis*, p.393. Newton, A. 1870. On Existing Remains of the Gare-fowl. *Ibis*, p.258. Selys Longchamps, E. 1870. Notes on various Birds observed in Italian Museums in 1866. *Ibis*, p.450. Blasius, W. 1884. Zur Geschichte der Ueberreste von *Alca impennis*. *Journal für Ornithologie*, pp.111-2. Grieve S. 1885. *The Great Auk or Garefowl*, p.81 *and* appendix p.24. Carruccio, A. 1902. Sovra un palmipede rarissimo e di gran valore...donato da S.M. il Re Vittorio Emanuele III al Museo Zoologico...Roma. *Bolletin Society Zoologico Italiane*, pp.1-15, pl.1 *and* pl.2. Salvadori, T. 1902. [Letter to the Editor] *Ibis*, p.523. Rothschild, W. 1907. *Extinct Birds*, p.156. Oddi, E. 1914. Notizie sull' Alca maggiore. *Revista Italiana di Ornitologia*, pp.1-3 *and* pl.1. Chigi, F. 1936. La morte delle specie animali. *Rassegna Faunistica*, III, no.3-4, pl. p.5. Hahn, P. 1963. *Where is that Vanished Bird?* pp.225-6. Violani, C. 1975. L'Alca impenne...nelle Collezioni Italiane. *Natura- Revista di Scienze Naturali*, p.20 *and* pl.1.

Obtained in 1832 by a Lutheran minister, Pastor C. L. Brehm, from the Royal Museum, Copenhagen, it may be assumed that this bird came from Iceland and is probably one of the individuals taken on Eldey during 1831.

Pastor Christian Ludwig Brehm (1787-1864) was a prolific writer on birds who assembled a huge collection of ornithological specimens at his house in Renthendorf near Leipzig. After his death they lay, largely undisturbed, in the attic of the house for almost 40 years. Eventually, they were bought by Walter Rothschild and most are now in the American Museum of Natural History along with much else of Rothschild's ornitho-

Pastor Brehm's Auk. *Courtesy of Fausto Barbagli.*

Victor Emmanuel, King of Italy (1820-78).

logical collection. Rothschild missed out on the Great Auk, however. It was long gone.

Probably in 1867 - although it may have been a year or so later - one of Brehm's sons, Dr. A. E. Brehm, sold it with the help of the well known German ornithologist Otto Finsch. The buyer was Victor Emmanuel, King of Italy. A fairly elaborate bargain seems to have been struck, which included a re-stuffing (by Schwerdfeger of Bremen) before the specimen left Germany. Victor Emmanuel is supposed to have paid an appropriately princely sum for his bird (given variously as 5,000 francs or 7,000 marks) and the money went according to Pastor Brehm's expressed wish to a son (he had six) who was particularly in need.

For many years this Great Auk was kept at the Veneria Reale in Turin but when the collection there was disbanded it passed to the museum at Rome.

THE TURIN AUK

Bird no.55 (Grieve no. 71; Hahn no. 45) - adult in summer plumage
Origin: Unknown but probably Iceland
Present location: Turin - Museo di Zoologia, Dipartimento di Biologia Animale dell' Università di Torino, Via Accademia Albertina 17, 10123 Torino, Italy.

Champley, R. 1864. The Great Auk. *Annals and Magazine of Natural History*, p.235. Fatio, V. 1868. Liste des divers représentants de l'*Alca impennis* en Europe. *Bullétin de la Société Ornithologique Suisse*, tome II, pt. 1, p.82. Selys Longchamps, E. 1870. Notes on various Birds observed in Italian Museums in 1866. *Ibis*, p.449. Blasius, W. 1884. Zur Geschichte der Ueberreste von *Alca impennis*. *Journal fur Ornithologie*, p.111. Grieve, S. 1885. *The Great Auk or Garefowl*, p.81 *and* appendix p.24. Hahn, P. 1963. *Where is that Vanished Bird?* p.226. Violani, C. 1975. L'Alca impenne...nelle Collezioni Italiane. *Natura- Revista di Scienze Naturali*, p.16 *and* pl.2. Violani, C. et al. 1984. Uccelli Estinti e Rari nei Musei Naturalistici. *Rivista Italiana di Ornitologia*, p.144.

The Turin Auk has a history only marginally less obscure than the Turin Shroud. Obtained from a Herr Vogt in 1832, it was listed as a bird from Iceland. Nothing else is known of it. Vogt seems to have been something of a trader in great Auks. In the same year that he sold this bird to Turin, he sold another to the museum at Neuchâtel and a year later obtained a third example from the Royal Museum, Copenhagen. It seems reasonable to suppose that the Turin Auk is one of the 24 birds taken on Eldey in 1831.

The Turin Auk. *Courtesy of Pietro Passerin d'Entrèves and the University of Turin.*

THE AMSTERDAM AUK

Bird no. 56 (Grieve no. 5; Hahn no. 46) - adult in summer plumage
Origin: unknown but probably from Eldey, Iceland

THE GREAT AUK

Present location: Amsterdam - Universiteit van Amsterdam Zoölogisch Museum, Mauritskade 61-57, 1090 GT Amsterdam, The Netherlands.

Preyer, W. 1862. Der Brillenalk in Europäischen Sammlungen. *Journal für* Ornithology, p.78. Champley, R. 1864. The Great Auk. *Annals and Magazine of Natural History*, p.235. Fatio, V. 1868. Liste des divers représentants de l'*Alca impennis* en Europe. *Bullétin de la Société Ornithologique Suisse*, tome II, pt.1, p.82. Blasius, W. 1884. Zur Geschichte der Ueberreste von *Alca impennis*. *Journal für Ornithologie*, p.72. Grieve, S. 1885. *The Great Auk or Garefowl*, p.80 *and* appendix p.6. Grieve, S. 1888. Recent information about the Great Auk. *Transactions of the Edinburgh Field Naturalists' and Microscopical Society*, p.108. Hahn, P. 1963. *Where is that Vanished Bird?* p.226.

This specimen was bought for the equivalent of eight guineas (100 florins) on May 18th 1840 by the Natura Artis Magistra, Amsterdam from G.A. Frank, who got it from Israel of Copenhagen. Its presence in the hands of Israel - a trader known for his good connections in Iceland - at such a late date is suggestive of a bird from Eldey.

From the museum of the Zoological Society, the Auk eventually passed to the University Museum where it remains to this day.

THE LEIDEN AUK

Bird no.57 (Grieve no. 38; Hahn no. 47) - adult in summer plumage
Origin: unknown but probably Iceland
Present location: Leiden - Nationaal Natuurhistorisch Museum, 2300 RA Leiden, The Netherlands.

Sclater, P. 1864. Notes on the Great Auk. *Annals and Magazine of Natural History*, p.320. Fatio, V. 1868. Liste des divers représentants de l'*Alca impennis* en Europe. *Bullétin de la Société Ornithologique Suisse*, tome II, pt.1, p.82. Schlegel, H. 1867. *Muséum d'Histoire Naturelle des Pays Bas*, vol.6, *Urinatores*, p.13. Rosenberg, H. 1874. De Reuzenalk. *Jaarboekje van het Koninklijk Zoologisch Genootschap Natura Artis Magistra*, p.142. Blasius, W. 1884. Zur Geschichte der Ueberreste von *Alca impennis*. *Journal für Ornithologie*, p.92. Grieve, S. 1885. *The Great Auk or Garefowl*, p.80 *and* appendix p.15. Hahn, P. 1963. *Where is that Vanished Bird?* p.226.

The magnificent collection of birds at Leiden may once have included three Great Auks but only one remains there today.

There is no doubt that the Leiden Museum sold a specimen during 1860 to Frank of Amsterdam - a bird now in Milan - but there is a possibility that Leiden had already parted with another. C.J. Temminck, first Director at Leiden, is said to have given a Great Auk to his fellow ornithologist Josse Hardy. Whether it actually belonged to Leiden or whether it was Temminck's private property is uncertain. Before the establishment of Leiden Museum (1820) Temminck had formed a large collection of stuffed birds at his home in Amsterdam and it is possible that he gave the bird

C.J. Temminck (1778-1858).

The Leiden Auk. *Courtesy of René Dekker and the Nationaal Natuurhistorisch Museum, Leiden.*

away during this period. On the other hand, a second duplicate specimen at Leiden might have seemed entirely surplus to requirements.

As far as the bird that remains at Leiden is concerned, it can be traced to two likely sources. Either it came from Professor Reinhardt of the Royal Museum, Copenhagen in 1835 (Reinhardt's son recorded - in a note sent to Alfred Newton - that a Great Auk was sent to Leiden during this year) or it is a bird bought from Frank two years earlier. The archives of the Leiden Museum hold a letter to Temminck dated April 19th 1833 in which Frank offers a Garefowl - described as male - for 150 Dutch florins. The offer appears to have been taken up but it is uncertain whether it is this Frank bird or the Reinhardt one that Leiden has retained.

The matter is of purely academic interest for, whether acquired in 1833 or 1835, it is likely that the Leiden Auk is one of the many individuals killed on Eldey during the early 1830's.

Whether any reliance can be placed on Frank's designation of sex is uncertain. It is not known what kind of relationship the merchants who bought the birds had with the fishermen who actually got them. Whether they received any accompanying information with the specimens they acquired is doubtful. Indeed Sigriður Thorláksdótter, the Icelandic woman who skinned many of the individuals taken on Eldey, said that she never bothered trying to sex the birds she handled.

THE OSLO AUK

Bird no.58 (Grieve no. 51; Hahn no. 48) - adult in summer plumage
Origin: Iceland (1831)
Present location: Oslo - Zoologisk Museum, Universitetet I Oslo, Sarsgate 1, 0562 Oslo, Norway.

Collet, R. 1866. Über *Alca impennis* in Norwegen. *Journal für Ornithologie*, p.70. Newton, A. 1870. On Existing Remains of the Gare-fowl. *Ibis*, p.258. Blasius, W. 1884. Zur Geschichte der Ueberreste von *Alca impennis*. *Journal für Ornithologie*, p.98. Collet, R. 1884. Ueber *Alca impennis* in Norwegen. *Mittheilungen des Ornithologischen Vereines in Wien*, no.5/6, p.84. Grieve, S. 1885. *The Great Auk or Garefowl*, p.81 *and* appendix p.17. Stejneger, L. 1886. Review of 'Grieve on the Great Auk.' *Auk*, p.265. Grieve, S. 1888. Recent information about the Great Auk. *Transactions of the Edinburgh Field Naturalists' and Microscopical Society*, p.108. Salomonsen, F. 1944-5. Gejrfuglens et hundredaars minde. *Dyr i Natur og Museum*, p.106. Hahn, P. 1963. *Where is that Vanished Bird?* p.226.

This specimen was once the property of Nicolai Aall of Nees, some 25 miles from Arendal, Norway. Aall, the proprietor of a steel works at Twedestrand, acquired it in 1845 from Professor Reinhardt of Copenhagen in exchange for the skin of a bear.

Paul Hahn (1963) states that the bird was killed in 1831. Probably, therefore, it was one of the 24 individuals killed on Eldey during that year. It was purchased by the Zoologisk Museum in 1884 at about which time it was remounted.

The Oslo Auk. *Courtesy of the University of Oslo.*

THE WROCLAW AUK

Bird no.59 (Grieve no. 11; Hahn no. 49) - adult in summer plumage, said to be male
Origin: unknown
Present location: Wroclaw (Breslau) - Wroclaw University Wladyslaw Rydzewski, Museum of Natural History, Sienkiewicza 21, 50-335 Wroclaw, Poland.

Homeyer, A. 1865. *Alca impennis* im Breslauer Museum. *Journal für Ornithology*, pp.151-2. Fatio, V. 1868. Liste des divers représentants de l'*Alca impennis* en Europe. *Bullétin de la Société Ornithologique Suisse*, tome II, pt.1, p.81. Blasius, W. 1884. Zur Geschichte der Ueberreste von *Alca impennis*. *Journal für Ornithologie*, p.74. Grieve, S. 1885. *The Great Auk or*

Garefowl, p.79 *and* appendix p.7. Blasius, W. 1900. Der Riesenalk in der Ornithologischen Litteratur der lekten funfzehn Jahr. *Ornithologische Monatsschrift*, p.434 *and* pl.27. Peters, D. 1960. Ausgestorbene und seltene Vögel in den Zoologischen Museen von Breslau und Warschau. *Bonner Zoologische Beiträge*, p.28. Hahn, P. 1963. *Where is that Vanished Bird?* p.226-7.

The museum at Wroclaw has owned two Great Auks for more than 150 years. According to an oral tradition preserved for posterity by Wilhelm Blasius (1884), they were acquired from an itinerant dealer between 1830 and 1840. Blasius identified this dealer as a certain Herr Platow, well known in Prussian towns as a travelling salesman specialising in zoological items. Because the Platow family came from Silesia, the Brunswick ornithologist speculated on the likelihood of Herr Platow returning from a lengthy buying tour with two Garefowls for the capital city of his home province. Blasius was acquainted with the son of the elder Platow and perhaps received specific information from him.

Although the two birds are generally considered a pair there is no direct evidence bearing on this. One is mounted with the wings rather less extended than the other and this is the specimen usually considered male.

The more recent history of these two birds is as romantic as their origin is obscure. Towards the end of World War II, during the siege of Wroclaw (then generally known as Breslau), a bomb destroyed one wing of the museum building and the entire contents lay under rubble. Somewhere in the middle of the devastation were the two Auks. Dr. Jan Lontkowski (*pers. comm.*, March 1st 1995) of the Wroclaw Museum has kindly provided details of the situation at the time:

> The Polish authorities took over the museum collection in a condition of utter devastation. Its preservation was entrusted to the re-established university which ... organised a ... Polish team of those specialists who had survived the war. Today it is difficult to believe that at that hard time, these few people, including only two with experience, managed to save the collection and ensure the continuity of the museum.

Two photographs of the Wroclaw Auks taken before World War II.

The Wroclaw Auks today. *Courtesy of Jan Lontkowski and Wroclaw University.*

It is not hard to imagine the conditions under which these people worked and the appalling hardships that they and their fellow citizens suffered. Somewhere among the rubble they found the Great Auks, however. Amazingly, they were unharmed apart from the grime and dirt that now stain their white fronts.

THE WROCLAW SPREAD-WINGED AUK

Bird no. 60 (Grieve no. 12; Hahn no. 50) - adult in summer plumage, said to be female
Origin: unknown
Present location: Wroclaw (Breslau) - Wroclaw University Wladyslaw Rydzewski, Museum of Natural History, Sienkiewicza 21, 50-335 Wroclaw, Poland.

All references are identical to those of the Wroclaw Auk

Both Great Auks in Wroclaw are stuffed with their wings more widely spread than is usual. One has the wings particularly open and it is this specimen that is kept on show to the public. Although traditionally regarded as female, there is probably no solid ground for this belief. The two birds have been together for more than 150 years and their known history is identical.

BLASIUS'S AUK

Bird no.61 (Grieve no. 9; Hahn no. 59) - adult in summer plumage
Origin: unknown probably Iceland
Present location: Moscow - Darvinskij Musej, Malaja Pirogovskaja, Moskova, Russia.

Blasius, W. 1881-3 Ueber die letzen Vorkommnisse des Riesen-Alks. *Jahrsberichte des Vereins für Naturwissenschaft zu Braunschweig für die Vereinsjahre 1881-2 und 1882-3*, pp.98-9 *and* p.113. Blasius, W. 1884. Zur Geschichte der Ueberreste von *Alca impennis. Journal für Ornithologie*, p.73. Grieve, S. 1885. *The Great Auk or Garefowl*, p.79 *and* appendix p.7. Stone, W. 1912. Notes and News. *Auk*, p.573. Oddi, E. 1914. Notizie sull' Alca maggiore. *Revista Italiana di Ornitologia*, ser.1, vol.3, pp.1-3 *and* pl.1. Hahn, P. 1963. *Where is that Vanished Bird?* p.228 *and* p.237. Luther, D. 1986. *Die Ausgestorbenen Vögel der Welt*, p.83.

The specimen now in Moscow was once the property of Wilhelm Blasius of Brunswick. Paul Hahn (1963) received letters to this effect from both the Brunswick and the Moscow Museums. Yet in his own writing Blasius was surprisingly coy about the fact that he owned a stuffed Garefowl. He describes two specimens at the Brunswick Museum and although recording that one of these was on loan from a private individual, he fails to mention that the private individual was none other than himself.

A drawing based on the Wroclaw Auks reproduced in *Ornithologische Monatsschrift* (1900).

Blasius's Auk before restoration (circa 1912).

According to Blasius's account the bird was previously owned by Baron von Pechlin, Danish Deputy to the old Diet of the German Confederation. He traces its history back no further than this but from the mode of preparation he judged its origin to be Icelandic.

After Blasius's death in May 1912 his Garefowl became available. A note in *The Auk* later in the year (*see* p.573) states:

(*see* p.573)

> We are informed that Mrs M. Blasius of Inselwall 13, Braunschweig, sister-in-law of the late Prof. W. Blasius, desires to dispose of the Great Auk which was contained in his collection.

Eventually the bird was sold by auction at Halle. According to Oddi (1914) it was purchased by an unknown Russian who carried it off to his own country. Later, it was given to the Darwin Museum by A. S. Chomyakov.

THE ST. PETERSBURG AUK

Bird no.62 (Grieve no. 66; Hahn no. 58) - adult in summer plumage
Origin: Iceland, probably Eldey
Present location: St. Petersburg - Zoological Institute, Russian Academy of Sciences, St. Petersburg 199034, Russia (catalogue no.138180).

Brandt, J. 1836 *and* 1837. Berichte über eine Wissenschaftliche Reise nach Deutschland *and* Rapport sur une Monographie de la famille des Alcadées. *Bullétin scientifique publié par l'Académie Impériale des Sciences de Saint-Petersburg,* tome 1, p.176 *and* tome 2, p.345. Preyer, W. 1862. Der Brillenalk in Europäischen Sammlungen. *Journal für Ornithologie,* p.78. Fatio, V. 1868. Liste des divers représentants de l'*Alca impennis* en Europe. *Bullétin de la*

Blasius's Auk after restoration (circa 1912).

The St. Petersburg Auk photographed in February, 1938.

Société Ornithologique Suisse, tome II, pt.1, p.83. Blasius, W. 1884. Zur Geschichte der Ueberreste von *Alca impennis*. *Journal für Ornithologie*, p.107. Grieve, S. 1885. *The Great Auk or Garefowl*, p.81, p.110 *and* appendix p.21. Stresemann, E. 1954. Ausgestorbene und Aussterbende Vogelarten, Vertreten im Zoologische Museum zu Berlin. *Mittheilungen aus dem Zoologischen Museum in Berlin*, p.40. Hahn, P. 1963. *Where is that Vanished Bird?* p.228. Cocker, M. 1989. *Richard Meinertzhagen- Soldier, Scientist and Spy*, plate between. pp.148-9.

In May of 1884 Wilhelm Blasius travelled to St. Petersburg to examine the Great Auk acquired by the Zoological Museum of the Imperial Academy during the 1830's. He found it labelled with the name Brandt and the location Iceland. This was hardly surprising to Blasius who already knew that the specimen received its first definite mention from J.F. Brandt in the Academy's *Bullétin* for 1837 and that it was probably referred to, in rather vaguer terms, by the same author in the *Bullétin* of the previous year.

The earlier reference details a trip to Hamburg in 1836 when J.F. Brandt purchased from his namesake, the Hamburg dealer J.G.W. Brandt, a number of bird skins. During the course of his career Hamburg Brandt handled several Great Auk skins. One, for instance, taken on Eldey in 1834 he offered to the Berlin Museum for 100 marks but the offer was turned down. Quite possibly, it is this same bird that St. Petersburg eventually acquired, although it could easily be a different one. Blasius calculated that eight Great Auks went to Hamburg from Iceland in the year 1834 alone. Whatever the precise truth, it is fairly safe to assume that this bird came from Eldey at some time during the early 1830's.

THE LUND AUK

Bird no.63 (Grieve no. 45; Hahn no. 53) - adult in summer plumage
Origin: unknown but probably Iceland
Present location: Lund - Museum of Zoology, Lund University, Helgonavägen 3, S-223 62 Lund, Sweden.

Newton, A. 1870. On Existing Remains of the Gare-fowl. *Ibis*, p.258. Blasius, W. 1881-3. Ueber die letzen Vorkommnisse des Riesen-Alks. *Jahresberichte des Vereins für Naturwissenschaft zu Braunschweig für die Vereinsjahre 1881-2 und 1882-3*, p.92, p.109 *and* p.114. Blasius, W. 1884. Zur Geschichte der Ueberreste von *Alca impennis*. *Journal für Ornithologie*, p.94. Grieve, S. 1885. *The Great Auk or Garefowl*, p.81 *and* appendix pp.15-6. Salomonsen, F. 1944-5 Gejrfuglens et hundredaars minde. *Dyr i Natur og Museum*, p.106. Hahn, P. 1963. *Where is that Vanished Bird?* p.227.

Although this bird was once labelled 'Greenland, 1835,' there is little doubt that its origin is Icelandic. It was presented to the University of Lund in 1835 by Reinhardt of Copenhagen at the request of Professor Nilsson. It can be assumed that it came from one of the early raids on Eldey.

PAYKULL'S STOCKHOLM AUK

Bird no.64 (Grieve no. 68; Hahn no. 54) - adult in summer plumage
Origin: unknown
Present location: Stockholm - Naturhistoriska Riksmuseet, Frescativägen 44, S-104 05 Stockholm, Sweden.

Newton, A. 1870. On Existing Remains of the Gare-fowl. *Ibis*, p.258. Blasius, W. 1884. Zur Geschichte der Ueberreste von *Alca impennis. Journal für Ornithologie*, p.108. Grieve, S. 1885. *The Great Auk or Garefowl*, p.81 *and* appendix p.22. Lönnberg, E. 1926. The Ornithological Collection of the Natural History Museum in Stockholm. *Auk*, p.438 *and* p.446. Salomonsen, F. 1944-5. Gejrfuglens et hundredaars minde. *Dyr i Natur og Museum*, p.106. Hahn, P. 1963. *Where is that Vanished Bird?* p.227.

For a short period during the nineteenth century there were two Great Auks in Stockholm. Due to the pressure of space and a genuine desire to increase the scope and variety of the collection many exchanges with other museum were arranged. In the museum records for 1835-6 one particular swap, organised by museum Superintendent Professor Fries, is detailed. A recently acquired Great Auk - considered an unnecessary duplicate - was sent to Florence. In exchange Stockholm received... a stuffed Platypus! Probably it seemed perfectly fair at the time.

For reasons that are now quite unclear, Alfred Newton became very confused about the Stockholm birds. He believed, and communicated this belief to his fellow Garefowl scholar Wilhelm Blasius, that the specimen remaining in Stockholm was taken in Iceland and arrived in Sweden - via Copenhagen - at some time during the early 1830's. This is wrong. The Stockholm Auk has a history that goes back at least as far as 1817 when it was recorded in the museum of Baron Gustaf von Paykull, Councillor of the Swedish Chancery. Paykull, anxious that a natural history museum belonging to the state should be established, offered to donate all his valuable zoological collection as a basis for such a museum. Indeed, he did just that and his Great Auk has stayed in Stockholm ever since. With a known history going back to 1817, this is one of the earliest examples in existence. Only a handful can claim equal or superior seniority. Probably the bird came from the Geirfuglasker but this is far from certain.

THE AARAU AUK

Bird no.65 (Grieve no. 2; Hahn no. 55) - adult in summer plumage
Origin: unknown but probably Iceland
Present location: Aarau - Aargauisches Naturmuseum, Aarau, Switzerland.

Michahelles, K. 1833. Geschichte von *Alca impennis. Isis*, p.650. Fatio, V. 1868. Quelques mots sur les exemplaires de l'*Alca impennis*, Oiseaux et Oeufs qui se trouvent en Suisse *and* Liste des divers représentants de l'*Alca impennis* en Europe. *Bullétin de la Société Ornithologique Suisse*, tome II, pt.1, pp.73-4 *and* p.83. Blasius, W. 1884. Zur Geschichte der Ueberreste von *Alca impennis. Journal für Ornithologie*, pp.70-1. Grieve, S. 1885. *The Great Auk or Garefowl*, p.81, p.91, appendix p.5, p.7 *and* p.19. Salomonsen, F. 1944-5. Gejrfuglen et hundredaars minde. *Dyr i Natur og Museum*, p.106. Hahn. P. 1963. *Where is that Vanished Bird?* p.227. Foelix, R. 1996. Seltsame Vögel im Aargauischen Naturmuseum. *Aarauer Neujahrsblätter, 1996*, p.55 *and* pl. p.53.

The accounts given of this bird by Wilhelm Blasius (1884) and Paul Hahn (1963) are particularly unclear and confusing, but some facts can be established. It was presented to the Aarau Museum about the year 1865 by Counsellor of State Frey Hérosée who'd purchased it for 80 florins (which Grieve, 1885, rather charmingly translates as six pounds, eleven shillings and eightpence) during 1842 or 1843 from a man named Michahelles.

Herr Michahelles was the father of the ornithological writer Dr. Karl Michahelles, a young man who interested himself greatly in *Alca impennis* - owning several stuffed examples - but who had died of dysentry at Nauplia, Greece, several years previously (1835). The Aarau Auk was the second of two birds that Frey Hérosée purchased from Michahelles senior and it came to him just a few days after the first. Because of the date discrepancy there is some doubt over whether this particular bird ever actually belonged to the younger Michahelles. If it did, it is probably one of the birds mentioned by him in a paper written for the German journal *Isis* in 1833.

Blasius believed that this was a specimen Karl Michahelles never saw, however. He considered it a bird sent to Michahelles on approval by a friend who was aware of his keen interest in Garefowls but was unaware of his death. According to Blasius, the bird arrived with the Michahelles family during 1840 or, perhaps, 1841. As the friend was expecting 80 florins for it, the father of the deceased simply sold it on for that price to Counsellor of State Hérosée.

Whatever the truth about the Michahelles's connection, it is very likely that Aarau's bird came from Iceland. All the specimens actually owned by the ill-fated doctor came from there and if, alternatively, this one was more recently taken then it is improbable that it originated elsewhere.

Soon after acquiring its Great Auk the town of Aarau refused an offer of 1,500 francs for it.

The Aarau Auk. *Courtesy of Dr. R.F. Foelix and the Aargauisches Naturmuseum.*

CAPTAIN VOUGA'S AUK

Bird no.66 (Grieve no. 19; Hahn no. 56) - adult in summer plumage
Origin: unknown
Present location: Lausanne - Musée Zoologique, Place Riponne 6, CH-1000 Lausanne 17, Switzerland.

Fatio, V. 1868. Quelques mots sur les exemplaires de l'*Alca impennis*, Oiseaux et Oeufs qui se trouvent en Suisse *and* Liste des divers représentants de l'*Alca impennis* en Europe. *Bullétin de la Société Ornithologique Suisse*, tome II, pt.1, pp.74-5, pl.1 *and* p.83. Blasius, W. 1884. Zur Geschichte der Ueberreste von *Alca impennis. Journal für Ornithologie*, pp.76-7. Grieve, S. 1885. *The Great Auk or Garefowl*, p.81 *and* appendix p.9. Grieve, S. 1888.

Captain Vouga's Auk. *Courtesy of Prof. P. Goeldlin and the Musée Zoologique, Lausanne.*

Chromolithograph in the *Bullétin de la Société Ornithologique Suisse* (1868) said to be drawn by Captain Vouga's son using the stuffed bird as a model.

Recent information about the Great Auk. *Transactions of the Edinburgh Field Naturalists' and Microscopical Society*, p.108. Saunders, H. 1891. Notes on Birds observed in Switzerland. *Ibis*, p.158. Grieve, S. 1896-7. Supplementary note on the Great Auk. *Transactions of the Edinburgh Field Naturalists' and Microscopical Society*, p.252. Duchaussoy, H. 1897-8. Le Grand Pingouin du Musée d'Histoire naturelle d'Amiens- notes additionnelles. *Mémoires de la Société Linnéenne du Nord de la France*, p.250. Hahn, P. 1963. *Where is that Vanished Bird?* p.228. Bárðarson, H. 1986. *Birds of Iceland*, pl. p.25.

Displayed in a glass case at the Musée Zoologique Lausanne, along with five other extinct birds, is Captain Vouga's Auk. Still in good condition, having been well cared for throughout its existence, the specimen was bought by the museum around the year 1886. Along with the rest of Captain Vouga's collection it cost 12,000 francs.

Although its origin is doubtful, a number of details concerning the specimen's history are well recorded. For much of the nineteenth century it belonged to Captain A. Vouga (and then to his son) of Cortaillod near

Neuchâtel, Switzerland. According to Vouga, his Great Auk was brought, preserved in salt, by whalers to a port in northern France. This information gave rise to speculation that the bird was brought back from the coasts of Newfoundland but the probable date of the event (first decades of the nineteenth century) makes it more likely that it came from one of the Icelandic skerries; these were certainly visited by French vessels from time to time. The salted bird was taken to Amiens where it was stuffed and sold to a friend of the taxidermist. This man happened also to be a friend of Captain Vouga who some years later (supposedly around the year 1838) was himself given the opportunity to buy it- an opportunity he eagerly took.

Vouga believed his bird to be male. Perhaps he had some grounds for this belief. Judging from the described circumstances of the bird's arrival in Amiens, however, the taxidermist received just the skin preserved in salt. From this alone there would be no reliable way of determining sex and it is doubtful if whalers would have troubled to discover it. Frank of London, who handled many skins, saw the specimen in 1886 and declared it a fine female. Like the reason for the Captain's belief, the basis of Frank's remains a mystery. The specimen shows the area of grey on the flanks that Rothschild (1907) stated to be the mark of the female but the logic behind Rothschild's statement is as obscure as the thinking of Frank and Vouga.

To accompany a paper in which he describes the three Great Auks in Swiss collections, Victor Fatio (1868) reproduced a charming portrait of this bird, the original of which is said to be the work of Captain Vouga's son. It hardly resembles the sitter and erroneously depicts the Auk among towering mountains of ice but nonetheless has enormous appeal. This son of Vouga showed considerable interest in Garefowls and travelled to Iceland in search of eggs and bones but was unable to find any sailor willing to take him out to the islands he most wished to visit.

THE NEUCHÂTEL AUK

Bird no.67 (Grieve no. 52; Hahn no. 57) - adult in summer plumage
Origin: Iceland, probably Eldey
Present location: Neuchâtel - Musee d'Histoire Naturelle, Rue des Terreaux 14, CH-2000 Neuchâtel, Switzerland.

Olphe-Galliard, L. 1862. [Letter to the Editor] *Ibis*, p.302. Fatio, V. 1868. Quelques mots sur les exemplaires de l'*Alca impennis*, Oiseaux et Oeufs qui se trouvent en Suisse *and* Liste des divers représentants de l'*Alca impennis* en Europe. *Bullétin de la Société Ornithologique Suisse*, tome II, pt.1, p.74 *and* p.83. Blasius, W. 1884. Zur Geschichte der Ueberreste von *Alca impennis. Journal für Ornithologie*, p.98. Grieve, S. 1885. *The Great Auk or Garefowl*, p.81, p.91 *and* appendix p.17. Saunders, H. 1891. Notes on Birds observed in Switzerland. *Ibis*, p.159. Salomonsen, F. 1944-5. Gejrfuglen et hundredaars minde. *Dyr i Natur og Museum*, p.106. Hahn, P. 1963. *Where is that Vanished Bird?* p.228.

This is probably another of the birds killed on Eldey following the sinking of the Geirfuglasker. Victor Fatio (1868) was able to state, on the authority of Louis Coulon (then Director of the Neuchâtel Museum), that it was bought at Mannheim during 1832 - for either 200 or 300 francs - from the natural history dealer Heinrich Vogt, a man who traded several Great Auks. He is on record as selling the Turin Auk in 1832 and is supposed to have received another bird from Reinhardt of Copenhagen during 1833.

CHAMPLEY'S AUK

Bird no.68 (Grieve no. 67; Parkin no. 4; Hahn no. 60 *and* lost bird no. 3) - adult in summer plumage
Origin: Iceland
Present location: Andover - Phillips Academy, Andover, Massachusetts, USA.

Kjærbölling, N. 1856. *Ornithologica Danica, Danmark's Fugle*, p.415. Champley, R. 1864. The Great Auk. *Annals and Magazine of Natural History*, p.235. Fatio, V. 1868. Liste des divers représentants de l'*Alca impennis* en Europe. *Bullétin de la Société Ornithologique Suisse*, tome II, pt.1, p.82. Newton, A. 1870. On Existing Remains of the Gare-fowl. *Ibis*, p.258. Blasius, W. 1884. Zur Geschichte der Ueberreste von *Alca impennis.Journal für Ornithologie*, pp.107-8. Grieve, S. 1885. *The Great Auk or Garefowl*, p.78, p.91, appendix p.21 *and* p.33. Grieve, S. 1888. Recent information about the Great Auk. *Transactions of the Edinburgh Field Naturalists' and Microscopical Society*, p.115. Duchaussoy, H. 1897-8. Le Grand Pingouin du Musée d'Histoire naturelle d'Amiens. *Mémoires de la Société Linnéenne du Nord de la France*, p.112. Lydekker, R. 1908. *The Sportsman's British Bird Book*, p.195. Parkin, T. 1911. The Great Auk. *Hastings and East Sussex Naturalist*, vol.1, pt.6 (extra paper), pp.24-5 *and* p.36. Allingham, E. 1924. *A Romance of the Rostrum*, p.166. Anon. 1929 (June 15th). The Story of a Great Auk- Some Rare Birds in Mr. J.B. Nichols' Collection. *The Field*, p.936 *and* pl. p.936. Chigi, F. 1936. La morte delle specie animali. *Rassegna Faunistica*, pl. p.31.

> I purchased the bird (said to have laid the egg) in 1861, from the Apothecary Mechlenburg, residing at Flensburg, Denmark.
>
> Robert Champley

The whereabouts of the Great Auk that once belonged to the indefatigable Auk collector Robert Champley went unrecognised for many years and his specimen was considered lost. Although it was widely known that there was a Great Auk at Phillips Academy, Andover, Mass., it was not correlated with the bird that Champley had owned. Even the staff at Phillips - while fully appreciative of their bird's importance - were quite unaware of its previous history. Only through the efforts of Thomas Hamilton (a member of the Academy staff) who unearthed a significant letter in the school's archives, has it been possible to correlate this Andover bird with the Auk that Robert Champley once kept at his home in Scarborough, Yorkshire.

Champley bought his Great Auk in 1861, together with its alleged egg (now at the Alexander Koenig Museum, Bonn), from Mechlenburg, the Apothecary of Flensburg (then Denmark, now Germany). The

Scarborough alderman always maintained that he paid £45 for the two items although the apothecary's papers suggested he paid the considerably larger sum of £120. The story of egg and bird belonging together - as told by Mechlenburg to Champley - is doubtful. According to this story, the bird was shot on its egg at Langenes, Iceland on June 20th 1820 but this sounds suspiciously like a dealer's tale aimed at making the items more desirable. A similarly romantic story is told- apparently on Mechlenburg's authority- by the Danish writer Nils Kjærbölling (1856). Kjærbölling recorded that among Mechlenburg's specimens were a pair killed in 1829 on the Geirfuglasker while courageously defending their egg. Whether or not Mechlenburg had any real idea about the origin of his birds is impossible to say. Through his hands no less than eight Great Auks are reputed to have passed; clearly, he knew exactly how to tantalise collectors into digging deep inside their pockets. There is no reason to doubt that all of Mechlenburg's birds - including this one - came from Iceland, however.

At some time during his period of ownership Champley considered having his bird re-stuffed. The skilled Newcastle taxidermist John Hancock offered to undertake the work and was confident of success. Although Hancock had successfully re-modelled his own stuffed bird (now at the Hancock Museum, Newcastle), Champley eventually decided not to go through with the project. Presumably, he felt the risks were too great and feared ending up with just a pile of black and white feathers.

When Champley died in 1895 his bird passed, along with nine eggs that he'd managed to amass, to his daughter. In 1902, on the 17th of April, it was sold at Stevens' Auction Rooms, London, the well-known venue for so many Great Auk sales. Here it fetched £315 and was knocked down to Mr. J. B. Nichols of Victoria Street, Westminster. The specimen stayed with Nichols until his death in 1929 when, in the month of June, it was sold - again by Stevens' (although not on their premises) - to the dealer W.H.F. Rosenberg for a sum of £660.

At this point all trace of the bird was lost - at least as far as Great Auk researchers were concerned. All that was known was that Rosenberg had purchased it on the 11th of June on behalf of an un-named client.

During 1963 Paul Hahn, a Toronto business man and naturalist, published the results of his survey of all known Great Auk specimens. Despite his extensive research he was unable to uncover the truth about this bird. He lists it twice, first as his 'bird no.60' (at Phillips Academy) and then as a lost specimen.

Late in 1994, an inquiry to Phillips, an independent boarding high school (more or less the equivalent of a British public school), was answered by Thomas Hamilton who regretted that the Academy had no knowledge of the bird's history. At Andover they knew only that it was donated by a generous benefactor called Thomas Cochran. However, Mr. Hamilton managed to find the only scrap of relevant information remaining at Andover: a letter from Thomas Cochran's secretary dated July 20th 1934. This was an answer to an inquiry from the Academy requesting details about the bird's origins. The answer, preserved by chance for all these years, held the clue needed to connect the Andover specimen with Champley's. Mr. Cochran, it seems, knew virtually nothing of his bird's

Champley's Auk (circa 1935).

prior history. What he did know, carefully explained by his secretary in the letter, was that it was purchased at auction in London on June 11th 1929 and that it cost him $3,529 and 46 cents. The mystery surrounding the whereabouts of Champley's Auk was a mystery no longer!

Quite why Mr. Cochran paid a large sum of money for a bird he knew nothing about and seems to have had little interest in - and then in a very short space of time presented to Phillips Academy - is something of a mystery in itself.

ROWLAND HILL'S AUK

Bird no.69 (Grieve no. 30; Hahn no. 61) - adult in summer plumage
Origin: unknown
Present location: Cambridge - Museum of Comparative Zoology, Harvard University, Cambridge, Massachusetts 02138, USA (catalogue no.171764).

Champley, R. 1864. The Great Auk. *Annals and Magazine of Natural History*, p.235. Fatio, V. 1868. Liste des divers représentants de l'*Alca impennis* en Europe. *Bullétin de la Société Ornithologique Suisse*, tome II, pt.1, p.82. Blasius, W. 1884. Zur Geschichte der Ueberreste von *Alca impennis*. *Journal für Ornithologie*, p.86. Grieve, S. 1885. *The Great Auk or Garefowl*, p.78 *and* appendix p.13. Grieve, S. 1896-7. Supplementary note on the Great Auk. *Transactions of the Edinburgh Field Naturalists' and Microscipical Society*, p.247. Thayer, J. 1905. The purchase of a Great Auk for the Thayer Museum at Lancaster, Mass. *Auk*, pp.300-3 *and* pl.XIII. Anon. 1905 (March 25th). Sale of a Great Auk. *The Field*, p.504. Evans, J. 1910 (December). *Country-side* [Magazine], pl.1. Ward, R. 1913. *A Naturalist's Life Study*, p.223. Hahn, P. 1963. *Where is that Vanished Bird?* p.229.

Rowland Hill (1795-1879), inventor of the postage stamp, owned a Great Auk that was kept at his family estate - Hawkstone - in Shropshire. It was purchased in 1838 for the sum of £9 from John Gould but it seems that its appearance did not at first satisfy Viscount Hill. An entry in the Hawkstone catalogue reads:

Henry Shaw (1812-87).

The Dodo and the Great Auk. Bromochrome after a painting by John Evans in *Country-side* Magazine (December 1910). The Great Auk is directly copied from a photograph of Lord Hill's Auk.

This bird was re-set up by H. Shaw [the Shrewsbury taxidermist] in
    1867 and is supposed to be the best in existence.
'Supposed' allows for all sorts of vagueness, of course. There are a surprising number of Great Auks 'supposed' to be the finest. Better examples than
this one certainly do exist.

Although it was probably unknown to Lord Hill, Henry Shaw removed
several bones from the extremities of the bird during the re-stuffing process.
Long after Hill's death, Shaw wrote to Walter Rothschild offering these
bones for £4. Rothschild bought them and they are now in the Natural
History Museum at Tring (catalogue no. 51972.1.156).

The entire Hawkstone collection, said by H.E. Forrest in *The Fauna of
Shropshire* (1899) to be particularly fine, was eventually sold to Mr. Beville
Stanier, a Shropshire gentleman who showed much interest in the birds of
the district where he lived. The Great Auk, not being exactly germane to
this interest, was quickly disposed of and at the end of 1904 the taxidermy
firm of Rowland Ward was entrusted with handling the sale. One of the
potential customers was Ernst Hartert, the German-born Curator of Birds
at Rothschild's Tring Museum, who tried to buy the specimen on behalf of
Alexander Koenig of Bonn. Dr. Goetz Rheinwald has kindly provided
translations of several letters concerning this Great Auk that are still kept
at the Alexander Koenig Museum. Commenting on a suggested price of
£400, Hartert advised Koenig as follows on December 16th 1904:

> Nobody will pay more than £300. This is the highest believable
> price ... Of course this ... can only be paid for a good one.

No-one understood the market better than Rowland Ward, however. On
February 16th 1905, Hartert was obliged to report back to Koenig:

> The Auk was unfortunately sold for £420 - to America ... I am sad
> that it left Europe ... In Spring another mounted Auk will be sold....

Lord Hill's Auk (circa 1900).

The Koenig Museum was, in fact, successful in buying this second bird; the
first was bagged by John E. Thayer the extremely enthusiastic collector
whose full Garefowl tally included ten eggs to go with his stuffed example.
All these specimens were presented to the Museum of Comparative
Zoology during the early 1930's and were added to the single egg and a
stuffed bird that the museum already possessed.

WILLIAM BARBOUR'S AUK

Bird no.70 (Grieve no. 7; Hahn no. 62) - adult in summer plumage
Origin: unknown
Present location: Cambridge - Museum of Comparative Zoology, Harvard
University, Cambridge, Massachusetts 02138, USA.

Bree, C. 1867 (Dec.14th). Birds in the collection of Mr. Hoy. *The Field*,
p.504. Newton, A. 1870. On Existing Remains of the Gare-fowl. *Ibis*,
p.258. Blasius, W. 1884. Zur Geschichte der Ueberreste von *Alca impennis*.
Journal für Ornithologie, p.73. Grieve, S. 1885. *The Great Auk or Garefowl*, p.77
and appendix p.7. Grieve, S. 1888. Recent information about the Great
Auk. *Transactions of the Edinburgh Field Naturalists' and Microscopical Society*,
p.107. Hahn, P. 1963. *Where is that Vanished Bird?* p.229.

William Barbour's Auk (circa 1920).

At Boyles Court (sometimes recorded as Bourne, sometimes as Boyne) in Essex there was once a splendid Great Auk. Very little is known about it. By the time Symington Grieve wrote *The Great Auk* (1885), this bird was the property of a Mrs Lescher. Some forty years previously it had belonged to her brother a Mr. J.D. Hoy but its origin cannot be reliably traced back further. In the Newton archive at Cambridge (England) there is a scribbled note recording that the bird came from a Mr. Tucker through the agency of a man by the name of Goodwin, but this is fairly meaningless. All that is known of Tucker is that he was a dealer from The Quadrant, Regent Street, London who played a small part in the history of the Great Auk that D.G. Elliot acquired for the American Museum of Natural History.

For most of the twentieth century the Auk from Boyle Court has been at Harvard. According to J.C. Greenway (*see* Hahn, 1963), once Curator of Birds at the Museum of Comparative Zoology:

It was purchased from Rowland Ward, London in 1923 ... [and was] the gift of William Barbour.

Barbour was, it seems, a considerable benefactor of the Museum. In 1905 he donated $5,000, most of which was spent on new storage cabinets. However, a small percentage was used to buy ... a Great Auk egg.

THE SELYS LONGCHAMPS AUK

Bird no.71 (Grieve no. 41; Hahn no. 7) - adult in summer plumage
Origin: Unknown
Presnt location: Chicago - Field Museum of Natural History, Roosevelt Road at Lake Shore Drive, Chicago, Illinois 60605-2496, USA (cat no. #285149).

Chromolithograph by E. de Maes from J. Fraipont's *Collections Zoologiques du Baron Edmund de Selys Longchamp - Oiseaux* (1910).

Newton, A. 1870. On Existing Remains of the Gare-fowl. *Ibis*, p.259. Selys Longchamps, E. 1870. Notes on various Birds observed in Italian Museums in 1866. *Ibis*, p.450. Blasius, W. 1884. Zur Geschichte der Ueberreste von *Alca impennis*. *Journal für Ornithologie*, p.92. Grieve, S. 1885. *The Great Auk or Garefowl*, p.77 *and* appendix p.15. Fraipont, J. 1910. *Collections Zoologiques du Baron Edmond de Selys Longchamps*, fasc.XXXI, *Oiseaux*, p.128 *and* pl.II. Hahn, P. 1963. *Where is that Vanished Bird?* p.211.

Nothing is known of the origin of this bird. Baron Edmund de Selys Longchamps obtained it in Turin in 1840. Although acquired from a man named Verany, he believed the actual owner to be the French naturalist Verreaux. For many years after its purchase the Great Auk was one of the centrepieces of the Baron's zoological collection at Longchamps near Waremme in Belgium and a coloured illustration of it is given in the collection catalogue. Eventually donated by Selys Longchamps to the Brussels Museum, the specimen joined the stuffed Garefowl already there. It stayed in Brussels until 1966 when an exchange was arranged with the Field Museum. Apparently feeling that holding two specimens of *Alca impennis* was excessive, Dr. A. Capart, then Director of the Institut Royal des Sciences Naturelles, decided to swap one of them. On December 27th 1966, the Selys Longchamps Auk was accessioned in Chicago in exchange for 2,000 North American bird skins. By arrangement only 1,902 skins were actually sent to Brussels. The odd two were skins of extinct birds (presumably a Passenger Pigeon *Ectopistes migratorius* and a Carolina Parakeet *Conuropsis carolinensis*) and by a rather bizarre mutual agreement these two were valued at a rate of 50 ordinary birds apiece. What the Baron would have made of the deal can only be imagined.

Fydell Rowley (1851-1933), who inherited two stuffed Auks and six eggs when his father died suddenly in 1878. *Courtesy of Peter Rowley.*

DAWSON ROWLEY'S CINCINNATI AUK

Bird no.72 (Grieve no. 14; Hahn no. 4) - adult in summer plumage, said to be male
Origin: Unknown
Present location: Cincinnati - Cincinnati Museum of Natural History, 1301 Western Avenue, Cincinnati, Ohio 45203, USA.

Newton, A. 1870. On Existing Remains of the Gare-fowl. *Ibis*, p.259. Blasius, W. 1884. Zur Geschichte der Ueberreste von *Alca impennis*. *Journal für Ornithologie*, pp.74-5 *and* p.127. Grieve, S. 1885. *The Great Auk or Garefowl*, p.77 *and* appendix p.8. Stevens' [auction catalogue],1934 (Nov.14th). *Important Ornithological Collections of the late George Dawson Rowley*, p.13. Anon. 1934 (Nov.15th). Rare Birds Bought at London Auction. *The Daily Sketch*, p.14. Jourdain, F. 1934 (Dec.1st). The Sale of G.D. Rowley's Collections. *Oologist's Record*, pp.75-9. Jourdain, F. 1935. The skins and eggs of the Great Auk. *British Birds*, p.233. Jourdain, F. 1935. Sale of the Skins and Eggs of the Great Auk. *Ibis*, p.245. Hahn, P. 1963. *Where is that Vanished Bird?* p.210. Spink and Son Ltd. (c1970). *Romantic Histories of Some Extinct Birds and Their Eggs*, p.11 *and* pl.5. Hywel, W. 1973. *Modest Millionaire*, pp.158-60 *and* pl.24. Knaggs, N. 1976 (Sept.). The Great Auk Expedition.

Explorer's Journal, pp.98-101. Bourne, W. 1993. The story of the Great Auk. *Archives of Natural History*, p.275. Laycock, G. 1994. *John A. Ruthven. In the Audubon Tradition*, p.50.

This Great Auk is one of two examples previously owned by the Victorian naturalist George Dawson Rowley, most famous for his beautifully illustrated *Ornithological Miscellany* (1875-8). When Rowley died in 1878 his two Auks, together with no less than six eggs, passed to his son Fydell. It was not until 1934 that his descendants sold off the magnificent Rowley collection of books and natural history items formerly housed at the family mansion- Chichester House, East Cliff, Brighton.

At Stevens' Auction Rooms King Street, Covent Garden, on November 14th 1934, one of the Great Auks was sold for 480 guineas and the other was knocked down for 500. Both went to the eccentric and fiercely wealthy Englishman, Captain Vivian Hewitt.

After Hewitt's death in 1965 his Auk collection, four stuffed birds and thirteen eggs, became the responsibility of the London art dealers Spink and Son Ltd., and nine years later the Cincinnati Museum paid $25,000 for the last of the stuffed birds (the other three being previously sold to museums in Birmingham, Cardiff and Los Angeles); included in the price was an egg. According to the catalogue that Spink produced to promote sales, Cincinnati had bought a mounted specimen of unknown provenance. In fact, the catalogue is not completely reliable and the museum had acquired the bird Hewitt bought from the Rowley collection for 500 guineas in 1934.

Very little is known of the specimen's early history. It is first recorded in the hands of the Paris dealer Lefevre during 1848. In that year the well known London taxidermist James Gardiner bought it and he kept it for twenty years before selling it to Rowley. An allegation that this Great Auk is male cannot be relied upon.

James Gardiner's receipt for the Great Auk.

The Great Auk exhibit at Cincinnati. *Courtesy of the Cincinnati Museum.*

DAWSON ROWLEY'S LOS ANGELES AUK

Bird no.73 (Grieve no. 13; Hahn no. 5) - adult in summer plumage, said to be female
Origin: Iceland, probably Eldey (perhaps June 1844)
Present location: Los Angeles - Los Angeles County Museum of Natural History, Exposition Park, 900 Exposition Boulevard, Los Angeles, California 90007, USA.

Altum, D. 1863. Veränderungen der Vogelfauna des Münsterlandes. *Journal für Ornithologie*, p.115. Gurney, J. 1869. Great Auks for Sale. *Zoologist*, p.1603. Newton, A. 1870. On Existing Remains of the Gare-fowl. *Ibis*, p.259. Blasius, W. 1884. Zur Geschichte der Ueberreste von *Alca impennis*. *Journal für Ornithologie*, pp.74-5. Grieve, S. 1885. *The Great Auk or Garefowl*, p.77, p.92 *and* appendix p.8. Stevens' [auction catalogue], 1934 (Nov.14th). *Important Ornithological Collections of the late George Dawson Rowley*, p.13. Anon. 1934 (Nov. 15th). Rare Birds Bought at a London Auction. *The Daily Sketch*, p.14. Jourdain, F. 1934 (Dec. 1st). The Sale of G.D. Rowley's Collections. *Oologist's Record*, pp.75-9. Jourdain, F. 1935. The skins and eggs of the Great Auk. *British Birds*, p.233. Jourdain, F. 1935. Sale of the Skins and Eggs of the Great Auk. *Ibis*, p.245. Hahn, P. 1963. *Where is that Vanished Bird?* p.210. Spink and Son Ltd. (c1970). *Romantic Histories of Some Extinct Birds and Their Eggs*, p.11 *and* pl.4. Hywel, W. 1973. *Modest Millionaire*, pp.158-60 *and* pl.24.

In 1965 the fabulously wealthy English eccentric Vivian Hewitt died leaving his affairs in chaos. Throughout his life he had collected furiously on

(Facing page, top). The Cincinnati (right) and Los Angeles Auks photographed by George Dawson Rowley in 1871. *Courtesy of the University Library, Cambridge.*

(Facing page, bottom). The Cincinnati (left) and Los Angeles Auks photographed in 1934 at the time of the Rowley sale. *Courtesy of David Wilson.*

(Below, left). The Cincinnati Auk (circa 1970).

(Below, right). The Los Angeles Auk (circa 1970).

a number of fronts and at his death his vast assemblage of possessions was left crammed into his house, many of the items - only partially unpacked - still in the boxes in which they'd been delivered. It is estimated that his egg collection consisted of at least half a million examples; these were accompanied by a huge quantity of bird skins and four stuffed Great Auks.

Due to the fact that Hewitt had collected coins as frantically as he'd collected birds, many of his effects were put into the charge of the London art dealer and coin specialist Spink to whom fell the task of selling as much of the material as possible. Quite understandably, Spink and Son Ltd. had no experience in the rather esoteric trade of selling stuffed specimens of extinct birds and perhaps it was due to this - as much as to the fact that Hewitt's affairs were left in such disorder - that the four Garefowls were mixed up and it became uncertain which was which. In the booklet Spink produced at the time of the sale, the identities of two of the specimens are hopelessly confused; the histories of the other two are not, therefore, to be necessarily relied upon. Despite this confusion over provenances, the four birds were sold to museums in Birmingham, Cardiff, Cincinnati and Los Angeles. From a comparison of nineteenth and early twentieth century photographs it is easy enough to determine the true identities of the birds in Cardiff and Birmingham. Those in Cincinnati and Los Angeles are more difficult to separate. Fortunately, there exist two annotated photographs in the Alfred Newton archive at Cambridge University that enable precise identifications to be made. These photos were taken in 1871 by George Dawson Rowley who was then the owner of both specimens.

It was from the Rowley estate that Hewitt acquired these two birds. On November 14th 1934 Rowley's descendants sold off many of his effects at Stevens' and Hewitt bought both stuffed Garefowls for sums that were huge by the standards of the day, paying 500 guineas for one and 480 for the other. The bird now in Los Angeles can be matched with the one sold for 480 guineas and it is described in the Stevens' catalogue as:

> A fine and perfect specimen of the Great Auk ... female in summer plumage, mounted under oval glass shade.

A certain amount is known about the earlier history of this bird and there is a strong likelihood that it is one of the two individuals killed during the notorious raid on Eldey of June 1844. The skins of these birds - generally supposed to be the last of their kind - vanished soon after they were taken (although the internal organs are preserved to this day in Copenhagen). Despite the efforts of nineteenth century researchers to trace and positively identify them, their whereabouts remain undetermined. The birds now in Los Angeles and Brussels, although hitherto unsuspected, seem the most likely candidates.

George Dawson Rowley obtained the Los Angeles specimen at the end of February 1869 from the Amsterdam based dealer Frank. Frank acquired it a little earlier in the year from the collection of the deceased Count Westerholt-Glikenberg whose home was near Münster, Germany. He paid ten Louis d'Or (which Grieve translates precisely: £9, 18 shillings and fivepence) to Westerholt-Glikenberg's heirs, but this was not the first time that Frank had owned this particular bird. He'd sold it to the Count in 1846, having received it, together with another, during the previous year from a Hamburg merchant by the name of Lintz. In a letter written from Amsterdam and dated March 15th 1869, Frank outlined to Dawson Rowley - in rather quaint English - some of the circumstances:

> The *Alca* you get from me was ... a skin ... I sold to the Count Westerholt which [he] let stuff by his [own] preparator. I found him very bad when I bought him [again] and have send him to the Museum at Leiden to make him better, which made him, I believe, very well.

Clearly the specimen was fairly fresh when Frank first got it from Lintz. Lintz had, in fact, just received both of the birds he sold to Frank from another dealer, a man known only as 'Israel of Copenhagen.' The date (1845) and the geographical implication are highly suggestive. The presence of two Great Auks in the Danish capital at this late period is indicative of specimens recently arrived from Iceland. Moreover, these two birds were in private hands (naturally, the Royal Museum still had several 'old' specimens) and have no known previous history. All that can relevantly be said of Israel is that he is a little known figure who forged good trading links with Iceland and who divided his time between Copenhagen and Amsterdam. Whether his two specimens were indeed the pair caught in 1844, there can be little doubt that both were Icelandic.

The Los Angeles bird has long been claimed to be female. The grounds for this claim are not known but the grey flank markings that Rothschild (1907) regarded as a female characteristic are evident. Despite Frank's good opinion of the specimen's state, Rowley had it re-stuffed by the Brighton taxidermist Swaysland who found it, 'full of pins, the wings cut off, the body filled with hay and pillaged of bones - save one leg bone.'

Los Angeles acquired the specimen in 1970 from Spink paying $12,600 for it. It has been on exhibition almost continuously ever since.

The Los Angeles Auk today. *Courtesy of Kimball Garrett and the Natural History Museum of Los Angeles County.*

ELLIOT'S AUK

Bird no.74 (Grieve no. 55; Parkin no. 2; Hahn no. 63) - adult in summer plumage
Origin: unknown
Present location: New York - American Museum of Natural History, Central Park West at 79th St., New York 10024-5192, USA (catalogue no.3934).

Champley, R. 1864. The Great Auk. *Annals and Magazine of Natural History*, p.235. Fatio, V. 1868. Liste des divers représentants de l'*Alca impennis* en Europe. *Bullétin de la Société Ornithologique Suisse*, tome II, pt.1, p.83. Gurney, J. 1869. Great Auks for Sale. *Zoologist*, p.1603. Blasius, W. 1884. Zur Geschichte der Ueberreste von *Alca impennis*. *Journal für Ornithologie*, p.101 *and* p.128. Grieve, S. 1885. *The Great Auk or Garefowl*, p.81, p.92, p.106, appendix p.19 *and* frontispiece. Stejneger, L. 1886. Grieve on the Great Auk. *Auk*, pp.263-4. Grieve, S. 1888. Recent information on the Great Auk. *Transactions of the Edinburgh Field Naturalists' and Microscopical Society*, p.108, p.119 *and* pl. p.92. Parkin, T. 1911. The Great Auk. *Hastings and East Sussex*

Naturalist, vol.1, pt.6 (extra paper), pp.8-9 *and* p.26. Allingham, E. 1924. *A Romance of the Rostrum*, pp.165-6. Eckert, A. 1963. *The Last Great Auk* (fig. on dustwrapper). Hahn, P. 1963. *Where is that Vanished Bird?* p.229. Bárðarson, H. 1986. *Birds of Iceland*, pl. p.18.

Elliot's Auk- an engraving produced as the frontispiece to S. Grieve's *The Great Auk* (1885).

Although the actual origin of this Great Auk is unknown, much of its rather complex history is recorded. It is first reported in the hands of an obscure dealer in natural history items by the name of Tucker. From Tucker, who operated from the Quadrant, Regent St., London, the bird passed, in 1837, to A.D. Bartlett, Superintendent of the Zoological Gardens in exchange for an assortment of specimens. Bartlett sold it, almost immediately, to Edmund Maunde but at a later date got it back again. During 1852 he sold it for a second time, on this occasion for £28 to Dr. Nathaniel Troughton Esq. Troughton possessed the bird for seventeen years and then, in April 1869, it was offered for sale at Stevens' Auction Rooms. Here it was knocked down - for a price of 90 guineas - to Thomas Cooke, another London dealer in natural history wares. Cooke, who traded in Oxford St., re-stuffed the skin then sold it to the distinguished American ornithological writer D.G. Elliot who was at that time in London making arrangements for the publication of his beautiful book on birds of paradise.

It is often recorded that Elliot purchased the specimen for the Central Park Museum (as the American Museum of Natural History was then known) but this does not quite seem to be the case. He was apparently operating on behalf of one Robert L. Stuart who, according to Symington Grieve (1885), bought the bird:

> For 625 dollars gold, which, calculating the value of each dollar at 4 shillings and 2 pence sterling, gives a total value of £130, 4 shillings and 2 pence.

Stuart's intention in buying the Great Auk was indeed to present it to the Central Park Museum and this he did. It has stayed there ever since.

During the nineteenth century there was some debate among European Auk scholars (most of whom were unlikely to visit the United States to examine the specimen) as to whether this was a repaired but once defective, footless skin that vanished without trace from the collection of the Apothecary Mechlenburg. This idea was thoroughly investigated by Stejneger (1886) who demonstrated fairly conclusively that it is not the case.

Symington Grieve used an illustration of Elliot's Auk as the frontispiece to his famous book.

COUNT DE RIOCOUR'S AUK

Bird no. 75 (Grieve no. 73; Hahn no. 65) - adult in summer plumage
Origin: unknown
Present location: New York - American Museum of Natural History, Central Park West at 79th St., New York 10024-5192, USA (catalogue no.747857).

Newton, A. 1870. On Existing Remains of the Gare-fowl. *Ibis*, p.258. Blasius, W. 1884. Zur Geschichte der Ueberreste von *Alca impennis*. *Journal für Ornithologie*, p.112. Grieve, S. 1885. *The Great Auk or Garefowl*, p.79 *and* appendix p.24. Grieve, S. 1896-8. Supplementary note on the Great Auk *and* Additional notes on the Great Auk. *Transactions of the Edinburgh Field Naturalists' and Microscopical Society*, pp.249-50 *and* p.329. Duchaussoy, H. 1897-8. Le Grand Pingouin du Musée d'Histoire naturelle d'Amiens *and* Notes additionnelles. *Mémoires de la Société Linnéenne du Nord de la France*, p.111 *and* p.247. Rothschild, W. 1907. *Extinct Birds*, p.155. Hahn, P. 1963. *Where is that Vanished Bird?* p.230.

The collection of the Count de Riocour of Vitry-le-François, Marne, France was a particularly fine and important one. Begun about 1820, it was built up over several decades and many of the type specimens of the famous naturalist Vieillot were incorporated. Vieillot was a close friend of the de Riocour family and at his death he willed them his unpublished manuscripts. Other naturalists of the day also showed considerable interest in the Count's museum. It is alleged that the natural history dealer Verreaux came every year to Vitry-le-François to study the rarities and varieties that were there. One of the fascinating features of the collection was a series of drawers that ran beneath the stands on which the specimens were placed. Housed in these drawers were manuscript notes, many written by celebrated naturalists, all relating to the specimens themselves. Among all these rare and precious items was a Great Auk.

In 1887 the grandson of the count who'd founded the collection decided to part with it. The price he asked was 40,000 francs. Unfortunately, perhaps, it didn't stay together although most of it, including the Great Auk, was sold by Alphonse Boucard, a French natural history dealer who operated from 225 High Holborn, London.

At around this time Boucard handled the sale of two Garefowls, both of which came from France and both of which were eventually acquired by Walter Rothschild for his museum at Tring. Naturally enough, this caused some confusion between the two. The de Riocour Auk was considered by Boucard to be superior as he offered it directly to Lord Rothschild for £1,000 whereas the other bird was priced at a mere £300. It is doubtful whether the French dealer actually received his asking price from the English lord but it is certain that Rothschild bought it along with much else of the de Riocour collection. Quite why Boucard rated this bird so highly is uncertain; perhaps it was simply the scent of Rothschild money. The specimen's condition did not greatly please Walter (apparently the feet were damaged) and he had it reduced to a cabinet skin.

De Riocour's Auk. Photograph by Peter Southon. *Courtesy of the American Museum of Natural History.*

The bird stayed with Rothschild until the early 1930's at which time, beset by financial worries, he decided to sell most of his bird collection. Although he retained one of his two Great Auks (and later willed it to the British people), the de Riocour Auk was sold to Mrs Harry Payne Whitney who bought it to present to the American Museum of Natural History.

The early history of this Auk is obscure, surprisingly so in view of the documentation that once existed at Vitry-le-François. According to Duchaussoy (1897-8) it was bought in Hamburg during 1828 for either 250 or 300 francs; certainly there is a record of its presence in the de Riocour collection in 1829. Probably this is a bird taken from the Geirfuglasker in the years just before it was destroyed by volcanic activity.

BONAPARTE'S AUK

(Below, left). Bonaparte's Auk before restoration (circa 1920).

(Below, right). Bonaparte's Auk after restoration (circa 1925).

Bird no. 76 (Grieve no. 36; Hahn no. 64) - adult in summer plumage
Origin: unknown
Present location: New York - American Museum of Natural History, Central Park West at 79th St., New York 10024-5192, USA (catalogue no.763826).

Newton, A. 1870. On Existing Remains of the Gare-fowl. *Ibis*, p.258. Blasius, W. 1884. Zur Geschichte der Ueberreste von *Alca impennis*. *Journal für Ornithologie*, p.91. Grieve, S. 1885. *The Great Auk or Garefowl*, p.78 *and* appendix p.14. Hahn, P. 1963. *Where is that Vanished Bird?* pp.229-30.

A Great Auk supposedly once the property of Prince Lucien Bonaparte was offered up for sale at Stevens' Rooms on Tuesday 23rd September 1919. Here it was bought by Rowland Ward Ltd. for £330 on behalf of the American ornithologist Leonard C. Sanford, the man who saved Audubon's Auk from beneath the Biology Laboratory sink at Vassar College, Poughkeepsie.

Unfortunately, the history of the specimen that Sanford acquired for himself can be traced back no further than to the time of the Prince's ownership. How it came to him is unrecorded. At an unknown date he sold it to the Paris natural history dealer Parzudaki from whom it was bought by the London taxidermy firm of Leadbeater.

During 1861 a friend of John Gould by the name of Naylor obtained the bird from Leadbeater for his collection at Leighton Hall, Montgomeryshire, Wales, where it remained until put into the hands of Stevens.'

On Sanford's death in 1950 his Great Auk became the property of the American Museum of Natural History.

THE RIVOLI AUK

Bird no.77 (Grieve no. 61; Hahn no. 66) - adult in summer plumage
Origin: unknown
Present location: Philadelphia - The Academy of Natural Sciences, 1900 Benjamin Franklin Parkway, Philadelphia, Pennsylvania 1903-1195, USA.

Anon. 1846. *Catalogue de la Magnificent Collection d'Oiseaux de M. Le Prince d'Essling, Duc de Rivoli, dont la vente aura lieu aux enchères publiques...le 8 Juin 1846*, p.37. Baird, S. et al. 1860. *The Birds of North America*, p.902. Newton, A. 1870. On Existing Remains of the Gare-fowl. *Ibis*, p.259. Blasius, W. 1884. Zur Geschichte der Ueberreste von *Alca impennis*. *Journal für Ornithologie*, p.105. Grieve, S. 1885. *The Great Auk or Garefowl*, p.81 *and* appendix p.20. de Schauensee, R. 1941. Rare and Extinct Birds in the Collection of the Academy of Natural Sciences, Philadelphia. *Proceedings of the Academy of Natural Sciences, Philadelphia*, p.291. Hahn, P. 1963 *Where is that Vanished Bird?* p.230.

André Masséna.

Marshall André Masséna, Napoleon's 'Enfant chèri de la Victoire,' distinguished himself gloriously at the Battle of Rivoli near Turin in 1796 and then at the Battle of Essling near Vienna in 1809. His son Victor - Le Prince d'Essling and Duc de Rivoli - collected stuffed birds. By the end of his collecting days he'd amassed well over 12,000 specimens but, sadly for the Prince, he ran short of money. In 1846 he made a rather abrupt decision to part with his collection and arranged to sell it by auction in the Rue de Lille, Paris on the 8th of June. The auction never took place, however. John Gray, Director of the British Museum, acting on behalf of the

The Rivoli Auk.

A page from *The Nidiologist* for December, 1893 showing the Smithsonian Auk before restoration. *The Nidiologist* was a short-lived journal that ran from 1893-1897 and afterwards became incorporated into the *Bulletin of the Cooper Ornithological Club*.

Academy of Natural Sciences, Philadelphia, arrived in town and sent an urgent message to the Prince. Not only did he offer to buy every bird at a rate of four francs per head, he offered to pay ... in ready money. The Prince sent back a message of his own. What exactly was meant by 'ready money' he required to know. And, could it be paid that very evening? On being told that no payment could be made until the banks opened on the following day, the Prince assured Gray that they opened early. Very early. At seven the next morning the two men met and the deal was concluded within a few minutes.

So it was that Philadelphia acquired a Great Auk for a mere trifle, for tucked in among the 12,000 four franc birds was a stuffed Garefowl. Naturally, the Academy of Natural Sciences has hung on to its bargain priced Auk ever since. Absolutely nothing is known of its origin.

THE SMITHSONIAN AUK

Bird no.78 (Grieve no. 74; Hahn no. 68) - adult in summer plumage
Origin: Eldey
Present location: Washington - National Museum of Natural History, The Smithsonian Institution, Washington DC 20560, USA (cat. no.57338).

Schlüter, W. 1868 (October). [Advertisement on inside front cover] *Ibis*. Gurney, J. 1869. Great Auks for Sale. *Zoologist*, p.1603. Newton, A. 1870. On Existing Remains of the Gare-fowl. *Ibis*, p.259. Blasius, W. 1884. Zur Geschichte der Ueberreste von *Alca impennis*. *Journal für Ornithologie*, p.112. Grieve, S. 1885. *The Great Auk or Garefowl*, p.81 *and* appendix p.24. Stejneger, L. 1886. Grieve on the Great Auk. *Auk*, p.263. Grieve, S. 1888. Recent information on the Great Auk. *Transactions of the Edinburgh Field Naturalists' and Microscopical Society*, p.109. Lucas, F. 1888. The Expedition to Funk Island. *Report of the U.S. National Museum*, pl.LXXII. Lucas, F. 1889. Animals Recently Extinct or Threatened...in...the U.S. National Museum. *Report of the U.S. National Museum*, p.638, pl.CIII. Lucas, F. 1890. General Notes. *Auk*, pp.203-4. Allen, J. 1891. Review of Recent Literature. *Auk*, p.306. Shufeldt, R. 1892. Scientific Taxidermy for Museums. *Report of the U.S. National Museum*, pp.406-7 *and* pl.XXXVII. Shufeldt, R. 1893. On remounting the specimen of the Great Auk. *The Nidiologist*, vol.1, no.4, pp.49-51, fig.1 *and* fig.2. Grieve, S. 1896-7. Supplementary note on the Great Auk. *Transactions of the Edinburgh Field Naturalists' and Microscopical Society*, p.253. Hahn, P. 1963. *Where is that Vanished Bird?* p.237.

It is safe to say that no Auk in the world ever stood in so <u>awkward</u> an attitude.

R.W. Shufeldt (1893)

Dr. Shufeldt's little pun was part of his description of the re-stuffing of the Great Auk belonging to the Smithsonian Institution. To prove his words were barely an exaggeration he published two photographs, one taken before and one taken after the restoration. The first shows the

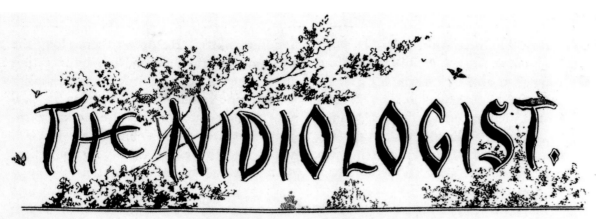

THE NIDIOLOGIST.

VOL. I. NO. 4.   ALAMEDA, CAL., DECEMBER, 1893.   { ONE DOLLAR
                                                       { PER YEAR.

ON RE-MOUNTING THE SPECIMEN OF THE GREAT AUK.

BY DR. R. W. SHUFELDT, M. A. O. U.

For years that priceless specimen, one of the greatest ornithological treasures of the Smithsonian Institution, the Great Auk (*Plautus impennis*), remained in the case where it belonged in that Museum, mounted as shown in Fig. 1. It is safe to say that no Auk in the world ever stood in so *awk*ward an attitude—the one that had been given it by the taxidermist who originally mounted the bird.

The neck was stretched out to rather more than its full length ; the back was nearly flat; but worse than all, the unfortunate fowl had been made to stand with its metatarsal joints nearly perpendicular to the ground, represented by the stand upon which it was placed. Those who are familiar with good drawings of the Great Auk, such as the one given by Audubon, or the one by the present writer, given in the *Century Magazine* (January 1886, p. 394), know full well that that species would never have commonly assumed any such posture. On the contrary, when on land, while the bird stood erect, it at the same time rested upon the back of the tarsi and the soles of the feet. Possibly, the posterior extremity of the body may also have come in contact with the ground.

As the years rolled by and taxidermy passing through its early crude stages to arrive at the dignity of an art, and, finally, to that of a science, a position now secured for it by its most advanced students, a piece of work like this could not, in a great museum, be daily viewed, and no thought entertained of its rectification.

Expert taxidermists came to be employed at the Smithsonian Institution and National Museum, and many of the gross errors seen in the specimens mounted by the old school "stuffers" were gradually taken in hand and re-modeled.

FIG. 1—THE GREAT AUK, AS FIRST MOUNTED IN THE SMITHSONIAN MUSEUM.

The Smithsonian Auk after restoration.

An advertisement placed in *The Ibis* for October, 1868.

specimen standing rather quaintly with its metatarsal joints arranged in a position almost perpendicular to the ground, a position virtually impossible for the living bird. As a more realistic interpretation of a live creature, the re-mounting - by a certain Nelson R. Wood - was an undoubted success.

This particular Great Auk was allegedly taken on Eldey during June of 1834, a date that is entirely likely as several Garefowls were killed in this year and were subsequently sent to Hamburg. Skinned in a traditional Icelandic manner with an incision beneath the right wing, it passed through the hands of Salmin the Hamburg dealer who obtained Great Auks fairly regularly from merchants in Reykjavik. He sold it to C.E. Götz of Dresden and Götz passed it on to a dealer from Halle, Herr Wilhelm Schlüter. It was Schlüter who sent the bird to Washington. Some years later he reported to Wilhelm Blasius that Salmin obtained the bird from Iceland around the year 1840 but this is probably incorrect. Salmin did indeed handle Garefowls killed on Eldey at around that time but it is rather more likely that this bird came from an earlier crop. Blasius (1884) calculated that eight birds passed through Hamburg during 1834.

What is certain is that the Great Auk arrived safely in America in 1869 and has stayed at the Smithsonian ever since.

THE GARE-FOWL OR GREAT AUK.

A beautiful stuffed *old male* specimen of ALCA IMPENNIS for sale.

Apply to Herr WILHELM SCHLÜTER, Landwehrstraesse, No. 17, Halle a. S., Prussia.

THE MAINZ AUK

Destroyed bird no.1 (Grieve no. 47) - adult in summer plumage, said to be female
Origin: Iceland

Preyer, W. 1862. Der Brillenalk in Europäischen Sammlungen. *Journal für Ornithologie*, p.78. Fatio, V. 1868. Liste des divers représentants de l'*Alca impennis* en Europe. *Bullétin de la Société Ornithologique Suisse*, tome II, pt.1, p.82. Blasius, W. 1884. Zur Geschichte der Ueberreste von *Alca impennis*. *Journal für Ornithologie*, p.95. Grieve, S. *The Great Auk or Garefowl*, p.79, appendix p.9 *and* p.16. Hahn, P. 1963. *Where is that Vanished Bird?* p.234.

Stuffed Great Auks seem remarkably resilient to the ravages of time and almost pre-ordained to escape disaster. The Strasbourg Auk endured a terrific bombardment during the Franco-Prussian War in circumstances described as 'miraculous.' The Dresden Auk was unaffected by the terrible fire-bombing of the city during 1945 and then survived being looted by the Red Army. At approximately the same time, the two specimens in Wroclaw (Breslau as it was then generally known) escaped relatively unscathed from the destruction of the museum wing that housed them. Of all the specimens known to researchers at the end of the nineteenth century only two have been destroyed. One of these is the Auk formerly in the Mainz Museum.

Destroyed during the Second World War this specimen was - for a time - the companion to the one now in Birmingham. The pair were purchased together by the Mainz Museum from G.A. Frank of Amsterdam in 1835 but during 1860 the second bird was disposed of. Both probably came from Iceland; indeed the destroyed specimen carried a label to this effect, a label that also described it as female. Apparently it showed the grey tinge to the flanks that Rothschild (1907) believed to be a female characteristic.

THE MUSEUM BOCAGE AUK

Destroyed bird no.2 (Grieve no. 40; Hahn no. 52) - adult in summer plumage, said to be female
Origin: unknown

Smith, A. 1868. A Sketch of the Birds of Portugal. *Ibis*, p.457. Newton, A. 1870. On Existing Remains of the Gare-fowl. *Ibis*, p.259. Selys Longchamps, E. 1870. Notes on various Birds observed in Italian Museums in 1866. *Ibis*, p.450. Blasius, W. 1884. Zur Geschichte der Ueberreste von

All that remains of the Museu Bocage Auk - a photograph of the head and a snapshot. *Courtesy of Dr. Carlos Almaça and the Museu Bocage.*

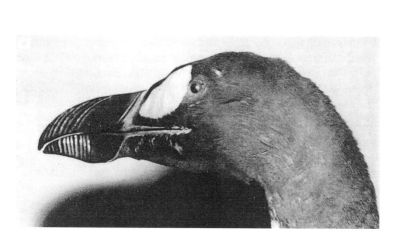

THE GREAT AUK

Alca impennis. Journal für Ornithologie, p.92. Sclater, P. 1884. The National Museum of Lisbon. *Ibis*, p.122. Grieve, S. 1885. *The Great Auk or Garefowl*, p.81 *and* appendix p.15. Hahn, P. 1963. *Where is that Vanished Bird?* p.227. Violani, C. 1975. L'Alca impenne...nelle Collezioni Italiane. *Natura- Revista di Scienze Naturali*, p.22 *and* pl.4.

On April 18th 1978 a fire engulfed the Museu Bocage, Lisbon and many treasures were lost. Among them were a Great Auk and an Auk's egg. All that remains of the bird is a snapshot of it and a rather clearer photograph of its head.

Nothing is known of its early history but it came to the ill-fated Museu Bocage from the collection of Luis I, King of Portugal. Luis is said to have 'coaxed' (*see* Smith, 1868) it from his father-in-law Victor Emmanuel, King of Italy. At about the time (1867) that King Luis got his bird, Victor Emmanuel acquired another (now in the museum at Rome) from the estate of Pastor C.L. Brehm. Whether the Italian King let his original specimen go because he'd got another or whether he got the new one as a replacement is not known. It is easy to confuse these two Great Auks. The one that still exists came to Victor Emmanuel from Pastor Brehm, the other was given to the King by the Marquis de Brême, Grand Master of the Royal Household.

THE STETTIN AUK

Alleged location: Stettin - Muzeum Noradowe, Staromlynska Nr 27, 70-561 Szczecin, Poland.

Hahn, P. 1963. *Where is that Vanished Bird?* p.227 (bird no. 51).

It is likely that this specimen doesn't exist and never did. The only reference to it in ornithological literature is in Paul Hahn's book *Where is that Vanished Bird?* (1963) and its listing by Hahn probably rests on a misunderstanding. While compiling his book Hahn sent out questionnaires to museums asking for details of any Great Auk holdings. The reply he received in 1959 from Professor W. Filipowiçk, Director of the Museum Pomorza Zachodniego (now the Muzeum Narodowe), caused him to assume the museum owned a specimen:

> We regret not to... give you precise details... in the inventory left by the Germans, the kinds of birds are treated in general and laconic terms.

Thirty-five years later, during late 1994, an inquiry to the museum brought the following reply from the Director, Professor Wladyslaw Filipowiçk - presumably the same gentleman who'd written so long ago to Paul Hahn:

> I am very sorry but I have to inform you that in our Museum we do not have [the] specimen of the Great Auk you mention in your letter. The specimen [has] not entered [our] list in our museum after 1945. I think that the bird that interests you disappeared during the last world war.

THE BIRDS

The Stettin Auk seems to be no more than a myth that grew up around a genuine linguistic misunderstanding.

THE EGGS

THE EGGS

It is an undeniable fact that eggs of the Great Auk became prized trophies in Victorian Britain. There were just enough of them to sustain a fairly constant interest but never too many to make them anything other than rare. They were also tantalisingly vulnerable.

Their emergence as symbols of power and wealth went hand in hand with the rise of egg collecting as a nationally approved pastime and a realisation that *Alca impennis* was actually extinct. Egg list after egg list was compiled, each attempting to supersede in completeness its immediate predecessor; names of owners were recorded and the whereabouts of examples disclosed. Often an estimation of fast-rising value was included. Like the similarly highly prized stuffed specimens, each egg has a history attached to it - often one redolent with intrigue, greed and what might simply be termed, 'human interest.'

Their heyday as objects of national interest and spectacular financial value passed long ago, however. Were an egg to be offered today it would doubtless fetch a good deal more than the £300 it might have brought around the year 1900. But £300 was a colossal sum at that time and it is extremely unlikely that any figure realised now would approach it in true value. When Captain Vivian Hewitt (the most formidable of all Great Auk egg collectors) died in the mid 1960's he left no fewer than thirteen examples, all of which were offered for sale. Five found new owners fairly quickly but it took almost 30 years to effect the sale of the remaining eight.

Where once the collecting of wild birds' eggs rivalled football or fishing in popularity, it has now all but vanished. Today the fuss over birds' eggs is long since over and the schoolboy passion that developed in so many men into a full-blown mania is largely a thing of the past. Despite ridiculous claims to the contrary, there is no lucrative black market in birds' eggs (apart, perhaps, from those taken to supply the aviaries of falconers) and the legions of active egg collectors feared by some bird protectionists are just a myth. Certainly, a few individuals still indulge themselves in this rather unendearing persuasion but these are isolated and secretive characters quite unlike the swaggering, self-assured 'eggers' of a century ago. The taking of eggs for collections is an activity that - in any meaningful sense - no longer exists.

Partly this is because egg collecting is now illegal. Probably more telling is the fact that it is now entirely unfashionable. Perhaps another factor is equally significant. The deliberate finding and taking of eggs is - notwithstanding its unpleasant overtones - something requiring enormous patience and skill, plus a remarkable degree of knowledge about the birds themselves. Few of today's ornithologists - priding themselves on their scientific approach - could match the knowledge of the old time egg collectors when it comes to understanding the ways of birds.

The nine eggs that once belonged to Robert Champley photographed circa 1890. *Courtesy of David Wilson.*

Yet all the frantic activity of these men was, perhaps, to no real point. Alfred Newton, who inherited John Wolley's fabled egg collection and cherished it for much of his life, believed that little of scientific value could be gleaned from the study of a collection. Once assembled, such collections were - he thought - simply things of aesthetic beauty.

The few Great Auk eggs that remain must be considered in the same way. What knowledge is to be drawn from them has - very likely - already been taken. Ornithologists may fuss over what they consider to be significant 'scientific' data, but none of this really matters. Whether a particular egg came from Newfoundland, Iceland, the Orkneys or anywhere else within the species' known range, is surely as irrelevant as whether it cost £50, £60 or £90. To regard such information as being in some way significant is an affectation. The living bird is gone and the eggs simply testify to the fact that it once lived.

As the very symbol of life itself, these eggs are objects that carry a terrible irony and a great wonder. Perhaps, also, they really are things of beauty. Ultimately, however, they represent a life that never was.

When a pair of hitherto unknown eggs were discovered in Edinburgh during 1880, the find caused much excitement among interested parties. A well known ornithologist tried to explain the significance (and outline the enormous financial value) to two perplexed citizens - a popular Edinburgh minister and a newspaper reporter. At the end of the short lecture these gentlemen gave their would-be teacher a look of profound pity and - together - uttered the same words:

But the eggs are of no use; they will never hatch.

The eggs will never hatch - three eggs and a skeleton.

HAY FENTON'S EGG

Egg no.1 (Grieve no. 20; Parkin no. XXIII; Tomkinson no. 58)
Origin: unknown
Colour: dirty white with brown and black blotches
Present location: Aberdeen - Museum of the Zoology Dept., University of Aberdeen, Tillydrone Avenue, Aberdeen, Scotland.

Champley, R. 1864. The Great Auk. *Annals and Magazine of Natural History*, p.236. Fatio, V. 1868. Liste des divers représentants de l'*Alca impennis* en Europe. *Bullétin de la Société Ornithologique Suisse*, tome II, pt.1, p.84. Blasius, W. 1884. Zur Geschichte der Ueberreste von *Alca impennis*. *Journal für Ornithologie*, p.155. Grieve, S. 1885. *The Great Auk or Garefowl*, p.89 *and* appendix p.27. Duchaussoy, H. 1897-8. Le Grand Pingouin du Musée d'Histoire naturelle d'Amiens. *Mémoires de la Société Linnéenne du Nord de la France*, p.101 *and* pp.108-10. Grieve, S. 1897-8. Additional notes on the Great Auk. *Transactions of the Edinburgh Field Naturalists' and Microscopical Society*, p.338. Parkin, T. 1911. The Great Auk. *Hastings and East Sussex Naturalist*, vol.1, pt.6 (extra paper), p.28. Allingham, E. 1924. *A Romance of the Rostrum*, p.163. Tomkinson, P. & J. 1966. *Eggs of the Great Auk*, p.117 *and* pl.58.

Edward Bidwell, dedicated gatherer of eggs and egg-lore, displayed this example before a meeting of the British Ornithologists' Club on June 17th 1908. Here he declared his belief that it was once in the French Royal Collection. He based this belief on evidence so slight that it is flattering to call it unconvincing. The word 'pingouin' is inscribed on the egg in large letters, letters that Bidwell claimed showed the hand of Monsieur Dufresne, Keeper of the Cabinet of Natural History belonging to the Empress Josephine. His reason for the association lay in the fact that rather similar inscriptions were present on other eggs believed to have come from Dufresne. The writing shows no particularly personal style, however. It is clear and straightforward printing. The notion that such a truly anonymous hand is referable to Dufresne and indicative of an illustrious provenance is romantic ornithology at its most creative; the lettering could be by Dufresne or it could belong to anyone else. Other than indicating that the egg was once in French hands, the 'pingouin' inscription reveals nothing.

This flight of fancy, full of wishful thinking and weak evidence, is fairly typical of nineteenth century attitudes to Great Auk eggs. The objects inspired wonder and only stories with the most splendid of associations could truly satisfy that emotion.

Leaving aside the question of a Dufresne connection, the egg's probable provenance is by no means without interest. Almost certainly it once belonged to Coenraad Temminck the ornithological writer who became first Director of the Leiden Museum. Temminck allegedly presented it to Monsieur Josse Hardy, the distinguished French ornithologist, in recognition of various acts of kindness. After Hardy's death the egg was loaned to the Dieppe Museum but was eventually retrieved by his granddaughter, Madame Ussel of Eu, who sent it for sale at Stevens' Rooms during February of 1909. Here its story takes another romantic turn.

10, Downing Street,
Whitehall, S.W.

February 6th, 1909.

Dear Sir,

I am directed by the Prime
Minister to acknowledge the receipt
of your letter of the 5th instant and
to say that he much regrets that, owing
to the very numerous calls upon him,
he is unable to contribute towards the
cost of procuring an egg of the extinct
Great Awk for the Natural History
Museum of Aberdeen University.

Yours faithfully,

Mark Sturgis

The reply (from Prime Minister Herbert
Asquith) to one of Hay Fenton's fund-raising let-
ters. *Courtesy of Robert Ralph and the University of
Aberdeen.*

Hay Fenton's Egg in its box at Aberdeen
University.

It attracted the attention of R. Hay Fenton, a man who had - with patience and endeavour - formed a magnificent collection of eggs that he intended to give to the University of his home town of Aberdeen. Due to relatively modest circumstances he'd never quite felt able to buy the one item needed to round out the collection. A Great Auk's egg was wanting ... and then this example became available. The moment seemed as good as any; several eggs had recently been sold and - with the market perhaps glutted - the last of these realised a mere £110. Knowing this, Hay Fenton felt he might stand a chance and he devised a plan. Just over a week before the sale he wrote to the Chancellor of Aberdeen University, Lord Strathcona, offering to make an immediate gift to the University of his entire collection (which he valued at £400) if Strathcona would buy the Auk's egg and add it to the rest. The Scottish baron answered straight-away. Yes, he would give Hay Fenton money for the egg but would only commit himself to the £110 that the previous egg had made. With five days to go before the sale Fenton sent back a brief thank-you note and concluded;

If I find the egg is all the catalogue claims for it, Aberdeen University must have it by hook or by crook.

Realising that the money put up would not necessarily be enough, he quickly wrote to other eminent Scots explaining his predicament. Another £75 was promised and Hay Fenton proceeded to the auction armed with little more than his pledges.

On February 9th he bought the egg for £199 and ten shillings, a sum £14 and ten shillings above the amount he had. Evidently impressed by the enthusiasm and determination shown, Lord Strathcona wrote to Fenton immediately.

Circumstances are such that it affords me much pleasure to hand you forthwith my cheque of even date, to your order for £124, ten shillings instead of the £110 I proposed giving when you approached me first on the subject.

Hay Fenton's egg collection, together with its star exhibit, was delivered to Aberdeen University and it has remained there ever since.

Nothing is known of the egg's origin.

DR. BOLTON'S EGG

Egg no.2 (Parkin no. XIX; Tomkinson no. 74 *but see also* no. 72)
Origin: unknown
Colour: yellowish white, blotched and streaked with black and brown
Present location: Bristol - Bristol Museum and Art Gallery, Queen's Rd., Bristol BS8 1RL.

Olphe-Galliard, L. 1862. [Letter to the Editor] *Ibis*, p.302. Parkin, T. 1911. The Great Auk. *Hastings and East Sussex Naturalist*, vol.1, pt.6 (extra paper), p.23. Heim de Balzac, H. 1929. Un nouvel œuf d'*Alca impennis*. *Alauda*, p.368. Tomkinson, P. & J. 1966. *Eggs of the Great Auk*, pp.121-2 *and* pl.74.

Towards the end of the nineteenth century Sir J.H. Greville Smythe of Ashton Court in Somerset - like others before him and after - became a serial egg collector. He managed to accumulate the very respectable total of four. One of these was an apparently unrecorded egg (possibly it is mentioned by Olphe-Galliard in *The Ibis* of 1862 but description and egg can't be matched with any certainty) sent from Lyons in France to Stevens' Auction Rooms. Here, on June 20th 1900, it was bought by the natural history dealer James Gardiner for 315 guineas. Almost immediately, Gardiner sold the egg - together with another - to Sir Greville but, sadly for the new owner, he didn't have long to enjoy his little purchases. He died during September of the following year and his collection was dispersed. This egg passed to his step-daughter The Hon. Esme Smythe. She kept it until 1945 - a year before her death - when she gave it to the Bristol Museum in honour of one of its former directors, Dr. H. Bolton.

Correspondence kept at the Bristol Museum indicates that the Tomkinsons - the husband and wife team who produced the excellent monograph *Eggs of the Great Auk* (1966) - had difficulty tracing the history of this egg and in their book they confused it with that of one now in Reykjavik. An annotated painting by the well known illustrator F.W. Frohawk (now the property of the Haslemere Museum), produced when the egg still belonged to Lady Smythe, confirms the correct identity.

Dr. Bolton's Egg (4/5 natural size). Watercolour by F.W. Frohawk. *Courtesy of Haslemere Museum.*

MRS WISE'S EGG

Egg no.3 (Grieve no. 55; Parkin no. XI; Tomkinson no. 43)
Origin: unknown
Colour: white with brown, grey and black blotches, spots and streaks
Present location: Bristol - Bristol Museum and Art Gallery, Queen's Rd. Bristol BS8 1RL.

Newton, A. 1870. On Existing Remains of the Gare-fowl. *Ibis*, p.261. Blasius, W. 1884. Zur Geschichte der Ueberreste von *Alca impennis*. *Journal für Ornithologie*, p.165. Grieve, S. 1885. *The Great Auk or Garefowl*, p.88 *and* appendix p.32. Grieve, S. 1888. Recent information about the Great Auk. *Transactions of the Edinburgh Field Naturalists' and Microscopical Society*, p.114. Tegetmeier, W. 1888 (March 17th). Sale of an egg of the Great Auk. *The Field*, p.387 *and* fig.2. Anon. 1888 (March 24th). *Punch*, p.153. Grieve, S. 1896-7. Supplementary note on the Great Auk. *Transactions of the Edinburgh Field Naturalists' and Microscopical Society*, pp.237-8 *and* p.258. Duchaussoy, H. 1897-8. Le Grand Pingouin du Musée d'Histoire naturelle d'Amiens- notes additionnelles. *Mémoires de la Société Linnéenne du Nord de la France*, p.242. Parkin, T. 1911. The Great Auk. *Hastings and East Sussex Naturalist*, vol.1, pt.6 (extra paper), p.11 *and* pl.3. Tomkinson, P. & J. 1966. *Eggs of the Great Auk*, p.112 *and* pl. 43. Chatfield, J. 1987. *F.W. Frohawk- his life and work*, fig. p.53.

> Mrs Wise, you were wise to keep open your eyes
> To the value of *Alca impennis*
> Few eggs gain by keeping, whatever their size,
> But *Alca*'s will keep you in pennies.
>
> R. Scot Skirving (1888)

In late nineteenth century Britain the sale of a Great Auk egg was an event surrounded by much publicity and a good deal of curiosity. Such occasions often inspired literary efforts of the kind produced by Mr. Skirving (*in* Grieve, 1888). Puns were particularly fashionable. An edition of Punch (March 24th 1888) carries the following anonymous notice:

> Another Great Auk's egg has just turned up, been put up to Aukshun, and knocked down again, without being smashed, fortunately, frail a curiosity as it was to come under the hammer... We hope it reached its destination in safety. An accident might happen from mere Aukwardness.

Nor were artistic efforts of the painterly kind neglected. When Mrs Wise offered up her egg for sale, the zoological illustrator F. W. Frohawk was commissioned to paint its portrait and the picture was reproduced in *The Field*. Later, a small lithographic print was issued.

Mrs Wise's Egg can be traced back to the possession of Monsieur Thiebaut de Berneaud from whom it passed to the Paris dealer Lefèvre, but the whiff of scandal hangs over its early history. It is declared, in a curious but anonymous note preserved in the Newton Archive at

EGG OF THE GREAT AUK, *(Alca impennis)*

The property of JAMES GARDNER, *29, late 426, Oxford Street, London.*

Recently bought at auction for £225.

12ᵗʰ March 1888.

F.W.Frohawk lith.

Cambridge, that this egg was stolen from the Paris Museum during the revolution of Louis Phillipe's time. According to the note, Mrs. Wise was herself well aware of the position. Perhaps because of this problematical background, Lefèvre was anxious to get the specimen out of Paris and he quickly sold it to the London taxidermist J. Williams. In 1851 Williams sold it, for £18, to one Lancelot Holland, the father of Mrs. Wise.

By the time Mrs Wise parted with her egg (March 12th 1888) its value had rocketed to £225. It was bought by another London taxidermist, James Gardiner, who kept it for a while but then sold it - apparently for 300 guineas - to Sir J.H. Greville Smythe. After Smythe's death his widow presented this egg to the Bristol Museum.

Mrs Wise's Egg (natural size). Hand-coloured lithograph by F.W. Frohawk. *Courtesy of Peter Blest.*

WOLLEY'S EGG

Egg no.4 (Grieve no. 15; Tomkinson no. 33)
Origin: Iceland, probably Eldey

Colour: whitish with chestnut brown and fawn streaks and blotches
Present location: Cambridge - University Museum of Zoology, Downing Street, Cambridge.

Hewitson, W. 1853-6 (3rd ed.). *Coloured Illustrations of the Eggs of British Birds*, p.470. Roberts, A. 1861. Skins and Eggs of the Great Auk. *Zoologist*, p.7353. Champley, R. 1864. The Great Auk. *Annals and Magazine of Natural History*, p.236. Fatio, V. 1868. Liste des divers représentants de l'*Alca impennis* en Europe. *Bullétin de la Société Ornithologique Suisse*, tome II, pt.1, p.84. Blasius, W. 1884. Zur Geschichte der Ueberreste von *Alca impennis*. *Journal für Ornithologie*, p.154. Grieve, S. 1885. *The Great Auk or Garefowl*, p.87 *and* appendix p.26. Newton, A. 1905. *Ootheca Wolleyana*, vol.2, pp.364-7 *and* pl.14. Tomkinson, P. & J. 1966. *Eggs of the Great Auk*, p.109 *and* pl.33.

On December 12th 1846 John Wolley bought from the Rev. D. Barclay Bevan for 28 shillings a Great Auk's egg. The only information that Wolley got concerning the history of his purchase was that it came from John Gould. Unfortunately Gould - who, it seems, owned the egg for a little over a year - had forgotten how he'd come by it; several years had passed since the time of his ownership, a period during which he'd handled hundreds of natural history specimens. Fortunately the egg carried a rather distinctive label and when Wolley visited Hamburg in April of 1856 this piece of paper enabled him to pin down the egg's origin. On the 22nd of the month

Wolley's Egg (natural size). Lithograph by H. Grönvold from J. Wolley's and A. Newton's *Ootheca Wolleyana* (1864-1907).

Ootheca Wolleyana Tab. XIV.

H. Grönvold. pinx.

Bale & Danielsson, L.ᵗᵈ Lith.

Wolley spoke to the dealer J.G.W. Brandt and showed him an exact copy of the label. From this Herr Brandt was immediately able to identify the egg as one that he'd bought and sold, and after consulting his books he confirmed - with Germanic thoroughness - that Gould had purchased the egg on September 6th 1835. From this information Alfred Newton (1905) felt able piece together the egg's earlier history:

> It was certainly... from Iceland, the probability is greatly in favour of its having been obtained... from... Carl Siemsen, of Reykjavik, who must... have received it more or less directly from the people of Kyrkjuvogr, being part of the spoil which they year after year in those days brought from Eldey. Indeed there is not the least possibility of its having been obtained elsewhere, though the year in which it was taken cannot be surely fixed ... the probability is that it is an egg of 1835, in which year it reached Herr Brandt.

Before Wolley died, at the early age of 36, one of his last expressed wishes was that his wonderful egg collection should go to his friend Alfred Newton and his close relatives saw that this wish was put into effect. That the gift was not misplaced is evidenced by the masterly catalogue to the collection that Newton completed towards the end of his life. Not only did Newton lovingly care for the collection, he added to it considerably. At his death in 1907 he bequeathed it - along with much else - to Cambridge University.

BREE'S EGG

Egg no.5 (Grieve no. 16; Tomkinson no. 34)
Origin: unknown
Colour: dirty white with streaks of black, grey and fawn
Present location: Cambridge - University Museum of Zoology, Downing St., Cambridge.

Hewitson, W. 1842-6 (2nd ed.). *Coloured Illustrations of the Eggs of British Birds*, p.413. Roberts, A. 1861. Skins and Eggs of the Great Auk. *Zoologist*, p.7353. Champley, R. 1864. The Great Auk. *Annals and Magazine of Natural History*, p.236. Fatio, V. 1868. Liste des divers représentants de l'*Alca impennis* en Europe. *Bullétin de la Société Ornithologique Suisse*, tome II, pt.1, p.84. Blasius, W. 1884. Zur Geschichte der Ueberreste von *Alca impennis*. *Journal für Ornithologie*, p.154. Grieve, S. 1885. *The Great Auk or Garefowl*, p.87 *and* appendix p.26. Newton, A. 1905. *Ootheca Wolleyana*, vol.2, pp.367-72 *and* pl.15. Tomkinson, P. & J. 1966. *Eggs of the Great Auk*, p.109 *and* pl.34.

> If you should desire to possess it, and I could ascertain that Wm. Bree would not take amiss my parting with it, I imagine that we should have no difficulty arranging the exchange.
>
> J.P. Wilmot (1856)

So wrote Mr. Wilmot of Leamington Spa to John Wolley when Wolley showed an interest in acquiring this egg. Wilmot, who at one time owned two other Great Auk eggs (nos. 11 and 20), was experiencing only one

(Overleaf, top). Bree's Egg. Watercolour by F.W. Frowhawk. *Courtesy of Haslemere Museum.*

(Overleaf, bottom). Bree's Egg (Natural size). Lithograph by H. Grönvold from J. Wolley's and A. Newton's *Ootheca Wolleyana* (1864-1907).

problem with relinquishing ownership of his egg. It having been given by a friend, William Bree, he felt bound to obtain Bree's sanction concerning any transaction.

On hearing of the wish to part with the present he'd given, Bree sent Wilmot a note (March 10th 1856) that shows a certain generosity of spirit:

> I have never for one moment repented parting with the Great Auk's egg to yourself, neither have I the least wish to avail myself of your kind offer to return it ... I have a strong notion of things going to their right places, and whether you or Mr. Wolley have the possession of the Great Auk's egg it will in either case be ... in its right place.

Wolley, doomed to an early death, enjoyed possession of the egg for only a year or two and bequeathed it to Alfred Newton. Some 50 years later, as part of Newton's own bequest, the egg became the property of the Zoological Museum at Cambridge.

Its origin was never determined although Wolley, with characteristic industry, researched the history as best he could. His research revealed that Bree obtained the egg as a gift from an uncle, Alfred Dudley. Dudley, it seems, was given it by a school chum, one Thomas Davies, who, in turn, received it from another school friend, Alfred Mason.

The history of this egg is, in contrast to the stories of some of the others, a long tale of gifts, good feeling and *proper* behaviour. Alfred Mason got

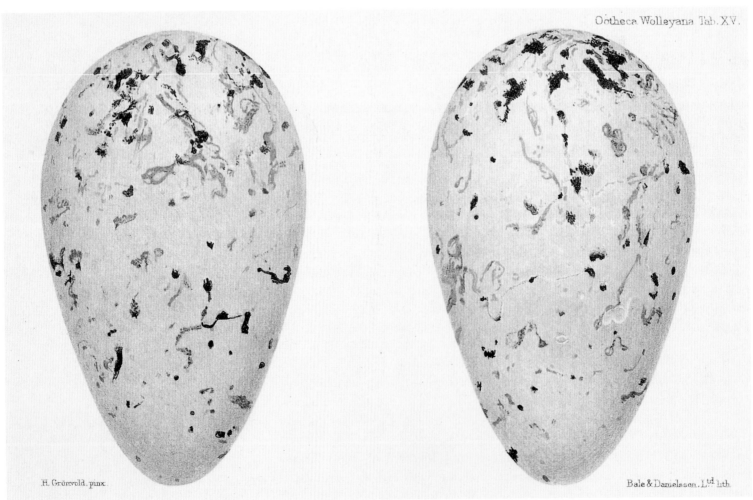

Ootheca Wolleyana Tab. XV.

H. Grönvold. pinx.

Bale & Danielsson. L.^{td} lith.

it, apparently, from his brother Augustus in among a small collection of other eggs, but at this point in his research Wolley could pursue his possession back through time no further. When he interviewed Augustus - who, it seems, distinctly remembered the egg in question - he learned nothing meaningful. Augustus thought back to his childhood and remembered only:

A kind lady coming in a carriage, with whom he thought the egg [was] connected.

NEWTON'S EGG

Egg no.6 (Grieve no. 17; Tomkinson no. 32)
Origin: Eldey
Colour: yellowish white sparsely spotted with black and brown
Present location: Cambridge - University Museum of Zoology, Downing St., Cambridge.

Hewitson, W. 1853-6 (3rd ed.). *Coloured Illustrations of the Eggs of British Birds,* p.470. Roberts, A. 1861. Skins and Eggs of the Great Auk. *Zoologist,* p.7375. Champley, R. 1864. The Great Auk. *Annals and Magazine of Natural History,* p.236. Fatio, V. 1868. Liste des divers représentants de l'*Alca impennis* en Europe. *Bullétin de la Société Ornithologique Suisse,* tome II, pt.1, p.84. Blasius,

Ootheca Wolleyana Tab. XVI.

H. Grönvold, pinx.   Bale & Danielsson, Ltᵈ lith.

W. 1884. Zur Geschichte der Ueberreste von *Alca impennis*. *Journal für Ornithologie*, p.154. Grieve, S. 1885. *The Great Auk or Garefowl*, p.87 *and* appendix p.26. Newton, A. 1905. *Ootheca Wolleyana*, pp.373-6 *and* pl.16. Wollaston, A. 1921. *Life of Alfred Newton*, pp.42-5. Tomkinson, P. & J. 1966. *Eggs of the Great Auk*, pp.108-9 *and* pl.32.

> It would be absurd of me to ignore the fact that persons there are, even among my friends, who have been inclined to think that I was guilty of some sharp practice in possessing myself of this egg.
>
> Alfred Newton (1905)

The story of human involvement with this egg is full of gossip, subterfuge and intrigue but quite apart from these considerations this is a particularly interesting example. It is one of the few eggs that can be reliably associated with specific locality data and, perhaps even more intriguingly, it has a companion. Another egg laid, it seems, by the same parent bird still exists. Referring to his own egg and to one now in the Museum of Comparative Zoology at Harvard University (egg no.60), Alfred Newton (1905) wrote:

> No one accustomed to eggs could possibly doubt that these ... were the produce of one and the same parent.

His grounds for this resolute statement were not just an overall similarity in patterning. He identified something rather more telling. At the pointed end of each egg was a semispiral depression:

> The effect no doubt of a sphincter muscle working upon the shell when in a soft and plastic condition.

That Newton was correct in his hypothesis is additionally borne out by the early histories of these two eggs, both - apparently - taken on Eldey, one probably in 1841, the other in 1840. These eggs came, therefore, from one of the very last Great Auks.

The one that Newton owned was probably - although not certainly - collected by Stephan Sveinsson when he raided the dwindling colony on Eldey in the summer of 1841 and procured an egg and two birds (another was probably taken during the same summer). These he sold at Keflavík to Carl Thaae and they passed through several hands in Iceland and Hamburg before reaching Robert Dunn of Hull during 1842.

In April of that year J.D. Salmon, a well known egg collector, bought the egg from Dunn (the birds may - or may not - have been returned to Hamburg) and it stayed in his collection until his death.

It is at this point in its history that the elements of intrigue and deception creep in. Salmon bequeathed his entire egg collection to the Linnean Society of London but in consequence of a condition attached to the bequest the Society was not immediately able to accept the gift. It was some time before the difficulty was cleared up, during which interval the Great Auk egg was surreptitiously removed from the collection. The Linnean Society lost it forever.

In the summer of 1860 it came to the attention of Alfred Newton that a London dealer by the name of Calvert had a Great Auk egg for sale. On investigating, Newton found that Calvert was not at his usual place of business but had to all appearances transferred to new premises. Newton recog-

Alfred Newton (1829-1907).

(Previous page, top). Newton's Egg. Watercolour by F.W. Frowhawk. *Courtesy of Haslemere Museum.*

(Previous page, bottom). Newton's Egg (natural size). Lithograph by H. Grönvold from J. Wolley's and A. Newton's *Ootheca Wolleyana* (1864-1907).

nised these new premises; they were, by curious co-incidence, the former offices of the deceased egg collector J.D. Salmon. Having, at this stage, no reason to suspect any improper conduct, Newton cheerfully proceeded to inspect the egg that was on offer. On making an inquiry as to its origin he was told that it had come from the recent sale of the Museum of the United Services Institution. That this might not be true was an idea that immediately occurred to Newton:

> I told him [Calvert] that I had learned from Mr. Leadbeater that there was no such thing in the collection, but he replied that the sale was so badly managed that whole boxes full of odds and ends were sold without examination.

This circumstance agreed with what Newton already knew of the sale and temporarily allayed any suspicions that might have been developing. Calvert then added a refinement to his story, one best interpreted as an additional protective smokescreen should anything approaching the truth ever be revealed. It also served to discourage any immediate supposition that he'd simply removed the egg from Salmon's collection while it was stored on the premises. The crafty dealer insisted that he'd not actually purchased the egg himself but had got it through an anonymous third party. Newton, still not completely happy with the story, allowed his sense of honour to come to an uneasy truce with his greed for eggs:

> It ended in my coming to terms with Mr. Calvert: I was to have the egg conditionally on his informing me whence he'd obtained it, and he was to keep it for me till my return from the Continent, whither I was intending to proceed that night - I paying a deposit upon it.

All did not quite go according to plan, however:

> On the 4th of September I called by appointment to redeem the egg, and ... Mr. Calvert ... informed me that he had it from one Westall, of Porchester or Portland Terrace, Bayswater - he could not recollect which. I complained that this was not according to our agreement, for that he had promised to give me the person's address, which he had plenty of opportunity to ascertain ... I lost no time, however, in writing to each of the places he named, but received no reply, though the fact of my letters not being returned to me by the Post Office showed me that they had been delivered to someone answering the designation.

As might be expected, Newton complained again to Calvert but he, of course, simply wished to wash his hands of the whole matter:

> When I told him that I had had no reply from "Westall," he said that he was not surprised, for he thought it was an assumed name, and that he had never seen the man since he bought the egg off him ... It was impossible to treat all that he said seriously, for at times he had a way of talking as though he were not in his right mind.

Clearly, Mr. Calvert was a master of prevarication and poor Newton was a little out of his depth when it came to understanding the wiles of unscrupulous dealers. For a year or so he tried to convince himself that the egg he'd bought had nothing to do with Salmon but finally was forced to face the inescapable fact. The legal difficulty over Mr. Salmon's bequest was eventually resolved and the Linnean Society took possession of his collection. In the place allocated to the Great Auk's egg was the egg of a Swan, rudely spotted and blotched with ink.

Establishing rightful ownership of the true egg might have presented a tricky legal problem. Whether or not the Linnean Society could have demonstrated a proper title to it is difficult to say. The fact is that the Society never tried and Professor Newton retained possession for the rest of his life. When he died he left the egg to the Zoology Museum of Cambridge University. Its companion, co-incidentally, is in Cambridge, Massachusetts.

MR. SMALL'S £100 EGG

Egg no.7 (Grieve no. 38 *or* 39; Parkin no. IX; Tomkinson no. 5)
Origin: unknown
Colour: dirty buff, spotted with black and brown
Present location: Cambridge - University Museum of Zoology, Downing St., Cambridge.

Gray, R. 1880. On two unrecorded Eggs of the Great Auk. *Proceedings of the Royal Society of Edinburgh*, p.668. Bree, C. 1880 (May 22nd and May 29th). Two Unrecorded Eggs of the Great Auk. *The Field*. Blasius, W. 1884. Zur Geschichte der Ueberreste von *Alca impennis*. *Journal für Ornithologie*, p.160. Grieve, S. 1885. *The Great Auk or Garefowl*, p.88, p.109, p.118 *and* appendix p.29. Grieve, S. 1888 *and* 1896-7. Recent information about the Great Auk *and* Supplementary note on the Great Auk. *Transactions of the Edinburgh Field Naturalists' and Microscopical Society*, p.113 (1888) *and* pp.259-60 (1896-7). Newton, A. 1905. *Ootheca Wolleyana*, p.377 *and* pl.17. Parkin, T. 1911. The Great Auk. *Hastings and East Sussex Naturalist*, vol.1, pt.6 (extra paper), pp.9-11 *and* pp.32-3. Tomkinson, P. & J. 1966. *Eggs of the Great Auk*, p.101 *and* pl.5.

The discovery of two hitherto unrecorded eggs in an obscure Edinburgh auction house during May of 1880 caused a considerable stir among Garefowl scholars. They were all too late to secure them at a bargain price, however. That piece of good fortune fell to Robert Small, an Edinburgh taxidermist and dealer in natural history specimens who bought them - virtually unopposed - for the remarkably low price of 32 shillings. Small sent the eggs - with almost indecent haste - to Stevens' Rooms in London where he expected to cash in on his lucky find. By early July he'd done just that. Both eggs were bought at Stevens' by Lord Lilford, the first for £100, the second, a rather darker example, for 102 guineas.

Their presence in Edinburgh - albeit unknown - and their rather poor condition gave rise to an ingenious theory concerning their history. It became the general opinion that they once formed part of the great consignment of natural history objects bought in 1819 by members of the Senatus of Edinburgh University from the French dealer Dufresne. On account of their defects (both are cracked and broken at the ends) and the fact that the Dufresne collection contained two better examples, this pair were, perhaps, soon disposed of. At the time of the Dufresne acquisition Great Auk eggs might not have seemed of any particular interest.

(Facing page, top). Mr. Small's £100 Egg (natural size). Lithograph by H. Grönvold from J. Wolley's and A. Newton's *Ootheca Wolleyana* (1864-1907).

(Facing page, bottom). Mr. Small's 102 Guinea Egg (natural size). Lithograph by H. Grönvold from J. Wolley's and A. Newton's *Ootheca Wolleyana* (1864-1907).

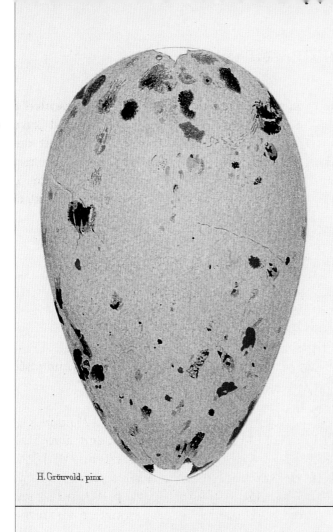

H.Grönvold, pinx.

Bale & Danielsson, L. td lith.

H.Grönvold, pinx.

Bale & Danielsson, L.td lith.

Additional, although very slight, evidence for this hypothesis came from inscriptions once visible on the eggs. The French word 'pingouin' was once present on both of them and so too was the inscription 'egal' or 'egale'- although the meaning of this remains unclear.

From this accumulation of evidence, and the implication of a French connection, Newton (1905) assumed that the eggs came originally from Newfoundland, a conjecture that may, or may not, be correct.

The eggs can be traced back with certainty to the possession of a Joseph Maul. It seems that this gentleman, once President of the Edinburgh Post Office, sold his eggs to a man named Lister (sometimes recorded as Little) who in turn sold them to a Mr. Cleghorn Murray. It was at the sale of Murray's effects that the naturalist Small purchased them.

Lord Lilford owned the eggs for several years and then gave them, along with two others, to Alfred Newton at Cambridge. Newton immediately presented them to the University he loved so much.

MR SMALL'S 102 GUINEA EGG

Egg no.8 (Grieve no. 38 *or* 39; Parkin no. X; Tomkinson no. 6)
Origin: unknown
Colour: yellowish brown, spotted and blotched - particularly at the blunt end- with black, grey and brown
Present location: Cambridge - University Museum of Zoology, Downing St., Cambridge.

All references are identical to those of egg no.7, except for:

Newton, A. 1905. *Ootheca Wolleyana*, pl.18 *and* Tomkinson P. & J. 1966. *Eggs of the Great Auk*, pl.6.

The known history of this egg is identical to that of the previous example. It is darker in colour than its companion but its condition is similar. Both eggs were given, along with two more, by Lord Lilford to Alfred Newton who immediately transferred them to the possession of his beloved university. In a display of gentlemanly courtesy Lilford corrected Symington Grieve (by letter on October 8th 1888) over the exact nature of this gift:

> I have not transferred my collection of eggs to the Cambridge Museum, but some time ago presented four of the five eggs of Garefowl in my possession to Professor Alfred Newton ... I distinctly wish to repudiate any claim to the gratitude of the Cambridge Museum, which is the sole and just due of Professor Alfred Newton.

Writing eleven years later, on March 16th 1897, Newton was equally gracious:

> The late Lord Lilford generously gave me four of the five eggs he possessed. These, with his full approval, I at once transferred to our Museum.

FRANK'S LAUSANNE EGG

Egg no.9 (Grieve no. 40; Tomkinson no. 7)
Origin: unknown
Colour: bright yellowish brown heavily spotted and streaked with black, brown and grey; according to Grieve this is the darkest of all known eggs.
Present location: Cambridge - University Museum of Zoology, Downing St., Cambridge.

Fatio,V. 1868. Quelques mots sur les exemplaires de l'*Alca impennis*, Oiseaux et Oeufs qui se trouvent en Suisse *and* Liste des divers représentants de l'*Alca impennis* en Europe. *Bullétin de la Société Ornithologique Suisse*, tome II, pt.1, pp.75-9 *and* p.85. Blasius, W. 1884. Zur Geschichte der Ueberreste von *Alca impennis*. *Journal für Ornithologie*, pp.157-8 *and* p.161. Grieve, S. 1885. *The Great Auk or Garefowl*, p.88, p.109 *and* appendix pp.29-30. Grieve, S. 1888. Recent information about the Great Auk. *Transactions of the Edinburgh Field Naturalists' and Microscopical Society*, p.113. Newton, A. 1905. *Ootheca Wolleyana*, p.376, p.378 *and* pl.19. Tomkinson, P. & J. 1966. *Eggs of the Great Auk*, pp.101-2 *and* pl.7.

> This egg has been disposed of by a great mistake ... They tell me that Frank has resold this interesting specimen at a very high price.
>
> Victor Fatio (1885)

During the 1880's this egg caused something of a scandal among those interested in Great Auks. The London dealer G.A. Frank (youngest of the Frank family) saw two eggs in the museum at Lausanne, Switzerland and, according to the tone of contemporary accounts, may be said to have *tricked* one of them out of the Curator. In exchange for the 'best' of the two, he gave a 'bad' gorilla skin, a swap that was considered a rather shady piece of business. Taking into account the immense novelty value of a gorilla to a nineteenth century public, not to mention the rarity of such a specimen at the time, it all seems perfectly fair and it is a little difficult to understand the chorus of disapproval that went up. Perhaps this is just another instance of the hold that the Great Auk has over certain men, its relics seeming to acquire an almost sacred mystique.

Victor Fatio, author of several papers on *Alca impennis*, wrote indignantly to Robert Champley on March 1st 1885:

> The Curator of the Museum, who now comprehends the great mistake he committed, is full of regret that he had not studied the subject more carefully.

Obviously the poor man had been nagged unrelentingly on 'the subject.' As for Frank, not only was every man's pen turned against him, he was also considerably beaten down on a price of £140 that he originally asked for the egg and he eventually settled for £110 offered by Lord Lilford. Frank was not a little stung by the accusations of sharp practice and Grieve (1888) provided him with a platform from which to defend his actions:

> In 1882 I went to see the hon. Curator of the Lausanne Museum, and I ... asked him if he would sell or make an exchange of one of the

two eggs. His reply was that he could not dispose of such a rare specimen without the full consent of the Museum Committee. The Committee ... decided, as they had two specimens, to let me have one ... for one stuffed gorilla (not a bad skin, as has been stated), a fine skull and several bones of *Alca impennis* ... and several other specimens which they selected... That his lordship obtained the finest egg [as it came to be afterwards regarded] was a mere chance, as Dr. Larguier [the Curator] wished to keep the most perfect for ... the Museum. I should have preferred the other one.

Whether Frank was as innocent as he suggests over the matter of the 'finest' egg is doubtful but Lausanne Museum seems to have got a very fair deal.

After Lord Lilford acquired the egg he included it in the parcel of four that he gave to Alfred Newton and which Newton passed on to the Zoological Museum of Cambridge University.

There is some doubt as to how Lausanne came by its pair of eggs, the second of which it retains to this day. They were allegedly part of a collection of natural history items acquired from a Professor D.A. Chavannes and lay unrecognised at the museum for twenty years or so. Before Chavannes they may have belonged to the ornithological writer François Levaillant but the provenance is really no more than a rumour.

This particular example is marked twice with the word 'pingouin,' once

Frank's Lausanne Egg (natural size). Lithograph by H. Grönvold from J. Wolley's and A. Newton's *Ootheca Wolleyana* (1864-1907).

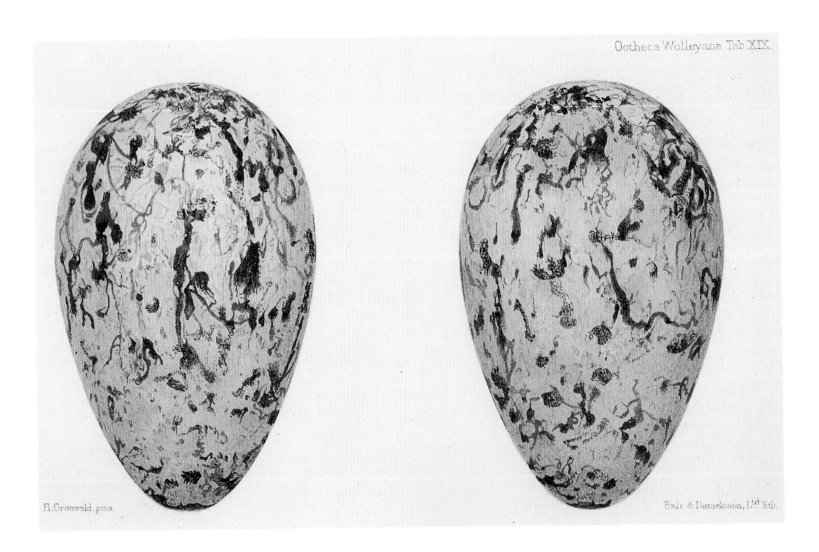

Ootheca Wolleyana Tab XIX

H. Grönvold, pinx.

Bale & Danielsson, Ltd lith.

in red and once in black. It is also inscribed with the number 61. Most nineteenth century commentators showed a considerable determination to give any egg with a French association a New World origin and they did so in this case. The origin is, in fact, entirely unknown.

THE PIMPERNE EGG

Egg no.10 (Tomkinson no. 8)
Origin: unknown
Colour: dirty white spotted and blotched with black, brown and grey.
Present location: Cambridge - University Museum of Zoology, Downing St., Cambridge.

Grieve, S. 1885. *The Great Auk or Garefowl*, p.88, p.110 *and* appendix p.30 (footnote). Grieve, S. 1888 *and* 1896-7. Recent information about the Great Auk *and* Supplementary note on the Great Auk. *Transactions of the Edinburgh Field Naturalists' and Microscopical Society*, pp.113-4 (1888), p.240 *and* pp.242-3 (1896-7). Newton, A. 1905. *Ootheca Wolleyana*, pp.378-80 *and* pl.20. Tomkinson, P. & J. 1966. *Eggs of the Great Auk*, p.102 *and* pl.8.

The Pimperne Egg (natural size). Lithograph by H. Grönvold from J. Wolley's and A. Newton's *Ootheca Wolleyana* (1864-1907).

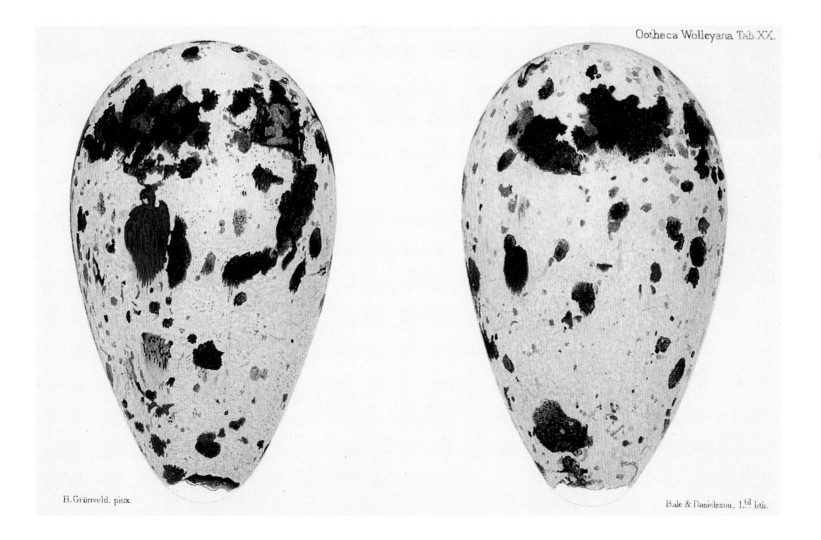

Ootheca Wolleyana Tab XX.

H.Grönvold, pinx.

Bale & Danielsson, Ltd lith.

On a spring morning during the year 1884 the Reverend S.A. Walker was shown into the drawing room of a Mr. and Mrs Philip Hill who lived at Pimperne near Blandford in Dorset. Lying snugly on a glass plate upon the mantelpiece he spotted what he immediately took to be the egg of a Great Auk. On asking why it was not better cared for the minister received an unexpected answer. Mr. Hill, a farmer, had several times been of a mind to get rid of it. Being broken at one end, he'd begun to see the egg as no ornament to his fireplace. Grieve (1888) records that Mr. Hill told Reverend Walker the object:

> Was a useless egg-shell, and it had often been intended to destroy it, but ... to settle the matter it would be thrown on the fire there and then.

Richard Bowdler Sharpe (1847-1909).

The Reverend Walker speedily outlined his suspicions and explained that if these proved correct the 'useless egg-shell' would be worth a good deal of money. The Hills - naturally - lost little time in finding out whether or not the minister was right and a day or two later took the egg to London where it was shown to Richard Bowdler Sharpe at the British Museum. Sharpe confirmed the minister's opinion and the Hills left their egg with him hoping he could find a buyer.

He quickly did so. On April 21st 1884 Lord Lilford became the new owner in exchange for £50 and this became one of the four eggs he later presented to Alfred Newton at Cambridge University.

Although the pointed end is indeed rather badly broken, Newton seems to have much liked this particular egg and he praised the boldness of its markings. Apparently it was very dirty when the Cambridge Professor received it and it was his careful washing that revealed the egg's full beauty.

As might be expected, Newton researched the origin as thoroughly as he was able. With his brother Edward he travelled to Pimperne where he interviewed Mrs Hill, but by his own admission he was able to uncover little that was meaningful. Mrs Hill could state only that the egg was given to her daughter Eliza in 1871 by a relative, Miss Betty Stone Way. She had had it from her brother James Way (who'd very much prized it) when he died in 1869. Before this it belonged to Way's father.

Miss Hill - or Mrs. Rose as she'd become by the time of Newton's inquiries - believed that the egg came to the Ways from a ship's captain but she could recall no details. Because Way lived so close to Poole, and because that port enjoyed a historic trade connection with Newfoundland, there was some conjecture that the egg had a New World origin. Perhaps it did, perhaps it didn't.

RUSSELL'S EGG

Egg no.11 (Grieve no. 41; Tomkinson no. 9)
Origin: unknown
Colour: brownish white spotted and blotched with black and brown
Present location: Cambridge - University Museum of Zoology, Downing St., Cambridge.

Hewitson, W. 1842-6 (2nd ed.) *and* 1853-6 (3rd ed.). *Coloured Illustrations of the Eggs of British Birds,* p.413 (2nd ed.), p.470 *and* pl.129 (3rd ed.). Roberts, A. 1861. Skins and Eggs of the Great Auk. *Zoologist,* p.7353. Champley, R. 1861. Additional Eggs of the Great Auk. *Zoologist,* p.7386. Wilmot, J. 1861. Additional Egg of the Great Auk. *Zoologist,* pp.7386-7. Roberts, A. 1861. Eggs of the Great Auk. *Zoologist,* pp.7438-9. Fatio, V. 1868. Liste des divers représentants de l'*Alca impennis* en Europe. *Bullétin de la Société Ornithologique Suisse,* tome II, pt.1, p.85. Blasius, W. 1884. Zur Geschichte der Ueberreste von *Alca impennis. Journal für Ornithologie,* p.181. Grieve, S. 1885. *The Great Auk or Garefowl,* p.88 *and* appendix p.30. Grieve, S. 1888. Recent information about the Great Auk. *Transactions of the Edinburgh Field Naturalists' and Microscopical Society,* p.114. Newton, A. 1905. *Ootheca Wolleyana,* p.365 *and* p.367. Wollaston, A. 1921. *Life of Alfred Newton,* p.47. Tomkinson, P. & J. 1966. *Eggs of the Great Auk,* p.98, p.102 *and* pl.9.

The most beautiful egg which I have seen is that... of Mr. Wilmot, which was purchased by my friend a few years ago from Mr. Leadbeater, at the cost of £5, a price which we both of us thought ... very extravagant, little dreaming that we should live to see the day when six times as much would be paid for one of them.

W.C. Hewitson (1856)

During 1861 a Mr. Alfred Roberts compiled a list of all the Great Auk eggs he knew of and he published it in *The Zoologist.* He hoped that readers would add to his knowledge by sending information concerning any examples he'd missed. Unfortunately for Mr. Roberts, he'd overlooked the beautiful egg belonging to J.P. Wilmot. This omission infuriated Wilmot who immediately fired off a very angry riposte to *The Zoologist* which was published in the next issue:

I have been quite amazed by the perusal of a list of ... great auk's eggs ...

Russell's Egg (natural size). Hand-coloured engraving from the third edition of W. Hewitson's *Coloured Illustrations of the Eggs of British Birds* (1853-6). *Courtesy of Peter Blest.*

It contains no reference to my egg figured by Mr. Hewitson in ... *Eggs of British Birds* ... Mr. Hewitson's figure of my egg is, unfortunately, placed across the plate, so ... there is not space to do it justice ... How Mr. Roberts could write his article without referring to Mr. Hewitson's work quite surpasses my comprehension. I am also surprised that both the editor and ... publisher of *The Zoologist* ... could allow Mr. Roberts' mis-statement to pass without comment.

If ever there was a case of wounded pride, then this was it. Not only was Wilmot disgusted at being missed off the list, he couldn't bear the way his prize possession was represented in Hewitson's well known book. Poor Mr. Roberts sent in a defensive note for publication in a later issue of *The Zoologist*. His defence was a simple re-affirmation that his intention in publishing was largely to elicit additional information thereby enabling the production of a more comprehensive list at a future date. So bitter had the whole affair become, however, that the Editor (Edward Newman) felt obliged to close the correspondence:

> I must request that any further observations... be confined to simple facts: as a matter of justice, not taste, I admit this note by Mr. Roberts... I must also say that I do not consider Mr. Wilmot's expression 'misstatement' at all called for by the circumstances of the case.

Such were the commotions caused by Great Auk eggs in nineteenth century Britain. Two years after this unrest Wilmot was dead and his beloved egg bequeathed to George Lake Russell. A quarter of a century later (1888) the egg was presented to Cambridge University by Russell's widow Lady Caroline and her son Cecil. The only condition attached to the gift was that mother and son be allowed to place on the cabinet housing the egg (and eggs of other birds presented at the same time) a plate explaining that the donation was made in Russell's memory. Sadly, perhaps, the egg is no longer kept in this cabinet.

Of the eight eggs now in Cambridge this is the only one not figured in Newton's *Ootheca Wolleyana* (1905). This is because it is the only one neither owned by Newton's great friend John Wolley nor added to the collection by Newton himself. By curious co-incidence another egg once belonging to Wilmot is at Cambridge but this specimen actually passed through Wolley's hands so it qualified for illustration in Newton's great œuvre.

VIVIAN HEWITT'S EGG

Egg no.12 (Grieve no. 6 *or* 7; Tomkinson no. 56)
Origin: unknown
Colour: whitish with small brown and black spots and streaks
Present location: Cardiff - National Museum of Wales, Cathays Park, Cardiff, Wales CF1 3NP.

Olphe-Galliard, L.1862. [Letter to the Editor] *Ibis*, p.302. Dubois, Ch.F. 1867. Note sur le *Plautus impennis*. *Archives Cosmologiques. Revue des Sciences Naturelles* (Bruxelles), no.2, pp.33-5 *and* pl.3 (lower fig.). Fatio, V. 1868. Liste des divers représentants de l'*Alca impennis* en Europe. *Bullétin de la Société Ornithologique Suisse*, tome II, pt.1, p.84. Dubois, Ch.F. 1872. *Les Oiseaux de l'Europe et leurs œufs*, vol.2, pl.27 (lower fig.). Blasius, W. 1884. Zur Geschichte der Ueberreste von *Alca impennis*. *Journal für Ornithologie*, p.152. Grieve, S. 1885. *The Great Auk or Garefowl*, p.89 *and* appendix p.25. Duchaussoy, H. 1897-8. Le Grand Pingouin du Musée d'Histoire naturelle d'Amiens. *Mémoires de la Société Linnéenne du Nord de la France*, pp.103-4 *and* p.106. Tomkinson, P. & J. 1966. *Eggs of the Great Auk*, pp.116-7 *and* pl.56. Spink and Son Ltd. (c.1970). *Romantic Histories of Some Extinct Birds and Their Eggs*, p.15 (specimen 17) *and* pl.8 (upper fig.). Bateman, J. 1971. Rare acquisitions by the Department of Zoology. *Amgueddfa*, 9, pp.4-5. Hywel, W. 1973. *Modest Millionaire*, appendix C. Bourne, W. 1993. The story of the Great Auk. *Archives of Natural History*, p.273 *and* p.275.

The number of men who have attempted to assemble a major collection of Great Auk eggs is surprisingly high. Some have managed two or three; several have acquired four or five. A few have put together altogether more spectacular groups. Dr. Jack Gibson obtained eight and Robert Champley, the Scarborough Alderman, got nine. It is possible that Dr. Dick, mysterious benefactor of the Royal College of Surgeons, went one better; J.E. Thayer of Lancaster, Massachusetts certainly did. But one collector stands out from the rest. Starting in the early 1930's, for a period of 25 years or so, Captain Vivian Hewitt hunted Great Auk eggs with a passion- and a fortune- that proved virtually unstoppable and he scooped up almost every egg that became available. At the end of this period of frantic spending he was the owner of no less than thirteen examples.

After his death in 1965 his matchless series was split up. All were sold by the London art and antique dealer Spink but, regrettably, no proper records were kept and the present whereabouts of several are unknown.

Vivian Hewitt's Egg (¾ natural size) - circa 1970.

Vivian Hewitt (front) in one of his racing cars (circa 1920).

This particular egg is not among the lost ones, however. It was sold to the National Museum of Wales along with a stuffed specimen and an egg of the Great Elephant Bird of Madagascar (*Aepyornis maximus*). The cost for the three was £11,000.

The Auk egg is supposed- with no solid ground for the supposition- to have come from Newfoundland. It is rumoured that it arrived in France (along with egg no.74) with the captain of a whaling vessel and was given - or, perhaps, sold - to a merchant in Bergues, near Dunkirk. This merchant apparently gave his two eggs to a young egg fancier whose entire collection was eventually bought by a local collector named de Méezemaker. In 1900 this egg was separated from its companion and sold to one Alfred Vaucher of Lausanne. Then, in 1937, the son of Vaucher sold it to a mysterious Englishman who he guessed might be Captain Hewitt. His guess was correct, of course.

DUFRESNE'S STREAKED EGG

Egg no.13 (Grieve no. 24; Tomkinson no. 3)
Origin: unknown
Colour: yellowish, heavily marked with brown blotches and streaks particularly at the blunt end
Present location: Edinburgh - National Museum of Scotland, Chambers St., Edinburgh, Scotland.

Newton, A. 1861. Abstract of Mr. J. Wolley's Researches in Iceland. *Ibis*, p.387. Fielden, H. 1869. [Letter to the Editor] *Ibis*, pp.358-60. Hügel, A. 1870. Great Auks Eggs in Edinburgh. *Zoologist*, p.1982. Blasius, W. 1884. Zur Geschichte der Ueberreste von *Alca impennis*. *Journal für Ornithologie*, p.157. Grieve, S. 1885. *The Great Auk or Garefowl*, p.87, pp.107-8, appendix p.28 *and* pl.2. Tomkinson, P. & J. 1966. *Eggs of the Great Auk*, p.100 *and* pl.3. Fuller, E. 1987. *Extinct Birds*, fig. p.97.

In 1819 Monsieur Dufresne, Keeper of the Cabinet of Natural History belonging to the Empress Josephine, Assistant Keeper of the Paris Museum and experienced dealer in natural history specimens, sold to some members of the Senatus of the University of Edinburgh an important collection of stuffed birds. Included were several specimens now perceived to be of great significance. A stuffed example of the extinct Pigeon Hollandais (*Alectroenas nitidissima*), a species known from only three preserved skins seems, for instance, to have been included in the sale. In among the stuffed birds were several miscellaneous items that were, apparently, quite disregarded at the time. At least two of these were Great Auk eggs so, naturally, the estimation of their worth has changed. The whole collection passed eventually to the Senatus as a body and in 1855 was transferred to the Edinburgh Museum of Science and Art.

How Dufresne came by his eggs is unknown. During a visit to Paris a certain Mr. Scales saw several in Dufresne's possession in 1816 or, perhaps, 1817 and, in fact, he bought one; this particular egg (no.76) was subsequently destroyed by fire. Presumably two of the others that Scales saw can

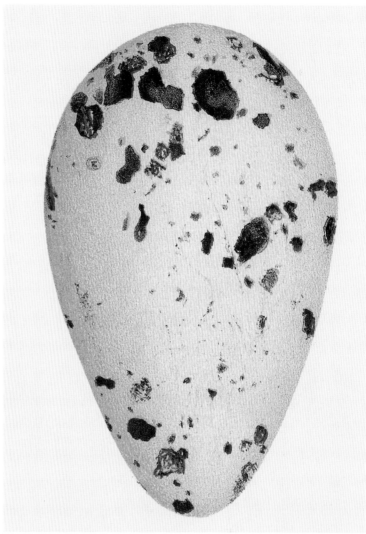

be correlated with the eggs now in Edinburgh.

DUFRESNE'S SPOTTED EGG

Egg no.14 (Grieve no. 23; Tomkinson no. 4)
Origin: unknown
Colour: yellowish with brown spots
Present location: Edinburgh - National Museum of Scotland, Chambers St., Edinburgh, Scotland.

All references are identical to those of egg no.13, except for:

Grieve, S. 1885. *The Great Auk or Garefowl*, pl.1. Tomkinson P. & J. 1966. *Eggs of the Great Auk*, pl.4.

This egg is the second of those supplied in 1819 to the Senatus of Edinburgh University by the French natural history specialist Dufresne. It is less heavily marked than its companion being decorated with a com-

(Above, left). Dufresne's Streaked Egg (natural size). Chromolithograph by Banks and Co. from S. Grieve's *The Great Auk* (1885).

(Above, right). Dufresne's Spotted Egg (natural size). Chromolithograph by Banks and Co. from S. Grieve's *The Great Auk* (1885).

paratively few large, brownish spots. Written on it is the inscription 'G. Pingouin.' This inscription and the known French provenance suggested to nineteenth century egg chroniclers only one possible origin - the egg came from Newfoundland. The rigid insistence on associating French eggs with a New World origin is mystifying. French ships certainly visited Newfoundland shores more frequently than they visited Icelandic ones, but Iceland was by no means neglected. In the very year, for instance, that Dufresne sold his eggs to Edinburgh a French vessel is rumoured to have chanced upon the Geirfuglasker; there are a number of such records for earlier and for later years. Nor should it be assumed that Dufresne, or any other French egg owner, acquired property only from French sources. It is quite possible (although less probable, of course) that they were bought from elsewhere in Europe. Newfoundland may very well be the most likely source for this particular egg but such an origin cannot be taken for granted.

ROYAL COLLEGE OF SURGEONS EGG NO.1 (ALFRED NEWTON'S HUNTERIAN EGG)

Egg no.15 (Grieve no. 33, 34 *or* 35; Tomkinson no. 11)
Origin: unknown
Colour: yellowish white, streaked and blotched with black, brown and grey.
Present location: Scottish Natural History Library (Dr. J.A. Gibson) - to be presented to the National Museum of Scotland, Edinburgh.

Alfred Newton's Hunterian Egg (¾ natural size) - circa 1970.

Champley, R. 1864. The Great Auk. *Annals and Magazine of Natural History*, p.236. Newton, A. 1865. The Gare-fowl and its Historians. *Natural History Review*, pp.483-4. Fatio, V. 1868. Liste des divers représentants de l'*Alca impennis* en Europe. *Bullétin de la Société Ornithologique Suisse*, tome II, pt.1, p.84. Blasius, W. 1884. Zur Geschichte der Ueberreste von *Alca impennis*. *Journal für Ornithologie*, p.159. Grieve, S. 1885. *The Great Auk or Garefowl*, p.87, pp.105-6 *and* appendix p.29. Tomkinson, P. & J. 1966. *Eggs of the Great Auk*, pp.102-3 *and* pl.11. Spink and Son Ltd. (c.1970). *Romantic Histories of Some Extinct Birds and Their Eggs*, p.12 (specimen 6) *and* pl.6 (lower fig.). Bourne, W. 1993. The story of the Great Auk. *Archives of Natural History*, p.273.

On the twelfth day of December 1861, Alfred Newton went to the Hunterian Museum of the Royal College of Surgeons to inspect a dissection of a Great Bustard (*Otis tarda*) made by the dour comparative anatomist Richard Owen. There, quite by accident, he made an extraordinary discovery. He found ten hitherto unrecorded Great Auk eggs!

Before finding the eggs, however, Newton spotted Darwin's great champion Thomas Huxley standing in the entrance hall and Huxley directed him to the place where the Bustard was likely to be found. Almost two weeks later, in a letter written on Christmas day, Alfred Newton described to his brother Edward what happened next:

Ascending to the topmost gallery of the innermost room, a glass case with birds' eggs met my eye. After looking at one or two grimy ostriches ... I saw, as I thought, a nice model of a Great Auk, next to a prickly

hen's, and then, on, on, on, as far as the eye could reach, Great Auks!!
To cut it short, there were <u>ten</u>, nearly all in excellent preservation ... As
soon as my first emotions were ... over I called out over the railing to
Huxley and told him what I had discovered; whereupon to the astonish-
ment of some grave-looking medical students in spectacles, he answered
back that I was like Saul who went out to seek his father's asses and
found a kingdom; to which I could only respond that I hoped I should,
like my illustrious prototype, succeed in gaining possession of my
discovery.

Unfortunately for Newton, he didn't manage to acquire any of the eggs
he'd discovered but perhaps he needn't be pitied too much. Lord Lilford
later gave him four eggs to go with the two he inherited from John Wolley
and the single example he bought for himself.

On discovering that such valuable specimens had remained overlooked, the College Trustees decided that seven of the eggs could be sold.
Three went to Robert Champley (nos. 56, 57 *and* 58) in 1864 and four
were sold by auction at Stevens' Rooms a year later (nos.18, 31, 37 *and*
66). The remaining three (this example and nos.16 *and* 17) were retained
and they stayed at the Hunterian Museum for almost 90 years longer.
Then, on June 19th 1946, they were sold through the agency of the dealer W.H.F. Rosenberg to Captain Vivian Hewitt. Hewitt paid £1,000 for
them, the money being put towards the cost of restoring the museum,
which was badly damaged by bombing during World War II.

This egg is one of the three that the College held back until the 1940's
and sold to Captain Hewitt. Hewitt kept them until his death in 1965 after
which they passed to Spink and Son Ltd. Not until 1993 were they sold.
During that year they were bought, along with five others, by Dr. J. A.
(Jack) Gibson on behalf of the Scottish Natural History Library, from where
Dr. Gibson intends to present all eight to the National Museum of Scotland.

Dr. Dick's Egg (¾ natural size) - circa 1970.

ROYAL COLLEGE OF SURGEONS EGG NO.2
(DR. DICK'S EGG)

Egg no.16 (Grieve no. 33, 34 *or* 35; Tomkinson no. 12)
Origin: unknown
Colour: fawnish brown with many fine streaks of brown and grey
Present location: Scottish Natural History Library (Dr. J.A. Gibson) - to be
presented to the National Museum of Scotland, Edinburgh.

All references are identical to those for egg no.15, except for:

Seebohm, H. 1896. *Coloured Figures of the Eggs of British Birds*, pl.28.
Tomkinson, P. & J. 1966. *Eggs of the Great Auk*, p.103 *and* pl.12. Spink and
Son Ltd. (c.1970). *Romantic Histories of Some Extinct Birds and Their Eggs*, p.12
(specimen 7) *and* pl.6 (upper fig.).

After Alfred Newton made his discovery of ten eggs in the Hunterian
Museum there was much speculation on how such a matchless series

came to be there without any informed person being aware of their existence.

Despite the best efforts of several researchers this is a mystery that persists to this day. The only real clue to origin is an inscription once existing on a box containing two of the eggs. It simply said, 'Penguin eggs from Dr. Dick.'

Dr. Dick has proved an elusive character, however, and his identity hasn't been pinned down. W.R.P. Bourne (1993) suggests that Assistant Ship's Surgeon Robert Dick, a man whose name occurs in the *Navy List* for 1809 until 1844 might fit the bill. This tentative association is as good, perhaps, as any. Certainly Alfred Newton was unable to track down Dr. Dick, despite trying hard to do so. He discovered, for instance, that someone with the surname 'Dick' bought 'a parcel of eggs' on June 11th 1806, the 32nd day of the celebrated Leverian Museum Sale, but such a vague piece of information leads nowhere. Tending to believe that the word 'penguin' automatically implied a French connection and a New World origin, Newton proposed Newfoundland as the source, but this is little more than a guess. Bourne (1993) points out that the eggs (eight of which may have had nothing to do with Dr. Dick and possibly came from a gift known as 'Mrs Robinson's donation') were perhaps obtained by John Hunter himself during a visit to Cape Breton in 1760 - in which case they would have been in the Hunterian Museum for a century when Newton found them.

This particular egg is one of the three kept back by the Royal College of Surgeons when it capitalised on its unexpected windfall by promptly selling most of it off. Its history since 1946 is identical to that of the previous egg.

ROYAL COLLEGE OF SURGEONS EGG NO.3 (DR. JACK GIBSON'S EGG)

Egg no.17 (Grieve no. 33, 34 *or* 35; Tomkinson no. 13)
Origin: unknown
Colour: cream with blotches and streaks of black and brown and grey
Present location: Scottish Natural History Library (Dr. J.A. Gibson) - to be presented to the National Museum of Scotland, Edinburgh.

All references are identical to those for egg no.15, except for:

Seebohm, H. 1896. *Coloured Figures of the Eggs of British Birds*, pl.27. Tomkinson, P. & J. 1966. *Eggs of the Great Auk*, p.103 *and* pl.13. Spink and Son Ltd. (c.1970). *Romantic Histories of Some Extinct Birds and Their Eggs*, p.12 (specimen 8) *and* pl.7 (lower fig.).

From the moment that he made his discovery of ten eggs in the Hunterian Museum, Alfred Newton hoped to acquire them. In a letter to his brother written a month or so after the discovery, he outlined the situation. As so often happens with Great Auk matters, the intrigue and plotting was intense and in this case it began with an unfortunate death:

THE EGGS

I may as well tell you that poor Stewart, who was acting Curator... went raving mad just about the time I made the discovery and died a few days later! Flower has since been chosen... Curator... and when I was in town last, I called on him and had some talk about them ... He told me he did not anticipate any objection on the part of the Museum Committee about parting with some - perhaps six or eight - and promised I should have the refusal of such as they were disposed ... of. I told him I would take as many as they had to spare either at a fair price for the lot- or exchanging for them some bones of the bird and a typical collection of other birds' eggs - which collection I would take care should equal the market value of those they would let me have - or that I was ready to ... offer... a combination of all these terms. I particularly told him there was no hurry - for I should much prefer waiting some time in order to give them every opportunity of learning the exact value of their property, and also that it might not be said that I had availed myself of their ignorance. Another reason why I should prefer this is that at present I have not many bones ... and I am in expectation of receiving no small number ... this summer from Funk.

At the end of February Newton wrote explaining that he'd done nothing more about getting the eggs, 'thinking it best not to appear too eager.' By late March, he was unable to restrain himself and wrote again to his brother:

I have now made a formal application for any number of Great Auk eggs ... and I hear that at present they do not intend to dispose of any ... Several applications have been made consequent on my discovery! I do not doubt that eventually we shall get some of them, and delay is all in our favour.

During April Newton was busy mounting a partial skeleton of a Great Auk:

It would, I really believe, almost buy the whole of the ten eggs.

He'd also discovered the names of those he believed to be his main rivals:

Flower tells me that the more urgent applications were from Tristram and Shaw the bird-stuffer at Shrewsbury. It is sharpish practice I think as regards the first, but I have not rebuked him for it.

By the autumn Newton was pinning his hopes on the irresistible temptation of an Aye-aye skeleton (*Daubentonia madagascariensis*) to break the developing deadlock but this ploy too was destined to fail.

Newton was never to realise his ambition. Nor were either of the rivals he mentioned. During 1864 the Scarborough collector Robert Champley coolly walked into Surgeons' Hall and carried off three of the eggs. His method was ruthlessly direct. He simply asked what the College needed, went out and purchased the £45 worth of anatomical specimens asked for, and effected a swap on the spot. With the sanctity of the collection breached, the Royal College sent up four more of the eggs for sale at Stevens' Rooms. Again poor Newton missed out and all four were sold to other parties.

The three remaining eggs, kept until June 19th 1946, eventually became the property of Vivian Hewitt and were among the thirteen that

Dr. Jack Gibson's Egg (¾ natural size) - circa 1970.

passed to Spink and Son. Five of these were sold during the early 1970's leaving eight (including the three College of Surgeons eggs) to languish precariously in a store cupboard at Spink's London headquarters. For complex legal reasons it had become impossible to sell them but, fortunately, Dr. Jack Gibson, Chairman of the Scottish Natural History Library, had taken an interest in their well-being. During 1976 he visited Spink and, fearing for the safety of the eggs, wrapped them carefully in cotton wool (they had previously been left to rattle around in a cardboard box) and replaced them securely in the cupboard. Over the next eighteen years, whenever he had occasion to visit London, Dr. Gibson called at Spink's to inspect the eggs and it is in large measure due to his efforts that they have survived. It was not until 1993 that it became possible for him to acquire them, which he then did on behalf of the Scottish Natural History Library. Dr. Gibson's intention is to present all eight eggs to the National Museum of Scotland thus making Scotland's national collection- ten eggs in all-unsurpassed by any other.

ROYAL COLLEGE OF SURGEONS EGG NO 4 (DAWSON ROWLEY'S EGG)

Egg no.18 (Grieve no. 9; Parkin no. V; Tomkinson no. 37)
Origin: unknown
Colour: cream with streaks, spots and blotches of black, brown and grey
Present location: Scottish Natural History Library (Dr. J.A. Gibson) - to be presented to the National Museum of Scotland, Edinburgh.

Champley, 1864. The Great Auk. *Annals and Magazine of Natural History*, p.236. Newton, A. 1865. The Gare-fowl and its Historians. *Natural History Review*, pp.483-4. Fatio, V. 1868. Liste des divers représentants de l'*Alca impennis* en Europe. *Bullétin de la Société Ornithologique Suisse*, tome II, pt.1, p.84. Newton, A. 1870. On Existing Remains of the Gare-fowl. *Ibis*, p.261. Blasius, W. 1884. Zur Geschichte der Ueberreste von *Alca impennis*. *Journal für Ornithologie*, p.153. Grieve, S. 1885. *The Great Auk or Garefowl*, p.87, pp.105-6 *and* appendix p.26. Parkin, T. 1911. The Great Auk. *Hastings and East Sussex Naturalist*, vol.1, pt.6 (extra paper), p.7. Jourdain, F. 1934 (Dec. 1st). The Sale of G.D. Rowley's Collections. *The Oologist's Record*, pp.75-9. Jourdain, F. 1935. Sale of Skins and Eggs of the Great Auk. *Ibis*, p.245. Jourdain, F. 1935. The skins and eggs of the Great Auk. *British Birds*, p.233. Carter, G. 1935 (March). The Eggs of the Great Auk. *The Oologist's Record*, pp.3-5. Tomkinson, P. & J. 1966. *Eggs of the Great Auk*, p.110 *and* pl.37. Spink and Son Ltd. (c.1970). *Romantic Histories of Some Extinct Birds and Their Eggs*, p.13 (specimen 11) *and* pl.7 (upper fig.). Bourne, W. 1993. The story of the Great Auk. *Archives of Natural History*, p.273.

This egg, one of the ten found in the Hunterian Museum, was sold - along with three others - at Stevens' Rooms, Covent Garden on July 11th 1865. According to the published record it was bought here by John Gould acting on behalf of his fellow writer George Dawson Rowley. According to

Rowley's own hand-written entry in his egg collection catalogue this was not the case. Rowley noted that a groom by the name of Napier acted on his behalf in the rooms. The cost was £33.

During his lifetime Rowley acquired two stuffed birds and six eggs and when he died in 1878 he left them to his son Fydell. It was not until 1934 that the Rowley family finally disposed of the collection and, once again, it was Stevens' who handled the sale. On November 14th this egg was bought by Captain Vivian Hewitt who was, of course, to prove an even more fanatical collector than Rowley. Hewitt paid £315 for it, a sum that - at the time - would have bought two modestly-sized houses.

Like the rest of the Captain's eggs, this one was offered for sale after his death by the London art and antique dealer Spink. It is one of the eight eggs acquired by Dr. Jack Gibson in 1993, all of which are to be presented to the National Museum of Scotland.

Dawson Rowley's Egg (¾ natural size) - circa 1970.

VICOMTE DE BARDE EGG NO.1 (LORD GARVAGH'S EGG)

Egg no.19 (Grieve no. 13 *or* 14; Parkin no. I; Tomkinson no. 41)
Origin: unknown
Colour: yellowish white, sparsely spotted and blotched with brown and black
Present location: Scottish Natural History Library (Dr. J.A. Gibson) - to be presented to the National Museum of Scotland, Edinburgh.

Roberts, A. 1861. Skins and Eggs of the Great Auk. *Zoologist,* p.7353. Champley, R. 1864. The Great Auk. *Annals and Magazine of Natural History,* p.236. Fatio, V. 1868. Liste des divers représentants de l'*Alca impennis* en Europe. *Bullétin de la Société Ornithologique Suisse,* tome II, pt.1, p.84. Blasius, W. 1884. Zur Geschichte der Ueberreste von *Alca impennis. Journal für Ornithologie,* p.154 *and* p.159. Grieve, S. 1885. *The Great Auk or Garefowl,* p.87, p.104 *and* appendix p.26. Newton, A. 1905. *Ootheca Wolleyana,* pp.383-4. Parkin, T. 1911. The Great Auk. *Hastings and East Sussex Naturalist,* vol.1, pt.6 (extra paper), p.6. Jourdain, F. 1934 (Dec. 1st). The Sale of G.D. Rowley's Collections. *The Oologist's Record,* pp.75-9. Jourdain, F. 1935. Sale of Skins and Eggs of the Great Auk. *Ibis,* p.246. Jourdain, F. 1935. The skins and eggs of the Great Auk. *British Birds,* p.234. Carter, G. 1935 (March). The Eggs of the Great Auk. *The Oologist's Record,* pp.3-5. Tomkinson, P. & J. 1966. *Eggs of the Great Auk,* pp.111-2 *and* pl.41. Spink and Son Ltd. (c.1970). *Romantic Histories of Some Extinct Birds and Their Eggs,* p.14 (specimen 15).

Obtained at an unknown date before 1795, this egg - along with egg no.40 and egg no.61 - was for 30 years or so in the collection of the Vicomte de Barde. In 1825 his entire egg collection passed to the Boulougne Museum and here it stayed for another 27 years. During 1852 the London taxidermist and natural history dealer James Gardiner received the three Auk eggs in exchange for an Ostrich skin.

T.H. Potts, destined to become one of the pioneers of ornithology in New Zealand, bought them from Gardiner but very soon afterwards took the decision to emigrate to the Antipodes. As a consequence, he offered two of his eggs for sale at Stevens' and on May 24th 1853 this one was sold to Lord Garvagh for £30. Almost twenty years later George Dawson Rowley bought it and after his death and that of his son Fydell the egg went back to Stevens' Rooms where, on November 14th 1934, it was knocked down to Vivian Hewitt. After Hewitt's death it passed to Spink and Son becoming one of the group of eight eggs acquired in 1993 by Dr. Jack Gibson for onward presentation to the National Museum of Scotland.

F.C.R. Jourdain.

BOURMAN LABREY'S EGG

Egg no.20 (Grieve no. 11; Tomkinson no. 39)
Origin: unknown
Colour: dirty yellowish white with brown and black spots and blotches
Present location: Scottish Natural History Library (Dr. J.A. Gibson) - to be presented to the National Museum of Scotland, Edinburgh.

Hewitson, W. 1853-6 (3rd ed.). *Coloured Illustrations of the Eggs of British Birds*, p.470. Roberts, A. 1861. Skins and Eggs of the Great Auk. *Zoologist*, p.7375. Champley, R. 1864. The Great Auk. *Annals and Magazine of Natural History*, p.236. Fatio, V. 1868. Liste des divers représentants de l'*Alca impennis* en Europe. *Bullétin de la Société Ornithologique Suisse*, tome II, pt.1, p.84. Blasius, W. 1884. Zur Geschichte der Ueberreste von *Alca impennis*. *Journal für Ornithologie*, pp.153-4. Grieve, S. 1885. *The Great Auk or Garefowl*, p.87 *and* appendix p.26. Jourdain, F. 1934 (Dec. 1st). The Sale of G.D. Rowley's Collections. *The Oologist's Record*, pp.75-9. Jourdain, F. 1935. Sale of the Skins and Eggs of the Great Auk. *Ibis*, p.246. Jourdain, F. 1935. The skins and eggs of the Great Auk. *British Birds*, p.234. Carter, G. 1935 (March). The Eggs of the Great Auk. *The Oologist's Record*, pp.3-5. Jourdain, F. 1935 (March 1933- Nov. 1935). The Egg of the Great Auk. *Bulletin of the British Oological Association*, vol.4, pp.128-9 *and* frontispiece. Tomkinson, P. & J. 1966. *Eggs of the Great Auk*, p.111 *and* pl.39. Spink and Son Ltd. (c.1970). *Romantic Histories of Some Extinct Birds and Their Eggs*, pp.13-4 (specimen 13).

The London taxidermist Leadbeater is the first recorded owner of this egg and he sold it to J.P. Wilmot of Leamington Spa. Being badly broken, it was crudely repaired by the naturalist William Yarrell and then - perhaps because he acquired two rather better examples - Wilmot gave it away to a resident of Manchester by the name of Bourman Labrey. During 1871 Labrey sold it to George Dawson Rowley and it soon passed to Rowley's son Fydell. It finally left the possession of the Rowley family on November 14th 1934 when it was sold at Stevens' Rooms. Here it was bought for 140 guineas by the inveterate egg collector the Reverend F.C.R. Jourdain. After Joudain's death the egg was acquired by the even more avid Captain Vivian Hewitt and eventually became the responsibility of Spink and Son Ltd. Doctor Jack Gibson bought it from Spink during 1993, as one of the group of eight eggs he is presenting to the National Museum of Scotland.

LADY CUST'S EGG

Egg no.21 (Grieve no. 12; Tomkinson no. 40)
Origin: unknown
Colour: white with one or two blackish streaks and spots at the larger end
Present location: Scottish Natural History Library (Dr. J.A. Gibson) - to be presented to the National Museum of Scotland, Edinburgh.

Newton, A. 1870. On Existing Remains of the Gare-fowl. *Ibis*, p.261. Blasius, W. 1884. Zur Geschichte der Ueberreste von *Alca impennis*. *Journal für Ornithologie*, p.154. Grieve, S. 1885. *The Great Auk or Garefowl*, p.87 *and* appendix p.26. Jourdain, F. 1934 (Dec. 1st). The Sale of G.D. Rowley's Collections. *Oologist's Record*, pp.75-9. Jourdain, F. 1935. Sale of Skins and Eggs of the Great Auk. *Ibis*, p.246. Jourdain, F. 1935. The skins and eggs of the Great Auk. *British Birds*, pp.233-4. Carter, G. 1935 (March). The Eggs of the Great Auk. *Oologist's Record*, pp.3-5. Tomkinson, P. & J. 1966. *Eggs of the Great Auk*, p.111 *and* pl.40. Spink and Son Ltd. (c.1970). *Romantic Histories of Some Extinct Birds and Their Eggs*, p.7 *and* p.14 (specimen 14).

William Yarrell, author of *A History of British Birds* (1837-43) is supposed to have bought this egg for Lady Cust in Paris. Nothing else is known of its early history. Because of the Paris origin and the Yarrell connection its history has sometimes been confused with that of an egg the ornithologist bought in France for a bargain price (egg no.22). This confusion seems to be a twentieth century addition to the story, however.

In 1878, the day before he died, George Dawson Rowley bought Lady Cust's Egg for £40 and it quickly passed into the possession of his son Fydell. After Fydell's own death the egg was sold at Stevens' to the Rev. F.C.R. Jourdain for 210 guineas. Jourdain disposed of it straightaway, exchanging it with F.G. Lupton for Yarrell's Egg (egg no.22) and from Lupton the egg was acquired by the firm of Gowland. Its next owner was Captain Hewitt and it eventually passed to Spink and Son. In 1993, after some eighteen years of negotiation, Dr. J.A. Gibson acquired it, along with seven other eggs at Spink's, for the Scottish Natural History Library, from where he intends to present all eight to the National Museum of Scotland.

YARRELL'S EGG

Egg no.22 (Grieve no. 43; Parkin no. III; Tomkinson no. 52)
Origin: unknown
Colour: yellowish white with black and brown blotches and streaks
Present location: Scottish Natural History Library (Dr. J.A. Gibson) - to be presented to the National Museum of Scotland, Edinburgh.

Hewitson, W. 1831-8 (1st ed.). *British Oology; being illustrations of the eggs of British Birds*, pl.145. Hewitson, W. 1842-6 (2nd ed.). *Coloured Illustrations of the Eggs of British Birds*, p.413. Roberts, A. 1861. Skins and Eggs of the Great Auk. *Zoologist*, p.7353. Champley, R. 1864. The Great Auk. *Annals*

William Yarrell (1784-1856).

Three pictures of Yarrell's Egg (all approx. natural size). (Right). Hand-coloured engraving from W. Hewitson's *British Oology* (1831-8). (Below, left). Chromolithograph from *Mémoires de la Société Zoologique de France* (1888). (Below, right). Watercolour by F.W. Frohawk (*courtesy of Haslemere Museum*)

and *Magazine of Natural History*, p.236. Fatio, V. 1868. Liste des divers représentants de l'*Alca impennis* en Europe. *Bullétin de la Société Ornithologique Suisse*, tome II, pt.1, p.84. Blasius, W. 1884. Zur Geschichte der Ueberreste von *Alca impennis*. *Journal für Ornithologie*, pp.161-2. Grieve, S. 1885. *The Great Auk or Garefowl*, p.89, p.105 *and* appendix p.30. Grieve, S. 1888. Recent information about the Great Auk. *Transactions of the Edinburgh Field Naturalists' and Microscopical Society*, pp.116-117. d'Hamonville, L. 1888. Note sur les quatre Oeufs d'*Alca impennis* appartenant a notre collection oologique. *Mémoires de la Société Zoologique de France*, p.225 *and* pl.5 (fig. A). d'Hamonville, L. 1891. Addition à une note sur les quatres œufs du *Pingouin brachyptère*. *Bullétin de la Société Zoologique de France*, p.34. Anon. 1894 (Feb.23rd). An Egg of Price. *Daily Graphic*. Anon. 1894 (Feb.23rd). The Eggs of the Great Auk. *The Times*. Newton, A. 1894 (March 3rd). The Great Auk. *The Athenaum*, no.3402. Grieve, S. 1896-7. Supplementary note on the Great Auk. *Transactions of the Edinburgh Field Naturalists' and Microscopical Society*, pp.258-9. Duchaussoy, H. 1897-8. Le Grand Pingouin du Musée d'Histoire naturelle d'Amiens. *Mémoires de la Société Linnéenne du Nord de la France*, p.102. Harting, J. 1901. *A Handbook of British Birds*, p.279. Newton, A. 1905. *Ootheca Wolleyana*, p.365. Parkin, T. 1911. The Great Auk. *Hastings and East Sussex Naturalist*, vol.1, pt.6 (extra paper), pp.12-4. Tomkinson, P. & J. 1966. *Eggs of the Great Auk*, p.115 *and* pl.52. Spink and Son Ltd. (c.1970). *Romantic Histories of Some Extinct Birds and Their Eggs*, pp.14-5 (specimen 16).

I would ask for the admission of a few lines in which to state what is known exactly of the origin of that specimen, which I well remember in the collection of the late Mr. Yarrell. He told me, as he told others of his friends, that he bought it in Paris; and to the best of my belief, not many years after the peace of 1815. In a little curiosity shop of mean appearance, he saw a number of eggs hanging on a string; he recognised one of them as... *Alca impennis*, and... was told that they were one franc apiece, except the large one, which from its size was worth two francs. He paid the money and walked away with the egg in his hat.

Alfred Newton (1894)

Aggravated by seemingly endless variations on this simple tale, and using the occasion of the egg's sale at Stevens' Rooms in 1894 as a pretext, Alfred Newton fired off this definitive letter to the editors of several British journals and newspapers. His intention was to put an end to the rumours and exaggerations concerning the acquisition of this egg, speculative stories that had persisted for more than three-quarters of a century.

Nineteenth century oologists seem never to have tired of telling and re-telling the tale of the finding of Yarrell's Egg and, as the decades passed, the story grew and grew in the telling. It held such an appeal for those who heard it that few could resist tampering with the sequence of events or leaving their own personal stamp on the basic plot. So many variations and amendments were added to the English ornithologist's original piece of good luck that the actual facts almost disappeared beneath them. Some place the scene of the action in a fisherwoman's hut close to the beach at Boulougne, others set it in a fashionable quarter of

AN EGG OF PRICE.

"I think I am justified in describing this egg," said Mr. Stevens to the audience of egg-collectors and curiosity-seekers who had assembled at Stevens' Auction Rooms, Covent Garden, yesterday, to see an egg of the great auk sold, "as the most interesting egg which it has ever been my privilege to sell. It is," continued Mr. Stevens, balancing delicately upon his palm a dirty looking egg with black spots at one end, and perhaps three and a half inches by two and a half at its greatest diameter—"It is—need I say?—a most beautiful egg. (A gentleman at the back of the crowd was understood to say that he thought such an explanation was distinctly necessary.) It has a story. It was bought by Yarrell, in whose collection originally it was, for a franc. He obtained it from a fisherwoman at Boulogne, to whom it had been given by a fisherman home from the North Sea. In 1856, upon Yarrell's death, the egg was sold here, at this very spot—not by me, gentlemen, for I was a school-boy at the time. It went," said Mr. Stevens, with a touch of sadness, "for twenty guineas. About twenty years later it again changed hands, and came into the possession of Baron Louis d'Hamonville, with whom it has remained till now. So I think I may say we know its history almost from the day it was laid. (No, sir, there is no known pedigree of the fisherwoman in existence.) There are but sixty-eight great auk's eggs in existence. There were only sixty-seven, but a short time ago a very beautiful specimen, a most exquisite egg, was found in a loft—a warning to us all to examine our lofts carefully. (The bird was *not* traced, sir.) Of these sixty-eight, two are in America. This one," concluded the auctioneer, with very creditable patriotism, "will, I hope, remain in Europe. Its proper place is the British Museum."

There was a long pause. Nobody knew quite what to offer for an egg of such unimpeachable connections. Mr. Stevens gave the bidders a hint. "Nothing less than a hundred guineas to begin with is to be thought of," he said. "I'd, well, I'd give that myself."

"Hundred and ten," said a responsive voice.

"Thank you, sir," said Mr. Stevens. "A hundred and ten. Thank you. Hundred and twenty, thirty, forty, fifty, thank *you*, sixty, can I say seventy for you, sir?" he added rapidly.

Another pause, during which the humorist in the back rows observes that auks' eggs are expensive eating.

"A hundred and seventy," says Mr. Stevens, suddenly, and then starts off again on another little run. "Eighty, ninety, two hundred guineas. Thank you! The most interesting auk's egg in the world for two hundred guineas. Won't do, gentlemen, won't do! Mustn't let the price go back. The last one sold here for two hundred and twenty-five. Two hundred and ten did I hear?"

It was two hundred and ten, and twenty and thirty and forty, at which it stopped for breath, and the auctioneer added that this was more promising. Could he (he asked) take two fifty from the

An account of the sale of Yarrell's Egg in *The Daily Graphic* for February 23 1894 (continued over).

(Cont. from previous page).

gentleman in the middle of the room? The gentleman in the middle of the room examined his boots with great attention and did not respond.

"Last one sold here for two hundred and twenty-five, was re-sold at a considerable profit," says Mr. Stevens, persuasively.

Still a pause, during which the "oldest attendant" tells his neighbour that the last but one sold here would never be re-sold at any profit. For as soon as its owner had succeeded in getting it he stamped on it—so as to make his other specimens rarer. "Salesman who told me," concludes the oldest attendant, "nearly wept." Meanwhile Mr. Stevens has been going over the advantages of the egg once more. "Did you say two-fifty, sir?" he asks ingenuously. "No? Well, you mustn't look at me, then. It's two-fifty if you do. Two-forty offered."

"Two hundred and fifty," says the voice in the middle of the room.

"Thank you," says Mr. Stevens; "two-sixty on my left."

"Two-seventy," says the voice.

"Two-eighty," says the nod.

"One more from you?" Mr. Stevens entreats the middle of the room; and adds to the other spectators (who avoid his eye), "He'll give it presently; he only wants to think of it."

He appears to want a very long time, and the auctioneer has almost resigned himself to taking two hundred and eighty guineas for "this unrivalled egg" when the middle of the room says manfully, "Two-ninety guineas!"

Cheers.

"You hear what he says," Mr. Stevens says, in another audible aside to the gentleman on his left; "now let me take another, just one more from you. I'll wait just as long for you as for him. Wait as long as you like for another ten guineas."

Thus adjured "the gentleman on my left" presently nods, and Mr. Stevens says, "Three hundred guineas! That's right. But I shall be happier still to take more from you, sir!" The "gentleman in the middle," however, digs his umbrella into his boot and says that he has quite finished, quite. So Mr. Stevens says, "Three hundred guineas once! Three hundred guineas twice! Reconsider, sir. No other chance. Three hundred guineas for the third time—for the third time—(crack)—gone! Purchaser, gentlemen, Sir Vauncey Crewe, Bart."

Paris; some said that Yarrell kept the egg for years, others believed he gave it away immediately. Some thought he'd broken it, others thought he hadn't. Even after Newton's clear statement of facts, the tale wouldn't rest. Eleven years later, in the *Ootheca Wolleyana* (1905), he felt obliged to try again:

> This story has been told with so many embellishments as to convey it into the realm of romance. Really the only variation it admits is to the price Mr. Yarrell paid. According to my recollection it was <u>two</u> francs, but I have known men who ought to remember put it as high as <u>five</u> ... The great points are that it was bought at Paris not long after 1815, and that the seller knew not what it was, or anything of its history.

The most interesting aspect of all this is the light it casts on the Victorian obsession with Great Auks. That grown men could attach such importance to the minutiæ of egg lore - without, it seems, a trace of a smile - is fairly remarkable. On the other hand, perhaps it's not. In a equally humourless spirit, quite typical of late twentieth century ornithological writing, here is the rest of the egg's story.

Yarrell kept the egg until his death in 1856 after which it was bought at Stevens' by the London taxidermist James Gardiner. Gardiner was acting on behalf of Mr. Frederick Bond of Kingsbury, Middlesex who paid £21 for it. During 1875 the egg was sold to Baron Louis d'Hamonville who took it to France but in 1894 it was back at Stevens' where it was sold to Sir Vauncey Crewe of Calke Abbey, Derbyshire for 300 guineas. In 1925 the egg visited Stevens' for a third time and was bought by a Mr. Hirch for five guineas more than it cost in 1894. It then passed through several more hands - Commander A.T. Wilson, F.G. Lupton and the Reverend F.C.R. Jourdain before being acquired by Captain Vivian Hewitt. Following Hewitt's death in 1965 the egg languished in a cupboard at the London headquarters of Spink and Son before being bought, in 1993, by Dr. Jack Gibson who intends to present it to the National Museum of Scotland.

MONSIEUR FAIRMAIRE'S EGG NO.1 (RODERICK STIRLING'S EGG)

Egg no.23 (Grieve no. 61; Parkin no. XXI; Tomkinson no. 22)
Origin: unknown
Colour: dirty white marked sparingly with brown and black spots
Present location: Inverness - Museum and Art Gallery, Castle Wynd, Inverness, Scotland.

Champley, R. The Great Auk. *Annals and Magazine of Natural History,* p.236. Fatio, V. 1868. Liste des divers représentants de l'*Alca impennis* en Europe. *Bullétin de la Société Ornithologique Suisse,* tome II, pt.1, p.84. Blasius, W. 1884. Zur Geschichte der Ueberreste von *Alca impennis. Journal für Ornithologie,* p.167. Grieve, S. 1885. *The Great Auk or Garefowl,* p.88 *and* appendix p.33. Grieve, S. 1888. Recent information about the Great Auk. *Transactions of the Edinburgh Field Naturalists' and Microscopical Society,* p.116.

Parkin, T. 1911. The Great Auk. *Hastings and East Sussex Naturalist*, vol.1, pt.6 (extra paper), p.24. Tomkinson, P. & J. 1966. *Eggs of the Great Auk*, p.106 *and* pl.22.

This egg, together with another (no.59), is first heard of in 1864 in the hands of the Parisian natural history dealer Monsieur E. Fairmaire of the *Magasin de Zoologie* at 56 Rue de l'Université. Fairmaire never really disclosed how he came by the pair, simply saying they were the property of

A letter from Monsieur Fairmaire to Robert Champley.

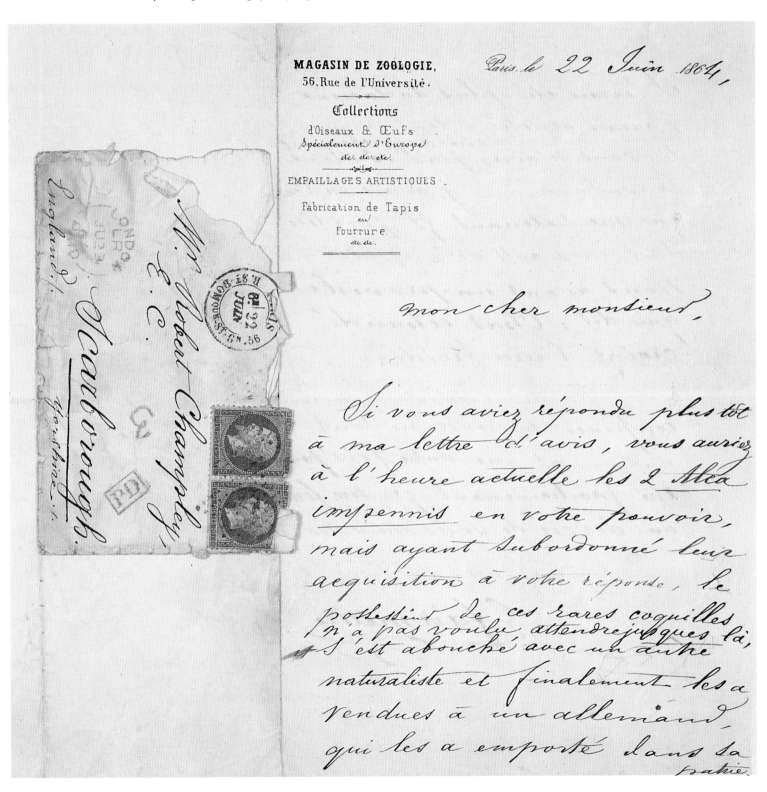

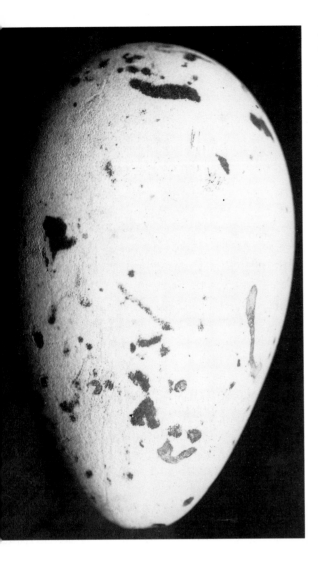

Roderick Stirling's Egg (natural size) - circa 1900.

an acquaintance who wished to sell them. Accordingly, he wrote to Robert Champley offering him the chance to buy.

The letter arrived at Champley's Scarborough home while the alderman collector was away and there was some delay in his responding. During this time another buyer was found and when, later, Champley complained at the haste with which the eggs had been sold, Fairmaire answered with a rather reproachful letter that can be translated as follows:

> If you had responded sooner to my bird letter, you would by now have ... the two *Alca impennis* in your power, but... acquisition was dependent on your response. The owner of the two rare egg shells did not wish to wait, entered into negotiations with another naturalist and finally sold them to a German who has taken them back to his country. I was not brave enough to buy them - at all risk to myself - without waiting for your agreement.

After mentioning a previous arrangement over which Champley had evidently let him down, Fairmaire finished his letter by writing:

> We have a proverb that says, 'A heated cat calls for cold water.'
> [Chat échaudé craint l'eau froide]

Champley's disappointment was short-lived, however. The eggs, it seems, did not go to Germany after all, or, if they did, they were there only very briefly. A few days after receiving Fairmaire's rebuke, Champley visited London and while there happened to call at the taxidermy shop of Henry Ward (father of the more famous Rowland) in Vere Street. Champley (*in* Grieve, 1888) described what happened:

> He showed me one egg, for which I gave him £25, and asked him if he had any more. He showed me another ... for which I paid him £30. I then asked him if he had any more, as I would take 20. He smiled. He would not say how he got them, but I afterwards found out they were the same as offered to me by Fairmaire. I called on Ward many times after, and he always regretted having parted with the eggs.

Champley kept both for the rest of his life but after his death the one that cost £25 was offered for sale at Stevens' Rooms. On April 17th 1902, it was knocked down to William Stirling of Fairburn, Ross-shire, Scotland for £252. It passed, by descent, to his son Major Sir John Stirling and then to his grandson Captain Roderick Stirling who, in 1983, donated it - along with another egg (no.24) acquired by his grandfather - to the Museum and Art Gallery of Inverness.

WILLIAM STIRLING'S EGG

Egg no.24 (Grieve no. 36; Parkin no. VIII; Tomkinson no. 71)
Origin: unknown, but probably Iceland
Colour: buff with spots and blotches of black and brown
Present location: Inverness - Museum and Art Gallery, Castle Wynd, Inverness, Scotland.

Roberts, A. 1861. Skins and Eggs of the Great Auk. *Zoologist*, p.7353. Champley, R. 1864. The Great Auk. *Annals and Magazine of Natural History*, p.236. Fatio, V. 1868. Liste des divers représentants de l'*Alca impennis* en Europe. *Bullétin de la Société Ornithologique Suisse*, tome II, pt.1, p.85. Blasius, W. 1884. Zur Geschichte der Ueberreste von *Alca impennis*. *Journal für Ornithologie*, pp.159-60. Grieve, S. 1885. *The Great Auk or Garefowl*, p.88, p.106 *and* appendix p.29. Grieve, S. 1897-8. Additional notes on the Great Auk. *Transactions of the Edinburgh Field Naturalists' and Microscopical Society*, pp.336-7. Parkin, T. 1911. The Great Auk. *Hastings and East Sussex Naturalist*, vol.1 pt.6 (extra paper), pp.8-9 *and* pp.26-7. Tomkinson, P. & J. 1966. *Eggs of the Great Auk*, p.121 *and* pl.71.

William Stirling's Egg (natural size). Water- colour by F.W. Frohawk. *Courtesy of Haslemere Museum.*

Several times during the nineteenth century this egg was supposed broken, lost or both. The rumours of its destruction were greatly exaggerated, however, and it still exists. Before becoming the property of the Stirling family in the early years of this century, it had many owners and its history is a rather tedious story of sales, dates, names and prices.

It is first recorded in the hands of either a Mr. Hoy or Mr. Robert Dunn of Hull, two men who were apparently related by marriage. Dunn was a zoological dealer with connections in Hamburg, from which information an Icelandic origin for this egg may be assumed. In 1838 Dunn sold the egg to A.D. Bartlett who re-sold it in 1842 to Mr. E. Maunde for £2. Nine years or so later Bartlett bought it back and promptly sold it again to Dr. Nathaniel Troughton. This time the price was £5. After Troughton's death his collection was sent up to Stevens' Rooms where, in April of 1869, Lord Garvagh gave £64 for the egg. Garvagh died in 1871 and the egg became the property of his wife before passing to her daughter, The Hon. Emmeline R. Canning. Evidently, Emmeline told no-one in the ornithological world about her possession and during the period of her ownership it was feared lost. After her death in 1890 the egg was found in a dusty attic at her home by J.E. Harting and it was subsequently sold to a Mr. Heatley Noble. In 1904 it was sent back to Stevens' Rooms to be sold yet again but on this occasion the reserve price wasn't reached. Six months later the Stevens' exercise was repeated - this time successfully - and Mr. William Stirling became the new owner in exchange for 200 guineas. This was Stirling's second Great Auk egg and both his examples passed to his son John and then eventually to his grandson Captain Roderick Stirling. During 1983 Captain Stirling presented them to the Inverness Museum.

THE EARL OF DERBY'S EGG

Egg no.25 (Grieve no. 30; Tomkinson no. 14)
Origin: unknown
Colour: cream, finely streaked and blotched with black, brown and grey
Present location: Liverpool - The Liverpool Museum, National Museums and Galleries on Merseyside, William Brown St., Liverpool.

Moore, T. 1861. Additional Egg of the Great Auk. *Zoologist*, p.7387. Champley, R. 1864. The Great Auk. *Annals and Magazine of Natural History* p.236. Fatio, V. 1868. Liste des divers représentants de l'*Alca impennis* en Europe. *Bullétin de la Société Ornithologique Suisse*, tome II, pt.1, p.84. Blasius, W. 1888. Zur Geschichte der Ueberreste von *Alca impennis*. *Journal für Ornithologie*, p.158. Grieve, S. 1885. *The Great Auk or Garefowl*, p.87 and appendix p.29. Seebohm, H. 1885. *History of British Birds*, vol.3, pl.41. Newton, A. 1905. *Ootheca Wolleyana*, p.376. Wollaston, A. 1921. *Life of Alfred Newton*, p.45. Tomkinson, P. & J. 1966. *Eggs of the Great Auk*, p.103 and pl.14.

Mr. Hancock tells me it is the most interesting specimen he has seen, and differs very much from any other egg that has passed through his hands. I have had it photographed, and shall be happy to exchange prints with any proprietor of these very rare eggs, as soon as the weather will permit their being taken with effect.

T.J. Moore (1861)

Two views of the Earl of Derby's Egg (natural size). Photographs by David Flower. *Courtesy of National Museums and Galleries on Merseyside.*

This was how Mr. T.J. Moore communicated to the world his discovery of a hitherto unknown egg in the collection of the 13th Earl of Derby. How

it got into the collection and whether or not the Earl was aware of its existence cannot be said. The Earl, who'd collected furiously, had died in 1851 and his material was presented to the City of Liverpool; it was here in the city museum that Mr. Moore made his discovery. Not only is this shell quite different in appearance to others, it also has the deserved reputation of being the most beautiful in existence.

HANCOCK'S EGG

Egg no.26 (Grieve no. 44; Tomkinson no. 15)
Origin: Iceland, probably Eldey
Colour: white with brown and black spots
Present location: Newcastle-upon-Tyne - The Hancock Museum, Barras Bridge, Claremont Rd., Newcastle-upon-Tyne NE2 4PT.

Hewitson, W. 1842-6 (2nd ed.) *Coloured Illustrations of the Eggs of British Birds*, p.413 *and* pl.115 (3rd ed., 1853-6, p.470 - no plate). Pässler, W. 1860. Die Eier der *Alca impennis* in deutschen Sammlungen. *Journal für Ornithologie*, p.59. Roberts, A. 1861. Skins and Eggs of the Great Auk. *Zoologist*, p.7353. Newton, A. 1861. Abstract of Mr. J. Wolley's Researches in Iceland. *Ibis*, p.392. Champley, R. 1864. The Great Auk. *Annals and Magazine of Natural History*, p.236. Fatio, V. 1868. Liste des divers représentants de l'*Alca impennis* en Europe. *Bullétin de la Société Ornithologique Suisse*, tome II, pt.1, p.84. Blasius, W. 1884. Zur Geschichte der Ueberreste von *Alca impennis*. *Journal für Ornithologie*, p.162. Grieve, S. 1885. *The Great Auk or Garefowl*, p.88, appendix p.18 *and* pp.30-1. Newton, A. 1905. *Ootheca Wolleyana*, p.381. Tomkinson, P. & J. 1966. *Eggs of the Great Auk*, pp.103-4 *and* pl.15.

Edward Stanley, 13th Earl of Derby (1775-1851).

Hancock's Egg (natural size). Hand-coloured engraving from the second edition of W. Hewitson's *Coloured Illustrations of the Eggs of British Birds* (1842-6). *Courtesy of Peter Blest.*

This egg is one of three once owned by Mechlenburg, the Apothecary of Flensburg. He sold it, together with a stuffed bird, to John Hancock the celebrated Newcastle taxidermist during April of 1844 through the agency of one John Sewell.

Mechlenburg claimed - in a letter to Hancock - that he'd received both skin and egg a year or two previously from an island off the north-east coast of Iceland but this, almost certainly, is an error. All serious researchers on the subject have reached the conclusion that the egg was taken on Eldey during the early 1830's.

Hancock gave both his bird and his egg to the Museum of the Natural History Society at Newcastle and both specimens now form part of the collection of the Hancock Museum.

James Reeve (1833-1920).

REEVE'S EGG

Egg no.27 (Grieve no. 49; Tomkinson no. 45)
Origin: unknown, but probably Eldey
Colour: yellowish white with large blotches and spots of black and brown
Present location: Norwich - The Castle Museum, Norwich NR1 3JU.

Hewitson, W. 1853-6 (3rd ed.). *Coloured Illustrations of the Eggs of British Birds*, p.470. Roberts, A. 1861. Skins and Eggs of the Great Auk. *Zoologist*, p.7353. Champley, R. 1864. The Great Auk. *Annals and Magazine of Natural History*, p.236. Fatio, V. 1868. Liste des divers représentants de l'*Alca impennis* en Europe. *Bullétin de la Société Ornithologique Suisse*, tome II, pt.1, p.85. Blasius, W. 1884. Zur Geschichte der Ueberreste von *Alca impennis*. *Journal für Ornithologie*, p.165. Grieve, S. 1885. *The Great Auk or Garefowl*, p.88 *and* appendix p.31. Whitaker, J. 1907. *Notes on the Birds of Nottinghamshire*, p.XVII. Gurney, J. 1911. The Great Auk and its egg in Norwich Museum. *Transactions of the Norfolk and Norwich Naturalists' Society*, pp.214-5 *and* pl.1. Anon. 1911 (Feb.18th). The Great Auk's Egg at Norwich. *The Field*, p.34. Tomkinson, P. & J. 1966. *Eggs of the Great Auk*, p.113 *and* pl.45.

In 1910 James Reeve, retiring Curator of the Castle Museum, purchased an egg to present to his museum and ruined the carefully laid plan of Auk egg specialist Edward Bidwell. It was Bidwell's intention that the Castle Museum should buy the egg *he* owned so that it would remain in Norwich (Bidwell's home town) for posterity. Whether the town appreciated (or was even aware of) the privilege Bidwell intended to confer upon it is not recorded but his annoyance concerning his thwarted wish is expressed in a letter to his fellow enthusiast Thomas Parkin, second on Bidwell's list of preferred buyers. Had he been as generous as Curator Reeve, the fate of his own egg (no.31) would not, of course, have presented a problem.

It seems that at one time Bidwell even hoped that Reeve might personally buy his egg but the Curator turned it down and bought this one instead. According to an unnamed correspondent of *The Field* (Feb.18th, 1911), the history of the egg Reeve chose is as follows:

It was bought ... from Mr. J.H. Walter, whose father had obtained it about the year 1850 from Dr. Pitman. In the opinion of the late Professor Newton this egg was one of those which came from... Brandt, the dealer at Hamburg ... it is thought that [it] was ... marked No.661 in his sale catalogue, priced 30 shillings.

Brandt's Icelandic trading connections and the time at which it passed through his hands make it virtually certain that this is an egg from Eldey.

BARCLAY'S EGG

Egg no.28 (Grieve no. 57; Parkin no. XXII; Tomkinson no. 18)
Origin: Iceland, probably Eldey
Colour: dirty white with brown spots and blotches
Present location: Norwich - The Castle Museum, Norwich NR1 3JU.

Barclay's Egg (natural size). Watercolour by F.W. Frohawk. *Courtesy of Haslemere Museum.*

Thienemann, F. 1845-54. *Einhundert Tafeln colorirter Abbildungen von Vogeleiern*, pl.96 (fig. lower right). Pässler, W. 1860. Die Eier der *Alca impennis* in deutschen Sammlungen. *Journal für Ornithologie*, p.59. Champley, R. 1864. The Great Auk. *Annals and Magazine of Natural History*, p.236. Fatio, V. 1868. Liste des divers représentants de l'*Alca impennis* en Europe. *Bullétin de la Société Ornithologique Suisse*, tome II, pt.1, p.84. Blasius, W. 1884. Zur Geschichte der Ueberreste von *Alca impennis*. *Journal für Ornithologie*, pp.166-7. Grieve, S. 1885. *The Great Auk or Garefowl*, p.86 *and* appendix p.33. Grieve, S. 1888. Recent information about the Great Auk. *Transactions of the Edinburgh Field Naturalists' and Microscopical Society*, pp.114-5. Parkin,T. 1911. The Great Auk. *Hastings and East Sussex Naturalist*, vol.1, pt.6 (extra paper), pp.25-6. Tomkinson, P. & J. 1966. *Eggs of the Great Auk*, pp.104-5 *and* pl.18.

The second of the eggs in the Norwich Museum was the first of nine acquired by the Scarborough collector Robert Champley and he got it almost by accident. Champley had written to the Editor of *The Zoologist* with an inquiry about Great Auk eggs and for an unknown reason his letter was printed on the journal's cover. This cover was seen by Herr G.H. Kunz, a Leipzig soap-maker, who wrote to Champley in July 1859 offering the egg for £18.

Champley believed this example, with its particularly pyriform shape, to be the finest egg known and, certainly, it was chosen by F.A.L. Thienemann to be illustrated for his rare book *Einhundert Tafeln colorirter Abbildungen von Vogeleiern* (1845-54).

Before Herr Kunz acquired it, the egg belonged to Theodore Schultz of Neuhaldensleben bei Magdeburg who'd obtained it, for approximately a guinea (7 thalers), from another Schultz - the dealer who traded out of Leipzig and Dresden. This Schultz was a major operator and his name is connected with several Great Auk specimens. He received this egg, in a consignment of seven, from Iceland at some time before 1835. Apparently Dresden Schultz sold it to his namesake with the following words:

I have had to keep the egg of the *Alca impennis* hidden, as I have several times been asked for it, and it is probably the last I shall get this year.

After Champley's death the egg was offered at Stevens' Rooms where it was knocked down to a Mr. Macgregor for £304 and ten shillings. In reality 'Mr. Macgregor' was fictional, the auctioneer having taken bids 'off the wall.' In other words, the reserve price wasn't reached. Eventually the egg was sold by Rowland Ward to Hugh Gurney Barclay of Colney Hall, Norwich. In 1936 his son Evelyn Barclay donated it to the local museum.

It is likely that it came from one of the early raids on Eldey.

THE OXFORD EGG

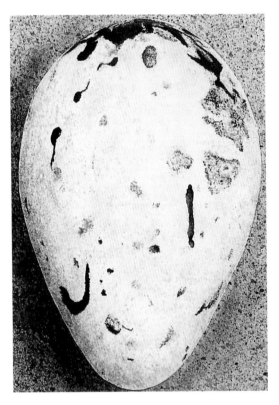

Egg no.29 (Grieve no. 48; Tomkinson no. 10)
Origin: unknown
Colour: whitish marked with brown and black spots and blotches
Present location: Oxford - The University Museum, Parks Rd., Oxford.

Roberts, A. 1861. Skins and Eggs of the Great Auk. *Zoologist*, p.7353. Champley, R. 1864. The Great Auk. *Annals and Magazine of Natural History*, p.236. Fatio, W. 1868. Liste des divers représentants de l'*Alca impennis* en Europe. *Bullétin de la Société Ornithologique Suisse*, tome II, pt.1, p.85. Blasius, W. 1884. Zur Geschichte der Ueberreste von *Alca impennis*. *Journal für Ornithologie*, p.163. Grieve, S. 1885. *The Great Auk or Garefowl*, p.88 *and* appendix p.31. Seebohm, H. 1885. *A History of British Birds*, vol.3, pl.40. Newton, A. 1905. *Ootheca Wolleyana*, vol.2, p.381. Tomkinson, P. & J. 1966. *Eggs of the Great Auk*, p.102 *and* pl.10.

This egg was given by Lady Wilson of Charlton House, Blackheath, London to her relative Sir Walter C. Trevelyan at a date prior to 1839. Before that, nothing is known of it. Sir Walter died in 1879 and bequeathed his egg to the University Museum at Oxford.

The Oxford Egg (¾ natural size). Watercolour by David Waring. *Courtesy of John Metcalf.*

THE SCARBOROUGH EGG

Egg no.30 (Grieve no. 56; Tomkinson no. 16)
Origin: unknown
Colour: white, sparingly marked with brown and black blotches and streaks
Present location: Scarborough - The Scarborough Museum, Scarborough, Yorkshire.

Blasius, W. 1884. Zur Geschichte der Ueberreste von *Alca impennis*. *Journal für Ornithologie*, p.165. Grieve, S. 1885. *The Great Auk or Garefowl*, p.88 *and* appendix p.33. Anon. 1905 (Jan.29th). *Daily Telegraph*. Tomkinson, P. & J. 1966. *Eggs of the Great Auk*, p.104 *and* pl.16.

On January 29th 1905 the *Daily Telegraph* carried the following rather sombre little report:

> The specimen of a Great Auk's egg owned by the Scarborough Museum authorities has been discovered lying on a chair in the building practically broken into two pieces. Our correspondent says it is authoritatively stated

that the value has depreciated £50 at least. Till recently the egg had not been exhibited in the usual way for fear it should be purloined but had been locked up in one of the officer's safes. It will now again be placed under lock and key.

The perpetrator of this deed was, it seems, never identified but more than 90 years later the egg - apparently repaired - is still in the safe.

The Yorkshire resort of Scarborough has something of an association with the Great Auk due to its one-time alderman, Robert Champley. The egg in the museum safe has nothing to do with Mr. Champley, however. Despite the fact that he brought back to Scarborough no less than nine eggs along with a single stuffed bird, all of his property has long since been dispersed and none of his material has stayed in the town.

The Scarborough Egg was bequeathed to the Scarborough Philosophical Society by a Mr. Alwin S. Bell. Little is known of its earlier history. Bell bought his egg from the London natural history dealer James Gardiner but there is some mystery as to how Gardiner came by it. He claimed to have purchased it from a collection somewhere in Derbyshire but always refused steadfastly to name the source.

ROYAL COLLEGE OF SURGEONS EGG NO.5 (PARKIN'S EGG)

Egg no.31 (Grieve no. 68; Parkin no. IV; Tomkinson no. 17)
Origin: unknown
Colour: cream with brown, black and grey spots, blotches and streaks
Present location: Spalding - Spalding Gentlemens' Society, The Museum, Broad St., Spalding, Lincolnshire PE11 1TB.

Newton, A. 1865. The Gare-fowl and its Historians. *Natural History Review*, pp.483-4. Fatio, V. 1868. Liste des divers représentants de l'*Alca impennis* en Europe. *Bullétin de la Société Ornithologique Suisse*, tome II, pt.1, p.84. Newton, A. 1870. On Existing Remains of the Gare-fowl. *Ibis*, p.261. Blasius, W. 1884. Zur Geschichte der Ueberreste von *Alca impennis*. *Journal für Ornithologie*, p.168. Grieve, S. 1885. *The Great Auk or Garefowl*, p.87, p.106, appendix p.29 *and* p.34. Parkin,T. 1911. The Great Auk. *Hastings and East Sussex Naturalist*, vol.1, pt.6 (extra paper), p.7. Frohawk, F. 1931 (May 13th). *£260 for a Bird's Egg. The Field*, p.752. Belt, A. 1933. Thomas Parkin- an obituary. *Hastings and East Sussex Naturalist*, vol.4, pp.126-7. Tomkinson, P. & J. 1966. *Eggs of the Great Auk*, p.104 *and* pl.17.

The Spalding Gentlemens' Society, one of the oldest learned societies in Britain, has a charming tradition. The members give presents to the Society's Museum and as a result - over several centuries - a quite amazing collection has been formed. Considering that men such as Sir Isaac Newton, Sir Joseph Banks, Alfred Lord Tennyson and Sir Hans Sloane were numbered among the members, the quality of the collection is scarcely surprising; in fact, the Museum of the Society is - with the exception of the Ashmolean at Oxford - the oldest museum in England.

Ashley K. Maples (1868-1950) was a comparatively modern, but exceptionally generous, benefactor of the Society. Among the many treasures he gave is a Great Auk's egg. Not only is this egg a particularly beautiful example, it is also an egg that was owned by two of the Garefowl's most devoted students, Edward Bidwell and Thomas Parkin. It was from Parkin that Maples got the egg, albeit via Stevens' Auction Rooms. Here on May 13th 1931 he paid what F.W. Frohawk (1931) described as, 'the very reasonable sum of £260,' for one of Parkin's most treasured possessions. Parkin, apparently, kept his egg in a secret cupboard situated behind a row of books. One of his great delights was to take it out and place it in the hands of a visitor or friend together with the words:

Now you can say you have had a Great Auk's egg in your hands!

It was only shortly before his death that he took the decision to part with his egg. His reason is hinted at in a letter written six years previously to his old friend Joseph Whitaker, author of *Notes on the Birds of Nottinghamshire* (1907):

Edward Bidwell and Thomas Parkin in the library at Parkin's home in Hastings.

A photograph of the library at Fairseat, High Wickham, Hastings. Mr. T. Parkin with a volume of his egg-registers, Mr. E. Bidwell perusing the history of the Great Auk's egg shown in the photograph. To the right of the egg is a model of it, and in the pamphlet is a plate of the egg showing the same markings. Next to the model are some bones of the Dodo

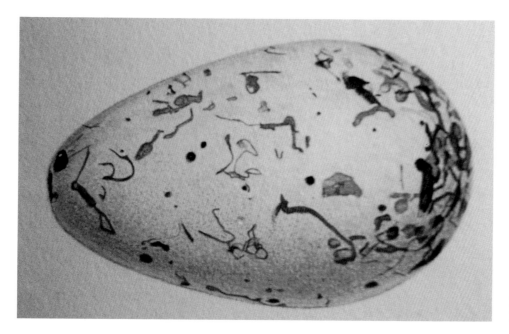

Parkin's Egg (natural size). Watercolour by F.W. Frohawk. *Courtesy of Haslemere Museum.*

> I hardly know what to do with my egg collection (about 5,000 eggs including a Great Auk's Egg) ... It is a pity that those who come after me care nothing about eggs or books!!! But so it is ... one collects and another scatters.

If Parkin was disappointed in having no-one to pass his egg on to, so too - after a fashion - was Edward Bidwell. It was from Bidwell, in April 1911, that Parkin obtained the egg, but Bidwell had always hoped it would find a different home. In July 1910 he wrote to Parkin:

> The late Curator to the Museum at Norwich has bought H. Walter's egg and presented it to the Museum. I knew he intended to give one and hoped he would buy mine, but he has gone past me so I am still open to let you have mine. It is a great nuisance. I should like to have mine in the Museum of my native city.

Perhaps it is the frequent use of the word *mine* that puts a reader out of sympathy with this note, or maybe it is the thought of poor Parkin waiting anxiously for the chance to buy the object he so much wanted.

Bidwell had owned the egg for 27 years, having bought it in 1884. Before that it was the property of the Reverend George W. Braikenridge of Clevedon in Somerset. Braikenridge obtained the egg - in exchange for £29 - at the well known sale of four Great Auk eggs that took place at Stevens' Rooms on July 11th 1865. These eggs were four of the ten found by Alfred Newton in the Hunterian Museum of the Royal College of Surgeons. As with the rest, nothing is known of its origin.

BULLOCK'S EGG

Egg no.32 (Grieve no. 31 *or* 32; Parkin no. XXV; Tomkinson no. 1)
Origin: unknown
Colour: white marked with black and brown spots and blotches
Present location: Tring - The Natural History Museum, Sub-department of

Ornithology, Tring, Hertfordshire.

Hewitson, W. 1853-6 (3rd ed.). *Coloured Illustrations of the Eggs of British Birds*, p.470. Roberts, A. 1861. Skins and Eggs of the Great Auk. *Zoologist*, p.7353. Champley, R. 1864. The Great Auk. *Annals and Magazine of Natural History*, p.236. Fatio, V. 1868. Liste des divers représentants de l'*Alca impennis* en Europe. *Bullétin de la Société Ornithologique Suisse*, tome II, pt.1, p.82. Blasius, W. 1884. Zur Geschichte der Ueberreste von *Alca impennis*. *Journal für Ornithologie*, pp.158-9. Grieve, S. 1885. *The Great Auk or Garefowl*, p.87 and appendix p.29. Oates, E. 1901-12. *Catalogue of the Collection of Birds' Eggs in the British Museum*, vol.1, p.165. Parkin, T. 1911. The Great Auk. *Hastings and East Sussex Naturalist*, vol.1, pt.6 (extra paper), pp.31-2. Tomkinson, P. & J. 1966. *Eggs of the Great Auk*, p.100 *and* pl.1. Knox, A. & Walters, M. 1994. *Extinct and Endangered Birds in the Collection of the Natural History Museum*, p.134.

This egg, now faded and badly broken, can be traced back to the ownership of William Bullock, goldsmith, jeweller and passionate collector of all manner of curiosities. Among Bullock's other possessions was the famous Papa Westray Auk and when he tired of his collection this egg

(Below, left). Bullock's Egg (natural size). Watercolour by F.W. Frohawk. *Courtesy of Haslemere Museum.*

(Below, right). Dr. Leach's '139' Egg (natural size). Watercolour by F.W. Frohawk. *Courtesy of Haslemere Museum.*

was packed in a box with the stuffed bird and the pair were lotted together for the monumental 'Bullock Sale.' This took place over many days during the early summer of 1819 and on May 16th egg and bird were knocked down to Dr. Leach - acting on behalf of the British Museum - for a sum usually recorded as £16, fifteen shillings and sixpence. Since that time both bird and egg - along with another egg that Bullock owned - have belonged to the Museum. For many years during the nineteenth century the egg was displayed glued down to a board and it is now badly spoiled. Today it is carefully stored at the Natural History Museum's Sub-department of Ornithology at Tring in Hertfordshire.

There is absolutely nothing to indicate where it originally came from. Suggestions of a New World provenance are speculation. So too is a rumour that the egg once belonged to Sir Joseph Banks.

DR. LEACH'S '139' EGG

Egg no. 33 (Grieve no. 31 *or* 32, Parkin no. XXVI; Tomkinson no. 2)
Origin: unknown
Colour: white marked with black and brown blotches
Present location: Tring - The Natural History Museum, Sub-department of Ornithology, Tring, Hertfordshire.

All references are identical to those of egg no.32, except for:

Tomkinson, P. & J. 1966. *Eggs of the Great Auk*, pl.2.

This egg, with the number 139 painted on its side, is the second of two owned by William Bullock. It was sold at Bullock's famous sale during which the eccentric jeweller and goldsmith himself took the chair to act as his own auctioneer. Like his other egg, this one was bought by Dr. Leach, Keeper of the Zoology Department of the British Museum. He bought it on June 3rd 1819, the twenty first day of the mammoth sale and its cost is recorded variously, sometimes as twelve shillings, sometimes as seventeen.

It has belonged to the British Museum ever since and, like its Bullock companion, is now faded and badly broken having been glued to a display board for many years. Nothing is known of its origin.

SPALLANZANI'S EGG

Egg no.34 (Grieve no. 60; Tomkinson no. 20)
Origin: unknown
Colour: dirty white marked with brown spots
Present location: Tring - The Natural History Museum, Sub-department of Ornithology, Tring, Hertfordshire.

Champley, R. 1864. The Great Auk. *Annals and Magazine of Natural History*, p.236. Fatio, V. 1868. Liste des divers représentants de l'*Alca impennis* en

THE GREAT AUK

Europe. *Bullétin de la Société Ornithologique Suisse*, tome II, pt.1, p.82. Blasius, W. 1884. Zur Geschichte der Ueberreste von *Alca impennis*. *Journal für Ornithologie*, p.167. Grieve, S. 1885. *The Great Auk or Garefowl*, p.88 *and* appendix p.33. Grieve, S. 1888. Recent information about the Great Auk. *Transactions of the Edinburgh Field Naturalists' and Microscopical Society*, p.115-6. Ward, R. 1913. *A Naturalist's Life Study*, p.221. Glegg, W. 1949. The History of a Great Auk's Egg. *Bulletin of the British Ornithologists' Club*, pp.77-80. Tomkinson, P. & J. 1966. *Eggs of the Great Auk*, p.105 *and* pl.20. Violani, C. 1975. L'Alca impenne...Nelle Collezioni Italiane. *Natura- Revista di Scienze Naturali*, p.13, p.22 *and* pl.4 (fig. B). Rothschild, M. 1983. *Dear Lord Rothschild*, p.105. Knox, A. & Walters, M. 1994. *Extinct and Endangered Birds in the Collection of the Natural History Museum*, p.134.

This egg was taken at an early date but its actual origin is unknown. It was found by Robert Champley during his Italian tour of 1861 and the Scarborough collector's notes on the circumstances surrounding his discovery were published by Symington Grieve in 1888:

> We ... drove on to Pavia ... and went over the Museum of Anatomy. I inquired if there were any eggs and birds and was answered in the affirmative ... Many eggs [were] stuck on wires on shelves, but all black ... with dust. I noticed... what I thought was an egg of a Great Auk [and] asked the attendant to open the case, but he had not the key. I told him to go for the sub-director. He returned with him and opened the case which was fastened with screws. I took down the egg, black over with dirt and rubbed it clean ... it was an *Alca impennis*. I told the sub-director I would exchange ... skins for it. He ... referred me to the chief director, and ... told me ... the collection was given by Professor Spallanzani one hundred years before, and that Spallanzani had been one of the lecturers in the University ... On my arrival at the director's residence, I told him that there was ... an egg of "Le Grand Penguin," and I should be glad if he would let me have it for an exchange. He accompanied me back to the museum...The sub-director told him I had offered five napoleons or an equivalent in exchange ... He said they would ... prefer the money. I therefore borrowed the amount from my Russian friend and, after packing the egg carefully, left the museum, they seeming sorry they had no more specimens, and considered they had got a good bargain.

When Robert Champley died, in 1895, his nine eggs passed to his daughter. This one was sold on November 1st 1901 by her husband, Mr. Rutter, to Walter Rothschild. Rothschild was, at this time, buying frantically for his museum at Tring but this purchase took place only after eight months of vacillation. Despite his enormous wealth, the English Baron felt a little guilty about lashing out a substantial sum on an egg, particularly since he already owned one. With a schoolboyish sense of shame the world's greatest ever natural history collector hid his new possession in a box containing one of his Great Auk skins. Only when he was safely away from his museum at Tring did he write to his curator Ernst Hartert to inform him of their most recent acquisition:

> I at last succumbed to the temptation though I was not going to tell you until my return.

Hartert was, of course, unlikely to complain too strongly; to Rothschild he owed his livelihood. On Walter's death in 1937, the egg was bequeathed to the British people along with much else of his collection.

(Above, left). Ernst Hartert (1860-1933). *Courtesy of The Hon. Miriam Rothschild.*

(Above, right). Walter Rothschild, 2nd Baron Rothschild of Tring (1868-1937). *Courtesy of The Hon. Miriam Rothschild.*

WALTER ROTHSCHILD'S EGG

Egg no.35 (Grieve no. 8; Tomkinson no. 36)
Origin: unknown, probably Iceland
Colour: white with large brown blotches and a few streaks
Present location: Tring - The Natural History Museum, Sub-department of Ornithology, Tring, Hertfordshire.

Pässler, W. 1860. Die Eier der *Alca impennis* in deutschen Sammlungen. *Journal für Ornithologie*, p.59. Fatio, V. 1868. Liste des divers représentants de l'*Alca impennis* en Europe. *Bullétin de la Société Ornithologique Suisse*, tome II, pt.1, p.84. Blasius, W. 1884. Zur Geschichte der Ueberreste von *Alca impennis*. *Journal für Ornithologie*, pp.152-3. Grieve, S. 1885. *The Great Auk or Garefowl*, p.89, p.103, p.108 *and* appendix pp.25-6. Grieve, S. 1888 *and* 1896-7. Recent information about the Great Auk *and* Supplementary note on the Great Auk. *Transactions of the Edinburgh Field Naturalists' and Microscopical Society*, p.117 (1888) *and* p.263 (1896-7). Blasius, W. (*in* Naumann) 1903. *Naturgeschichte der Vögel Mitteleuropas*, bd.12, pl.17b (fig.1). Rey, E. 1905. *Die Eier der Vögel Mitteleuropas*, pl.118. Newton, A. 1905. *Ootheca Wolleyana*, p.382. Rothschild, W. 1928 (March 28th). Eggs of the Great Auk. *The Field*, p.526. Glegg, W. 1949. The History of a Great

Walter Rothschild's Egg (2/3 natural size). Watercolour by F.W. Frohawk. *Courtesy of Haslemere Museum.*

Tristram's Egg (2/3 natural size). Watercolour by F.W. Frohawk. *Courtesy of Haslemere Museum.*

Auk's Egg. *Bulletin of the British Ornithologists' Club*, pp.77-80. Tomkinson, P. & J. 1966. *Eggs of the Great Auk*, p.110 *and* pl.36. Knox, A. & Walters, M. 1994. *Extinct and Endangered Birds in the Collection of the Natural History Museum*, p.134.

This egg is first recorded with certainty in the hands of the dealer Schultz who seems to have operated out of both Leipzig and Dresden. He is supposed to have bought it from a wealthy citizen of either Leipzig or, perhaps, Hamburg who'd obtained it from the Hamburg dealer Brandt. If this is the case it can be assumed that the egg has an Icelandic origin.

Schultz sold it to a Leipzig barber by the name of Hühnel for 7 thalers (about a guinea). Shortly before his death Barber Hühnel capitalised on his investment by selling the egg to Count Rödern of Wroclaw (Breslau) for 200 thalers.

In March of 1889, fairly near the beginning of his great acquisitorial career, Walter Rothschild bought the Count's collection and in so doing got the egg. Eventually, it formed part of his bequest to the British people.

During the nineteenth century it was several times stated that this egg is figured in Thienemann's celebrated book *Einhundert Tafeln colorirter Abbildungen von Vogeleiern* (1845-54) but this is not so.

TRISTRAM'S EGG

Egg no.36 (Grieve no. 19; Tominson no. 27)
Origin: Iceland, probably Eldey
Colour: whitish, spotted with brown and black
Present location: Tring - The Natural History Museum, Sub-department of Ornithology, Tring, Hertfordshire.

Roberts, A. 1861. Skins and Eggs of the Great Auk. *Zoologist*, p.7353. Champley, R. 1864. The Great Auk. *Annals and Magazine of Natural History*, p.236. Fatio, V. 1868. Liste des divers représentants de l'*Alca impennis* en Europe. *Bullétin de la Société Ornithologique Suisse*, tome II, pt.1, p.85. Blasius, W. 1884. Zur Geschichte der Ueberreste von *Alca impennis*. *Journal für Ornithologie*, pp.154-5. Grieve, S. 1885. *The Great Auk or Garefowl*, p.87 *and* appendix p.27. Grieve, S. 1888 *and* 1896-7. Recent information about the Great Auk *and* Supplementary note on the Great Auk. *Transactions of the Edinburgh Field Naturalists' and Microscopical Society*, p.114 (1888) *and* p.260 (1896-7). Newton, A. 1905. *Ootheca Wolleyana*, pp.381-2. Glegg, W. 1949. The History of a Great Auk's Egg. *Bulletin of the British Ornithologists' Club*, pp.77-80. Tomkinson P. & J. 1966. *Eggs of the Great Auk*, p.107 *and* pl.27. Knox, A. & Walters, M. 1994. *Extinct and Endangered Birds in the Collection of the Natural History Museum*, p.133.

Symington Grieve (1885) and the Tomkinsons (1966) relate that this egg was taken in Iceland possibly as late as 1844 but this is certainly a mistake. According to Alfred Newton's manuscript notes (at Cambridge

University Library) it was obtained in 1834 - which is, of course, altogether more likely.

It was brought from Iceland to Copenhagen where it stayed for several years and then, in 1851, was purchased by J. de Capel Wise from the Danish ornithologist Nils Kjærbölling. Shortly afterwards the egg arrived in Britain where the London taxidermist J. Williams acquired it, selling it during 1853 to the well known ornithological collector Canon Tristram for £35. Eventually Mr. Philip Crowley of Waddon House, Croydon purchased all of Tristram's egg collection, the Great Auk egg being included. In 1937 it went to the British Museum under the terms of the Crowley bequest.

ROYAL COLLEGE OF SURGEONS EGG NO.6 (LORD LILFORD'S EGG)

Egg no.37 (Grieve no. 37; Parkin no. VII; Tomkinson no. 29)
Origin: unknown
Colour: yellowish white with brown and black spots and blotches
Present location: Tring - The Natural History Museum, Sub-department of Ornithology, Tring, Hertfordshire.

Champley, R. 1864. The Great Auk. *Annals and Magazine of Natural History*, p.236. Newton, A. 1865. The Gare-fowl and its Historians. *Natural History Revue*, pp.483-4. Fatio, V. 1868. Liste des divers représentants de l'*Alca impennis* en Europe. *Bullétin de la Société Ornithologique Suisse*, tome II, pt.1, p.84. Blasius, W. 1884. Zur Geschichte der Ueberreste von *Alca impennis*. *Journal für Ornithologie*, p.160. Grieve, S. 1885. *The Great Auk or Garefowl*, p.88, pp.105-6 *and* appendix p.29. Parkin, T. 1911. The Great Auk. *Hastings and East Sussex Naturalist*, vol.1, pt.6 (extra paper), p.8. Glegg, W. 1949. The History of a Great Auk's Egg presented to the British Museum by Lord Lilford. *Bulletin of the British Ornithologists' Club*, pp.77-80. Tomkinson, P. & J. 1966. *Eggs of the Great Auk*, pp.107-8 *and* pl.29. Knox, A. & Walters, M. 1994. *Extinct and Endangered Birds in the Collection of the The Natural History Museum*, p.134.

Henry Baker Tristram (1822-1906).

Sold at Stevens' on July 11th 1865 when it was found surplus to the requirements of the Royal College of Surgeons, this is one of the ten eggs discovered by Alfred Newton in the Hunterian Museum. It was bought as 'Lot 143' for £29 by Mr. A.W. Crichton, the brother-in-law of Lord Lilford. At Crichton's death the egg passed to Lilford who kept it for the rest of his life; Lilford gave his other four eggs to Alfred Newton at Cambridge. Eventually - in 1949 - the egg he kept passed into the Natural History Museum's collection as part of the Lilford bequest.

THE TOMKINSON EGG

A life size model Great Auk made for David Evans by French sculptor Mario di-Ma'io. The egg is patterned after Tomkinson's Egg.

Egg no.38 (Grieve no. 3,4 *or* 5; Parkin no. XX; Tomkinson no. 53)
Origin: unknown
Colour: cream with some light brown staining, blackish blotches and streaks
Present location: private collection

Newton, A. 1870. On Existing Remains of the Gare-fowl. *Ibis*, p.261. Blasius, W. 1884. Zur Geschichte der Ueberreste von *Alca impennis*. *Journal für Ornithologie*, p.152. Grieve, S. 1885. *The Great Auk or Garefowl*, pp.104-5 *and* appendix p.25. Grieve, S. 1888. Recent information about the Great Auk. *Transactions of the Edinburgh Field Naturalists' and Microscopical Society*, pp.116-7. d'Hamonville, L. 1888. Note sur les quatre Oeufs d'*Alca impennis* appartenant a notre collection oologique. *Mémoires de la Société Zoologique de France*, p.224, p.227 *and* pl.5 (fig. D). Duchaussoy, H. 1897-8. Le Grand Pingouin du Musée d'Histoire naturelle d'Amiens. *Mémoires de la Société Linnéenne du Nord de la France*, p.108. Bidwell, E. 1901. Exhibition of a Great Auk's Egg. *Bulletin of the British Ornithologists' Club*, p.49. Parkin,T. 1911. The Great Auk. *Hastings and East Sussex Naturalist*, vol.1, pt.6, (extra paper), p.24. Tomkinson, P. & J. 1966. *Eggs of the Great Auk*, p.115 *and* pl.53.

During 1966 J.W. Tomkinson and his French wife P.M.L. Tomkinson produced a catalogue to the eggs of *Alca impennis*, giving it the splendidly straightforward name *Eggs of the Great Auk*. It lists every known example, gives brief historical details and illustrates each with two black and white photographs. The Tomkinsons were spurred to publish this interesting lit-

tle booklet by a death in the family. They inherited from Mr Tomkinson's father an egg along with which came a set of photographs, taken by Edward Bidwell, of almost every single specimen known to exist.

Their own egg had passed through many hands before it came to them but its actual origin is unknown. There seems no record of it earlier than 1855 at which time it was sold by the French dealer Parzudaki to the Baron Henri de Vèze for 500 francs. Three years later another Parisian dealer, Monsieur Fairmaire, sold it to le Comte Raoul de Baracé of Angers, owner of two other Great Auk eggs. When the Comte died in 1887, his collection passed to another French Baron, Louis Charles d'Hamonville. Baron d'Hamonville acquired one more egg to go with the three he got from the Comte de Baracé but eventually he disposed of all four. This one was sold at Stevens' Rooms on October 29th 1901 and was knocked down to Mr. Herbert Massey for £252. Massey was another serial egg collector but his final tally of three fell one short of Monsieur le Baron's.

During December of 1939 Mr. Gerald Tomkinson, father of J.W., was invited to choose and purchase one of Massey's eggs by Massey's executor G.H. Lings. The cost was £400 and the set of Bidwell photographs were included in the price. Massey's home, Ivy Lea, Didsbury, Manchester, was situated near to a munitions works - giving rise to fears of bombing raids - and the eggs had been evacuated. Mr. Tomkinson was obliged to travel to Windermere in the English Lake District to choose his egg.

After inheriting the egg and completing their research for *Eggs of the Great Auk* Mr. and Mrs Tomkinson retired to France and sold their egg to the well known bookseller David Evans who specialises in fine antiquarian ornithological volumes. He also acquired the set of Bidwell photographs.

Tomkinson's Egg (natural size). Chromolithograph from *Mémoires de la Société Zoologique de France* (1888).

Three views of the Tomkinson Egg (natural size). Photographs by David Evans. *Reproduced with his kind permission.*

THE COPENHAGEN EGG

Egg no.39 (Grieve no. 26; Tomkinson no. 47)
Origin: Iceland
Colour: dirty yellowish white with brown patches, spots and streaks
Present location: Copenhagen - Universitetets Zoologiske Museum, Universitetsparken 15, DK-2100 København Ø, Denmark.

Champley, R. 1861. Additional eggs of the Great Auk. *Zoologist*, p.7386. Champley, R. 1864. The Great Auk. *Annals and Magazine of Natural History*, p.236. Fatio, V. 1868. Liste des divers représentants de l'*Alca impennis* en Europe. *Bullétin de la Société Ornithologique Suisse*, tome II, pt.1, p.84. Blasius, W. 1884. Zur Geschichte der Ueberreste von *Alca impennis*. *Journal für Ornithologie*, p.157. Grieve, S. 1885. *The Great Auk or Garefowl*, p.88 *and* appendix p.28. Salomonsen, F. 1944-5. Gejrfuglen et hundredaars minde. *Dyr i Natur og Museum*, p.105. Tomkinson, P. & J. 1966. *Eggs of the Great Auk*, p.114 *and* pl.47.

This egg came from Iceland and probably arrived at the Royal Museum, Copenhagen in the 1830's. In the museum archives there is a record of an egg being exhibited during late 1830, a record that may well relate to this specimen. If so, then this is an egg collected on, 'the most distant Garefowl island,' by M.V. Moltke, probably earlier in the same year.

VICOMTE DE BARDE EGG NO.2 (LORD GARVAGH'S FOOTMAN'S EGG)

Lord Garvagh (1826-71).

Egg no.40 (Grieve no. 13 *or* 14; Parkin no. II; Tomkinson no. 42)
Origin: unknown
Colour: yellowish white streaked with brown and black
Present location: Helsinki - Finnish Museum of Natural History, University of Helsinki, P. Rautatiekatu 13, FIN- 00014, Helsinki, Finland.

Hewitson, W. 1853-6 (3rd ed.) *Coloured Illustrations of the Eggs of British Birds*, p.470. Roberts, A. 1861. Skins and Eggs of the Great Auk. *Zoologist*, p.7353. Champley, R. 1864. The Great Auk. *Annals and Magazine of Natural History*, p.236. Fatio, V. 1868. Liste des divers représentants de l'*Alca impennis* en Europe. *Bullétin de la Société Ornithologique Suisse*, tome II, pt.1, p.84. Blasius, W. 1884. Zur Geschichte der Ueberreste von *Alca impennis*. *Journal für Ornithologie*, p.154 *and* p.159. Grieve, S. 1885. *The Great Auk or Garefowl*, p.87, p.104, p.106 *and* appendix p.26. Grieve, S. 1897-8. Additional notes on the Great Auk. *Transactions of the Edinburgh Field Naturalists' and Microscopical Society*, pp.336-7. Newton, A. 1905. *Ootheca Wolleyana*, pp.383-4. Parkin, T. 1911. The Great Auk. *Hastings and East Sussex Naturalist*, vol.1, pt.6 (extra paper), p.6 *and* p.21. Jourdain, F. 1934 (Dec.1st). The Sale of G.D. Rowley's Collections. *The Oologist's Record*, pp.75-9. Jourdain, F. 1935. Sale of Skins and Eggs of the Great Auk. *Ibis*, p.246. Jourdain, F. 1935. The skins and eggs of the Great Auk. *British*

Birds, p.234. Carter, G. 1935 (March). The Eggs of the Great Auk. *The Oologist's Record*, pp.3-4. Tomkinson, P. & J. 1966. *Eggs of the Great Auk*, p.112 *and* pl.42.

The most dramatic event in the known tale of this egg came when it was accidentally dropped by one of Lord Garvagh's footmen. According to a letter written by George Dawson Rowley to Alfred Newton, the poor footman was instantly dismissed from his post; the egg, badly damaged at both ends, was immediately repaired, albeit rather clumsily.

Such were the whispers, rumours and general secrecy surrounding the sorry fate of this egg that the story of breakage was wrongly transferred to another of Garvagh's eggs and it is not always easy to separate the two in nineteenth century accounts (*see* Parkin, 1911, p.21).

The egg that the footman broke was collected at an unknown date before 1795. Like numbers 19 and 61, it was owned by the Vicomte de Barde until about 1825 at which time all three were presented to the Boulougne Museum. Although their origin is unknown, the early French connection makes a Newfoundland provenance likely. At a date given variously as 1848, 1849 or (as is most likely) 1852, the three were swapped for an Ostrich skin with the London dealer James Gardiner who subsequently sold them to T.H. Potts. Soon afterwards Potts took the decision to emigrate to New Zealand (where he produced several important papers and books on the natural history of that country) and so, rather quickly, he parted with two of his eggs. Both went to Lord Garvagh and after his death to George Dawson Rowley. They stayed in the possession of the Rowley family until 1934 when they were sold by auction at Stevens' Rooms.

This egg was bought by G.N. Carter of Wolseley Place, Manchester. Being so badly damaged its price was a mere 100 guineas. In 1956, after Carter's death, his widow sold it for £175 to Ragnar Kreuger of Helsinki. Kreuger owned an important collection of birds' eggs acquired from all over the world; some 3,200 species were represented. During the early 1960's he donated this collection to the Zoological Museum of the University of Helsinki but kept back two eggs. One was that of an *Aepyornis* and the other was his Garefowl egg. In 1986 Mr Kreuger finally parted with his two prize possessions and added them to his gift to the University.

To make good the damage caused by the footman (and the original inadequate restoration) the egg has twice subsequently been repaired, once in England in 1935 and then 21 years later by John Grönvall in Finland.

Ragnar Kreuger holding the egg of an *Aepyornis*.

THE ANGERS EGG

Egg no.41 (Grieve no. 2; Tomkinson no. 48)
Origin: unknown
Colour: dirty white with brown and black spots and streaks
Present location: Angers - Museum d'Histoire Naturelle, 43 Rue Jules Guitton, 49100 Angers, France.

The Angers Egg (2/3 natural size) - circa 1900.

Blasius, W. 1884. Zur Geschichte der Ueberreste von *Alca impennis*. *Journal für Ornithologie*, p.152. Grieve, S. 1885. *The Great Auk or Garefowl*, p.88 *and* appendix p.25. Duchaussoy, H. 1897-8. Le Grand Pingouin du Musée d'Histoire naturelle d'Amiens *and* Notes additionnelles. *Mémoires de la Société Linnéenne du Nord de la France*, p.106 *and* p.242. Grieve, S. 1897-8. Additional notes on the Great Auk. *Transactions of the Edinburgh Field Naturalists' and Microscopical Society*, pp.337-8. Tomkinson, P. & J. 1966. *Eggs of the Great Auk*, p.114 *and* pl.48.

Little is known about this badly damaged and crudely repaired egg. It was found in 1859 hanging from a piece of string in the window of a shop in Brest. Others, it seems, were with it. On May 12th 1862 the Curator of the Angers Museum, Monsieur Boreau, bought it and it has remained in Angers ever since.

THE ABBÉ MANESSE'S EGG

Egg no.42 (Grieve no. 50; Tomkinson no. 49)
Origin: unknown
Colour: creamy white blotched sparsely at the larger end with brown
Present location: Paris - Muséum National d'Histoire Naturelle, 55 Rue de Buffon, 75005 Paris, France.

Champley, R. 1861. Additional eggs of the Great Auk. *Zoologist*, p.7386. Des Murs, O. 1863. Notice sur l'œuf de l'*Alca impennis*. *Revue et Magasin de Zoologie*, p.4. Champley, R. 1864. The Great Auk. *Annals and Magazine of Natural History*, p.236. Fatio, V. 1868. Liste des divers représentants de l'*Alca impennis* en Europe. *Bullétin de la Société Ornithologique Suisse*, tome II, pt.1, p.84. Blasius, W. 1884. Zur Geschichte der Ueberreste von *Alca impennis*. *Journal für Ornithologie*, p.163. Grieve, S. 1885. *The Great Auk or Garefowl*, p.89 *and* appendix p.31. Milne Edwards, A. & Oustalet, E. 1893. Notice sur quelques espèces d'oiseaux...éteintes. *Centenaire de la Fondation du Muséum d'Histoire Naturelle*, pp.59-60. Duchaussoy, H. Le Grand Pingouin du Musée d'Histoire naturelle d'Amiens *and* Notes addition-nelles. *Mémoires de la Société Linnéenne du Nord de la France*, p.103 *and* p.242. Didier, R. 1934 (Jan.). Le Grand Pingouin. *La Terre et La Vie*, p.20 *and* fig.6. Berlioz, J. 1935. Notice sur les spécimens...d'oiseaux éteints. *Archives du Muséum d'Histoire Naturelle*, p.487. Tomkinson, P. & J. 1966. *Eggs of the Great Auk*, p.114 *and* pl.49.

Apart from the fact that at some time during the eighteenth century this egg belonged to the Abbé Manesse, nothing is known of it. For more than a century and a half it has been in the Muséum d'Histoire Naturelle, Paris.

THE BLOTCHED VERSAILLES EGG

Egg no.43 (Grieve no. 51; Tomkinson no. 50)
Origin: probably Newfoundland

Colour: whitish with brown and black blotches and spots at the larger end
Present location: Paris - Muséum National d'Histoire Naturelle, 55 Rue de Buffon, 75005 Paris, France.

Blasius, W. 1884. Zur Geschichte der Ueberreste von *Alca impennis*. *Journal für Ornithologie*, pp.163-4. Grieve, S. 1885. *The Great Auk or Garefowl*, p.89 *and* appendix pp.31-2. Milne Edwards, A. & Oustalet, E. 1893. Notice sur Quelques espèces d'oiseaux...éteintes. *Centenaire de la Fondation du Muséum d'Histoire Naturelle*, pp.59-60. Duchaussoy, H. Le Grand Pingouin du Musée d'Histoire naturelle d'Amiens. *Mémoires de la Société Linnéenne du Nord de la France*, pp.103-4. Berlioz, J. 1935. Notice sur les spécimens...d'oiseaux éteints. *Archives du Muséum d'Histoire Naturelle*, p.487. Tomkinson, P. & J. 1966. *Eggs of the Great Auk*, p.114 *and* pl.50.

This particularly long and pointed egg was discovered with another (more typically shaped) during December 1873 in the Lycée of Versailles and both were transferred to the Muséum d'Histoire Naturelle, Paris.

Each was inscribed with the words, 'St Pierre, Miquelon,' from which it may be inferred that they came from Newfoundland.

THE STREAKED VERSAILLES EGG

Egg no.44 (Grieve no. 52; Tomkinson no. 51)
Origin: probably Newfoundland
Colour: yellowish white with fine brown squiggles and spots
Present location: Paris - Muséum National d'Histoire Naturelle, 55 Rue de Buffon, 75005 Paris, France.

All references are identical to those of egg no.43, except for:

Didier, R. 1934 (Jan.). Le Grand Pingouin. *La Terre et la Vie*, p.20 *and* fig.7 *and* Tomkinson, P. & J. 1966. *Eggs of the Great Auk*, pl.51.

This is the second of the two eggs found in the Lycée at Versailles and quickly transferred to the Muséum d'Histoire Naturelle in Paris. Like its companion it was inscribed with the words, 'St Pierre, Miquelon,' indicating - presumably - that it came from Newfoundland.

MECHLENBURG'S EGG

Egg no.45 (Grieve no. 58; Tomkinson no. 19)
Origin: Iceland
Colour: yellowish white with brown and black spots and streaks
Present location: Bonn - Zoologisches Forschungsinstitut und Museum Alexander Koenig, Adenauerallee 160, 53113 Bonn, Germany.

Baedeker, F. 1855-63. *Die Eier der Europaeischen Vögel*, pl.70 (upper fig.). Champley, 1864. The Great Auk. *Annals and Magazine of Natural History*,

Alexander Koenig.

p.236. Fatio, V. 1868. Liste des divers représentants de l'*Alca impennis* en Europe. *Bullétin de la Société Ornithologique Suisse*, tome II, pt.1, p.84. Blasius, W. 1884. Zur Geschichte der Ueberreste von *Alca impennis*. *Journal für Ornithologie*, p.167. Grieve, S. 1885. *The Great Auk or Garefowl*, p.88 *and* appendix p.33. Grieve, S. 1888 *and* 1896-7. Recent information about the Great Auk *and* Supplementary note on the Great Auk. *Transactions of the Edinburgh Field Naturalists' and Microscopical Society*, p.115 (1888) *and* p.272 (1896-7). Koenig, A. 1931-2. *Katalog der Nido-Oologischen Sammlung im Museum Alexander Koenig*, vol.3, pp.811-2, p.899 *and* vol.4, pl.17 (fig.1). Tomkinson, P. & J. 1966. *Eggs of the Great Auk*, p.105 *and* pl.19. Luther, D. 1986. *Die Ausgestorbenen Vögel der Welt*, p.83.

This egg once belonged to Mechlenburg, the Apothecary of Flensburg who claimed it was taken, together with the bird that laid it (*see* bird no.68), at Langenes, Iceland on June 20th 1820. Whether this information can be relied upon is uncertain. Mechlenburg may have invented the data or he may have received it in good faith from the merchants who supplied him with specimens. Perhaps it is correct, perhaps it isn't; it is virtually certain that the egg came from Iceland, however.

Shortly before his death in 1861, Mechlenburg sold this egg, along with the bird said to have laid it, to the Scarborough collector Robert Champley. There was some controversy about the price paid - Champley said £45, Mechlenburg's papers suggest £120.

When Champley died in 1895 he left to his daughter his collection of birds' eggs including nine of the Great Auk. She kept them for several years but finally decided to sell. This presented something of a problem. Nine eggs, coming on to the market together, were quite enough to flood it and some failed to sell at auction. The majority were eventually sold by Rowland Ward. This one went to the German egg collector Alexander Koenig and today is in Bonn at the museum that bears his name.

ALEXANDER KOENIG'S EGG

Egg no.46 (Grieve no. 59; Tomkinson no. 21)
Origin: unknown
Colour: yellowish white marked with black and brown spots and streaks
Present location: Bonn - Zoologisches Forschungsinstitut und Museum Alexander Koenig, Adenauerallee 160, 53113 Bonn, Germany.

Champley, R. 1864. The Great Auk. *Annals and Magazine of Natural History*, p.236. Fatio, V. 1868. Liste des divers représentants de l'*Alca impennis* en Europe. *Bullétin de la Société Ornithologique Suisse*, tome II, pt.1, p.84. Blasius, W. 1884. Zur Geschichte der Ueberreste von *Alca impennis*. *Journal für Ornithologie*, p.167. Grieve, S. 1885. *The Great Auk or Garefowl*, p.88 *and* appendix p.33. Grieve, S. 1888. Recent information about the Great Auk. *Transactions of the Edinburgh Field Naturalists' and Microscopical Society*, p.116. Koenig, A. 1931-2. *Katalog der Nido-Oologischen Sammlung im Museum Alexander Koenig*, vol.3, pp.810-1, p.899 *and* vol.4, pl.17 (fig.2). Tomkinson,

P. & J. 1966. *Eggs of the Great Auk*, p.105 *and* pl.21. Luther, D. 1986. *Die Ausgestorbenen Vögel der Welt*, p.83.

This egg was obtained by the Abbé de la Motte of Abbeville from French whalers around 1820. During 1861 it was purchased, for £23, by the French naturalist Parzudaki on behalf of Robert Champley. Champley kept it, along with his eight others, until his sudden death in 1895 after which his collection was sold off. Due to the sudden glut of available specimens sales didn't prove easy and some of the eggs were marketed - over a period of several years - by the taxidermy firm of Rowland Ward. This egg was sold in March 1904 for £280 to Alexander Koenig and along with two others it is now in the Koenig Museum at Bonn.

Mechlenburg's Egg (left) and Alexander Koenig's Egg (right) - both natural size. Watercolour by Paul Preiss reproduced in A. Koenig's *Katalog der Nido-Oologischen Sammlung im Museum Alexander Koenig* (1931-2). *Courtesy of Peter Blest.*

THE CLUNGUNFORD EGG

Egg no.47 (Grieve no. 18; Tomkinson no. 35)
Origin: unknown
Colour: buff, heavily blotched and spotted with brown
Present location: Bonn - Zoologisches Forschungsinstitut und Museum Alexander Koenig, Adenauerallee 160, 53113 Bonn, Germany.

Newton, A. 1870. On Existing Remains of the Great Auk. *Ibis*, p.259. Blasius, W. 1884. Zur Geschichte der Ueberreste von *Alca impennis. Journal für Ornithologie*, p.154. Grieve, S. 1885. *The Great Auk or Garefowl*, p.87 *and* appendix pp.26-7. Watkins, M. 1890 (May). The collection of birds at Clungunford House. *Transactions of the Woolhope Naturalists' Field Club*, p.32 *and* facing plate. Rothschild, W. 1928. Exhibition of a mounted Great Auk and an Egg. *Bulletin of the British Ornithologists' Club*, p.9. Koenig, A. 1931-2. *Katalog de Nido-Oologischen Sammlung im Museum Alexander Koenig*, vol.3, pp.812-3, p.990 *and* vol.4, pl.18 (fig.1). Tomkinson, P. & J. 1966. *Eggs of the Great Auk*, pp.109-10 *and* pl.35. Luther, D. 1986. *Die Ausgestorbenen Vögel der Welt*, p.83.

This egg was purchased by John Rocke of Clungunford House, Shropshire during 1869 as a companion to the mounted Great Auk that he already owned. He obtained it from a Mr. E. Burgh who assured Rocke that the egg had been in the possession of his family for upwards of 70 years. Nothing else is known of its early history. After Rocke's death his bird and his egg were kept for several decades by his wife and descendants but during 1928 both were acquired by the Rowland Ward Company. During 1929 Messrs. Ward sold the egg to Professor Alexander Koenig for £200 and it has stayed in his museum ever since.

WOOLHOPE CLUB,
FIELD DAY, May 30: 1890.

GREAT AUK'S EGG REAL SIZE
In the collection of J.C. L. ROCKE Esq.
CLUNGUNFORD.

Three paintings of the Clungunford Egg. (Facing page). Watercolour (natural size) by Paul Preiss reproduced in A. Koenig's *Katalog der Nido-Oologischen Sammlung im Museum Alexander Koenig*. (1931-2). (This page, above). Chromolithograph by R. Clarke from *Transactions of the Woolhope Naturalists' Field Club* (1890). (This page, left). Watercolour (natural size) by F.W. Frohawk *(Courtesy of Haslemere Museum)*.

THE DRESDEN EGG

Egg no.48 (Grieve no. 21; Tomkinson no. 59)
Origin: Felseninsel, south-west Iceland
Colour: yellowish white with black and brown spots and streaks
Present location: Dresden - Staatliches Museum für Tierkunde, Augustusstraße 2, 01067 Dresden, Germany.

Thienemann, F. 1845-54. *Einhundert Tafeln coloriter Abbildungen von Vogeleiern,* pl.96 (fig. lower left). Pässler, W. 1860. Die Eier der *Alca impennis* in deutschen Sammlungen. *Journal für Ornithologie,* p.59. Champley, R. 1864. The Great Auk. *Annals and Magazine of Natural History,* p.236. Fatio, V. 1868. Liste des divers représentants de l'*Alca impennis* en Europe. *Bullétin de la Société Ornithologique Suisse,* tome II, pt.1, p.84. Blasius, W. 1884. Zur Geschichte der Ueberreste von *Alca impennis. Journal für Ornithologie,* pp.155-6. Grieve, S. 1885. *The Great Auk or Garefowl,* p.89 and appendix p.27. Blasius, W. (*in* Naumann) 1903. *Naturgeschichte der Vögel Mitteleuropas,* bd.12, pl.17b (fig.4). Rey, E. 1905. *Die Eier der Vögel Mitteleuropas,* pl.119. Jourdain, F. 1906. *The Eggs of European Birds,* pl.122. Tomkinson, P. & J. 1966. *Eggs of the Great Auk,* p.117 and pl.59. Luther, D. 1986. *Die Ausgestorbenen Vögel der Welt,* p.82.

The Dresden Egg (natural size). Photograph by H. Höhler. *Courtesy of Siegfried Eck and the Staatliches Museum für Tierkunde, Dresden.*

Many of the most precious objects belonging to the Dresden Museum escaped the terrible bombing that destroyed the city at the end of World War II. It is said that twelve enormous packing cases were filled with treasures and stored in the historic fortress of Königheim.

Although these cases escaped the bombs, they didn't avoid the attention of the victorious Red Army as it swept across Germany from the east. Russian soldiers carried them off as spoils of war and for many years the whereabouts of many important natural history specimens were unknown- at least as far as the western world was concerned. It was generally believed they'd been destroyed in the chaos that followed the fall of The Third Reich. This was not the case, however.

Many of the greatest treasures were carried back to Leningrad (as it was then called) where they remained for upwards of 30 years. Among the most precious of these relics were a stuffed Great Auk and an egg, both of which had belonged to Dresden since the first half of the nineteenth century.

During 1982 Siegfried Eck, Curator of Birds at the Staatliches Museum, was invited to Leningrad to reclaim for Dresden its ornithological treasures and among the items he took back to Germany were the Auk and the egg.

This egg was once the property of Friedrich Thienemann, Dresden resident and author of the rare and beautifully illustrated egg book, *Einhundert Tafeln coloriter Abbildungen von Vogeleiern* (1845-54). Indeed one of the plates (pl.96) shows this egg along with two others. From Thienemann the egg passed to what was then the Royal Zoological Museum of Dresden and, apart from its Russian sojourn, has remained in the city ever since.

It originated in Iceland and is listed in an old museum register as coming from Felseninsel, south-west Iceland during 1828. Presumably, it is an egg from the Geirfuglasker.

(Overleaf). *Eggs* (approx. 2/3 natural size) *and heads*. (Top left). Walter Rothschild's Egg (no.35), after a painting by F.W. Frohawk. (Top right). The Oldenburg Egg (no.51), after a painting by R. Dieck. (Bottom, left). Löbbecke's Egg (no.49), after a painting by H. Klönne. (Bottom, right). The Dresden Egg (no.48), after a painting by Bruno Geisler. Also shown (clockwise from the top) are the heads of birds no's. 22, 15, 9 and 50. Chromolithograph from J. Naumann's *Naturgeschichte der Vögel Mitteleuropas* (1896-1905).

LÖBBECKE'S EGG

Egg no.49 (Grieve no. 22; Tomkinson no. 61)
Origin: unknown
Colour: brownish yellow with a greenish cast and black, brown and grey spots and blotches, particularly at the larger end
Present location: Düsseldorf - Löbbecke Museum & Aquazoo, Kaiserswerther Strasse 380, D-40002 Düsseldorf, Germany.

Blasius, W. 1884. Zur Geschichte der Ueberreste von *Alca impennis*. *Journal für Ornithologie*, pp.156-7. Grieve, S. 1885. *The Great Auk or Garefowl*, p.89, p.104 *and* appendix pp.27-8. Blasius, W. (*in* Naumann) 1903. *Naturgeschichte der Vögel Mitteleuropas*, bd.12, pl.17b (fig.3). Rey, E. 1905. *Die Eier der Vögel Mitteleuropas*, pl.119. Jourdain, F. 1906. *The Eggs of European Birds*, pl.122. Tomkinson P. & J. 1966. *Eggs of the Great Auk*, p.118 *and* pl.61.

Thomas Löbbecke, until 1873 an apothecary in Duisberg, assembled a large collection of natural history specimens during his lifetime, a collection that forms the basis of today's Löbbecke Museum at Düsseldorf. Among the treasures of the collection is an Auk egg.

Löbbecke inherited this egg from his uncle, Friedrich Löbbecke, a Rotterdam merchant, but its origin is unknown. Nineteenth century

Eier natürliche Grösse,

Köpfe etwa ¼ natürl. Grösse.

Alca impennis L., Riesenalk.

1. Ei im Museum Rothschild zu Tring (früher im Besitz des Grafen Rödern in Breslau); 2. Ei im Grossherzogl. Naturhist. Museum zu Oldenburg; 3. Ei im Museum Löbbeckeanum zu Düsseldorf; 4. Ei im Königl. Zoolog. und Anthropolog. Museum zu Dresden; 5. Kopf des jungen Exemplars im Winterkleide im Museum zu Dublin; 6. Kopf des jungen Exemplars im Winterkleide im Museum zu Prag; 7. Kopf des Exemplars im Übergangskleide im Museum Rothschild zu Tring; 8. Kopf des sehr jungen Exemplars im Übergangskleide im Museum zu Newcastle-upon-Tyne.

researchers were able to trace it back to a pile of rubbish in a Paris street where it was found by the French dealer Perrot. Understandably, they could pursue its history no further. In 1846 Perrot sold it to Friedrich Thienemann, an intimate friend of the older Löbbecke, and at an unknown date the oological writer passed on the egg, the least perfect of two that he then owned.

It was once suggested by Wilhelm Blasius that the rather generalised portrait forming the lower figure of plate 70 in F. Baedeker's *Die Eier der Europaeischen Vögel* (1855-63) is intended as an illustration of this egg. There seems little reason to connect the two and Blasius later changed his mind. An attempt at photographing the egg during the period when the younger Löbbecke owned it resulted in something of a disaster. The photographer's clumsiness caused a breakage but, fortunately, a skilful repair was effected and the damage kept to a minimum.

THE ERLANGEN - NÜRNBERG EGG

Egg no.50
Origin: unknown
Colour: unknown
Present location: Nürnberg - Naturkundehaus, Tiergarten Nürnberg, Am Tiergarten, Nürnberg, Germany.

Luther, D. 1986. *Die Ausgestorbenen Vögel der Welt*, p.83.

This egg is the property of the Naturhistorische Gesellschaft Nürnberg (Gewerbemuseumsplatz 4, D-90403 Nürnberg) but is currently on loan to the Nürnberg Zoo. It was apparently first listed in an inventory compiled during January of 1970 but, other than this, nothing is known of its history or origin.

THE OLDENBURG EGG

Egg no.51 (Grieve no. 47; Tomkinson no. 60)
Origin: unknown
Colour: white with black and brown blotches and streaks
Present location: Oldenburg - Staatliches Museum für Naturkunde und Vorgeschichte, Damm 40-44, D-26135 Oldenburg, Germany.

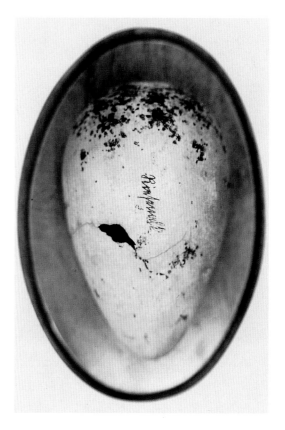

Two views of the Erlangen - Nürnberg egg (2/3 natural size). Photographs by Bernd Weidemann of the Universität Erlangen - Nürnberg. *Reproduced by kind permission of Bernd Weidemann.*

Preyer, W. 1862. Der Brillenalk in Europäischen Sammlungen. *Journal für Ornithologie*, p.78. Sclater, P. 1864. Notes on the Great Auk. *Annals and Magazine of Natural History*, p.320. Fatio, V. 1868. Liste des divers représentants de l'*Alca impennis* en Europe. *Bullétin de la Société Ornithologique Suisse*, tome II, pt.1, p.84. Blasius, W. 1884. Zur Geschichte der Ueberreste von *Alca impennis*. *Journal für Ornithologie*, pp.162-3. Grieve, S. 1885. *The Great Auk or Garefowl*, p.89 and appendix p.31. Blasius, W. (*in* Naumann) 1903. *Naturgeschichte der Vögel Mitteleuropas*, bd.12, pl.17b (fig.2). Rey, E. 1905. *Die Eier der Vögel Mitteleuropas*, pl.118. Jourdain, F. 1906. *The Eggs of European*

Birds, pl.121. Tomkinson, P. & J. 1966. *Eggs of the Great Auk*, p.118 *and* pl.60. Luther, D. 1986. *Die Ausgestorbenen Vögel der Welt*, p.84.

In 1840 the Grand Ducal Museum at Oldenburg bought a collection of 150 eggs from a Dr. Graba of Kiel and the prize of the collection was a Great Auk's egg. Its origin is unknown although, given the geographical location of Kiel, it is likely that it came from Iceland via Hamburg or Copenhagen.

GREVILLE SMYTHE'S REYKJAVIK EGG

Egg no.52 (Tomkinson no. 72 *but see also* no. 74)
Origin: unknown
Colour: white with black and grey spots and streaks
Present location: Reykjavik - Náttúrufraedistofnun Íslands (Icelandic Museum of Natural History), IS-125 Reykjavík, Iceland.

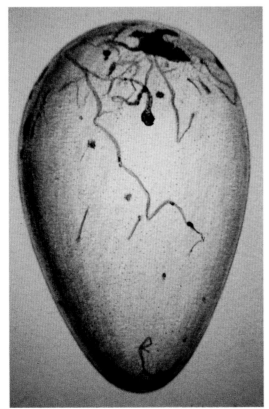

Greville Smythe's Reykjavik Egg (¾ natural size). Watercolour by F.W. Frohawk. *Courtesy of Haslemere Museum.*

Anon. 1912 (Thurs. April 18th). Sale of Great Auks' Eggs. *The Times.* Tomkinson, P. & J. 1966. *Eggs of the Great Auk*, pp.121-2 *and* pl.72. Bárðarson, H. 1986. *Birds of Iceland*, pp.20-1. Petersen, A. 1995. Brot úr sögu Geirfuglsins. *Náttúrufræðingurinn*, pp.61-2, p.66 *and* pl.6.

Sir J.H. Greville Smythe was a serial Great Auk egg collector who managed to acquire four examples towards the end of his life. Three of his eggs had a certain amount of history attached to them but the fourth is something of a mystery. All that is known is that Sir Greville got it among a collection of seabird eggs. He died during 1901 but his widow kept this egg until April 17th 1912 when it was sold - together with another - at Stevens' Rooms. It realised 140 guineas, being sold to Rowland Ward who was acting on behalf of an even more spectacular egg collector than Sir Greville: John Thayer of Lancaster, Massachusetts. Thayer managed to acquire no less than ten eggs and he gave all of these to the Museum of Comparative Zoology at the end of 1931. With eleven eggs in the collection (William Barbour had presented the extra example), the museum authorities felt able to make one available to the National Museum of Iceland and, during 1954, this egg re-crossed the Atlantic. Whether or not it was originally an Icelandic egg can only be guessed at but the probability is fairly strong.

In *Eggs of the Great Auk* (Tomkinson & Tomkinson, 1966), the authors confuse the identity of this egg with that of Dr. Bolton's Egg (egg no.2).

THE AMSTERDAM EGG

Egg no.53 (Grieve no. 1; Tomkinson no. 62)
Origin: unknown
Colour: cream with spots, blotches and streaks of brown, black and grey
Present location: Amsterdam - Universiteit van Amsterdam, Zoölogisch Museum, Mauritskade 61-57, 1090 GT Amsterdam, The Netherlands.

Thienemann, F. 1833. *Systematische Darstellung der Fortpflanzung Vögel Europás*, pt.5, pp.57-8. Champley, R. 1864. The Great Auk. *Annals and Magazine of Natural History*, p.236. Fatio, V. 1868. Liste des divers représentants de l'*Alca impennis* en Europe. *Bullétin de la Société Ornithologique Suisse*, tome II, pt.1, p.84. Blasius, W. 1884. Zur Geschichte der Ueberreste von *Alca impennis*. *Journal für Ornithologie*, pp.151-2. Grieve, S. 1885. *The Great Auk or Garefowl*, p.89 *and* appendix p.24. Tomkinson, P. & J. 1966. *Eggs of the Great Auk*, p.118 *and* pl.62.

According to a letter written to Robert Champley in February 1860 by H. Schlegel, then Director of the Leiden Museum, this egg, like one still at Leiden, was brought to Europe by a French whaling vessel at the start of the nineteenth century. Both eggs may have passed through the hands of the older Frank (the Leipzig dealer) although this is by no means certain. By 1807 they belonged to Coenraad Temminck and by the early 1840's were part of the collection of Leiden Museum. During 1859 the pair were separated when Leiden presented this egg to the Royal Zoological Society of Amsterdam; it has subsequently been incorporated into the collection of the University. Both eggs are said to be of Newfoundland origin but there is no real evidence for this.

THE LEIDEN EGG

Egg no.54 (Grieve no. 28; Tomkinson no. 63)
Origin: unknown
Colour: yellowish white with streaks and blotches of black, brown and grey
Present location: Leiden - Nationaal Natuurhistorisch Museum, 2300 RA Leiden, The Netherlands.

Thienemann, F. 1833. *Systematische Darstellung de Fortpflanzung Vögel Europás*, pt.5, pp.57-8. Thienemann, F. 1845-54. *Einhundert Tafeln coloriter Abbildungen von Vogeleiern*, pl.96 (upper fig.). Champley, R. 1864. The Great Auk. *Annals and Magazine of Natural History*, p.236. Fatio, V. 1868. Liste des divers représentants de l'*Alca impennis* en Europe. *Bullétin de la Société Ornithologique Suisse*, tome II, pt.1, p.84. Blasius, W. 1884. Zur Geschichte der Ueberreste von *Alca impennis*. *Journal für Ornithologie*, p.158. Grieve, S. 1885. *The Great Auk or Garefowl*, p.39 *and* appendix p.29. Grieve, S. 1888. Recent information about the Great Auk. *Transactions of the Edinburgh Field Naturalists' and Microscopical Society*, p.117. Tomkinson, P. & J. 1966. *Eggs of the Great Auk*, pp.118-9 *and* pl.63.

The illustrations in F.A.L. Thienemann's celebrated egg book *Einhundert Tafeln coloriter Abbildungen von Vogeleiern* (1845-54) are among the most beautiful of all egg pictures. Three Great Auk eggs are shown on plate 96 but only the second and third figures have been previously identified, the identity of the model for the upper figure remaining something of a mystery to Auk specialists. It is certainly a portrait of the egg at Leiden Museum, however. Despite some misleading inaccuracies in the drawing, René Dekker, Curator of Birds at the museum, has been able to correlate many of the markings on Leiden's egg with details shown in the picture.

Tab. IV C.

Alca impennis

This egg has been at Leiden for more than a century and a half and its known early history is identical to that of the example now in Amsterdam. An assumed Newfoundland origin, although probably correct, is not necessarily reliable.

DR. DEPIERRE'S LAUSANNE EGG

Egg no.55 (Grieve no. 27; Tomkinson no. 65)
Origin: unknown
Colour: yellowish, spotted and streaked with black and brown
Present location: Lausanne - Musée Zoologique, Place Riponne 6, CH-1000 Lausanne 17, Switzerland.

Fatio, V. 1868. Quelques mots sur les exemplaires de l'*Alca impennis*, Oiseaux et Oeufs qui se trouvent en Suisse *and* Liste des divers représentants de l'*Alca impennis* en Europe. *Bullétin de la Société Ornithologique Suisse*, tome II, pt.1, pp.75-9 *and* p.85. Blasius, W. 1884. Zur Geschichte der Ueberreste von *Alca impennis*. *Journal für Ornithologie*, pp.157-8 *and* p.161. Grieve, S. 1885. *The Great Auk or Garefowl*, p.89 *and* appendix pp.28-30. Grieve, S. 1888. Recent information about the Great Auk. *Transactions of the Edinburgh Field Naturalists' and Microscopical Society*, p.113. Newton, A. 1905. *Ootheca Wolleyana*, p.376 *and* p.378. Tomkinson, P. & J. 1966. *Eggs of the Great Auk*, pp.101-2, p.119 *and* pl.65.

This egg is one of two (*see* egg no.9) discovered by Dr. Depierre during the 1860's in the Lausanne Museum. How they came to be there is uncertain but they are supposed to have lain unrecognised at the museum for upwards of twenty years and to have formed part of a collection of zoological specimens obtained from a Professor D.A. Chavannes, Before that they may, or may not, have been the property of François Levaillant, author of many beautifully illustrated ornithological books during the early nineteenth century. This suggestion, made by several Garefowl historians, is really no more than a rumour.

One of these eggs was included in an exchange with G.A. Frank during the 1880's but this example has stayed in Lausanne.

ROYAL COLLEGE OF SURGEONS EGG NO.7 (CHAMPLEY'S EGG)

Egg no.56 (Grieve no. 63; Tomkinson no. 24)
Origin: unknown
Colour: whitish with brown steaks and blotches at the larger end
Present location: Cambridge - Museum of Comparative Zoology, Harvard University, Cambridge, Massachusetts 02138, USA.

Champley, R. 1864. The Great Auk. *Annals and Magazine of Natural History*, p.236. Newton, A. 1865. The Gare-fowl and its Historians. *Natural History Review*, pp.483-4. Fatio, V. 1868. Liste des divers représen-

(Facing page). Three eggs (natural size). (Top). The Leiden Egg (no.54). (Bottom, left). The Dresden Egg (no.48). (Bottom, right). Barclay's Egg (no.28). Hand-coloured lithograph by an unknown artist from F. Thienemann's *Einhundert Tafeln Colorirter Abbildungen von Vögeleiern* (1845-56). *Courtesy of Peter Blest.*

(Left). Ward's Egg (no. 59). (Right). Champley's Egg (no.56). Both natural size. Photograph from H.E. Dresser's *Eggs of the Birds of Europe* (1910).

tants de l'*Alca impennis* en Europe. *Bullétin de la Société Ornithologique Suisse*, tome II. pt.1, p.84. Blasius, W. 1884. Zur Geschichte der Ueberreste von *Alca impennis*. *Journal für Ornithologie*, p.167. Grieve, S. 1885. *The Great Auk or Garefowl*, p.88, pp.105-6 *and* appendix p.33. Grieve, S. 1888. Recent information about the Great Auk. *Transactions of the Edinburgh Field Naturalists' and Microscopical Society*, p.116. Thayer, J. 1905. The purchase of a Great Auk for the Thayer Museum. *Auk*, pp.300-2 *and* pl.14 (upper fig.). Dresser, H. 1910. *Eggs of the Birds of Europe*, vol.1, p.790 *and* vol.2, pl.102 (fig.2). Tomkinson, P. & J. 1966. *Eggs of the Great Auk*, p.99, p.106 *and* pl.24. Bourne, W. 1993. The story of the Great Auk. *Archives of Natural History*, p.273.

Bought in 1905 by Colonel J.E. Thayer of Lancaster, Mass from Rowland Ward for £315, this egg is one of the ten discovered by Alfred Newton at the Royal College of Surgeons during December of 1861. Considered surplus to the Royal College's requirements three of these were disposed of, in 1864, to the Scarborough collector Robert Champley who greedily described their acquisition (*in* Grieve, 1888):

THE EGGS

I had difficulty in getting them, as at the time they would not take money. I got over the difficulty by purchasing a collection of anatomical specimens for £45, which the museum was anxious to possess, and then exchanged it for the ... eggs, all very fine specimens.

After Champley's death most of his eggs were sent for sale to Rowland Ward and it took the famous taxidermist several years to find buyers for all that he received. J.E. Thayer proved a very good customer and purchased three of the Champley eggs and also several others from Ward. In fact Thayer eventually surpassed Champley's score of nine eggs by acquiring a grand total of ten.

Robert Champley (1830-95) with four of his nine eggs.

John Eliot Thayer (1862 -1933).

In Thayer's own account of the acquisition of three of his examples (*Auk*, 1905, pp.300-2 *and* pl.14) his description of this egg does not match the plate, such confusion being, perhaps, the inevitable consequence of greedy egg buying!

Shortly before his death Thayer presented his ten eggs to the Museum of Comparative Zoology at Harvard and these were added to an example that the museum already had. Perhaps believing a collection of eleven eggs excessive, the authorities at Harvard sold one of Thayer's specimens to the National Museum of Iceland in 1954. The ten remaining eggs (nine of them, including this example, once Thayer's) are still at Cambridge.

ROYAL COLLEGE OF SURGEONS EGG NO.8 (THAYER'S EGG)

Egg no.57 (Grieve no. 64; Tomkinson no. 25)
Origin: unknown
Colour: yellowish, streaked and spotted with brown and black
Present location: Cambridge - Museum of Comparative Zoology, Harvard University, Cambridge, Massachusetts 02138, USA.

All references are identical to those of egg no.56, except for:

Thayer, J. 1905. The purchase of a Great Auk for the Thayer Museum. *Auk*, pp.300-2 *but* no plate. Dresser, H. 1910. *Eggs of the Birds of Europe* (no reference *and* no plate). Tomkinson, P. & J. 1966. *Eggs of the Great Auk*, pl.25.

The known history of this egg is identical to that of the previous example

ROYAL COLLEGE OF SURGEONS EGG NO.9 (WILLIAM BARBOUR'S EGG)

Egg no.58 (Grieve no. 65; Tomkinson no. 26)
Origin: unknown
Colour: yellow, streaked all over with brown, grey and black
Present location: Cambridge - Museum of Comparative Zoology, Harvard University, Cambridge, Massachusetts 02138, USA.

All references are identical to those of egg no.57, except for:

Thayer, J. 1905. The purchase of a Great Auk for the Thayer Museum. *Auk* (no reference). *Report of the Museum of Comparative Zoology, 1906*, pl.2. Tomkinson, P. & J. 1966. *Eggs of the Great Auk*, pl.26.

In 1905, with part of a donation of $5,000 given by William Barbour (the rest was, apparently, spent on storage units) the Museum of Comparative Zoology bought this egg from Rowland Ward of London. Apart from this small episode in its tale, the known history of this egg is identical to the histories of eggs 56 and 57.

William Barbour was the father of Thomas Barbour, ornithological writer and one-time Director of the Museum at Harvard. Nearly twenty years after supplying funds for the egg, he bought a mounted Auk for his son's museum. Once again, Messrs Rowland Ward were the vendors.

MONSIEUR FAIRMAIRE'S EGG NO.2 (WARD'S EGG)

Egg no.59 (Grieve no. 62; Tomkinson no. 23)
Origin: unknown
Colour: whitish, spotted with black and brown particularly at the larger end
Present location: Cambridge - Museum of Comparative Zoology, Harvard University, Cambridge, Massachusetts 02138, USA.

Champley, R. 1864. The Great Auk. *Annals and Magazine of Natural History*, p.236. Fatio, V. 1868. Liste des divers représentants de l'*Alca impennis* en Europe. *Bullétin de la Société Ornithologique Suisse*, tome II, pt.1, p.84. Blasius, W, 1884. Zur Geschichte der Ueberreste von *Alca impennis*. *Journal für Ornithologie*, p.167. Grieve, S. 1885. *The Great Auk or Garefowl*, p.88 *and* appendix p.33. Grieve, S. 1888. Recent information about the Great Auk. *Transactions of the Edinburgh Field Naturalists' and Microscopical Society*, p.116. Thayer, J. 1905. The purchase of a Great Auk for the Thayer Museum. *Auk*, pp.300-2 *and* pl.14 (lower fig.). Dresser, H. 1910. *Eggs of the Birds of Europe*, vol.1, p.790 *and* vol.2, pl.102 (fig.1). Tomkinson, P. & J. 1966. *Eggs of the Great Auk*, p.99, p.106 *and* pl.23.

This is the second of the two eggs of unknown origin (*see* egg no.23) that Henry Ward - father of the rather more famous Rowland - got from the French natural history dealer Fairmaire during 1864 and sold to Robert Champley. This one cost Champley £30 but 41 years later realised £200 when sold to J.E. Thayer of Lancaster, Mass.

This second transaction - just like the first - was associated with the Ward family, the sale being handled by Rowland, Henry being long since dead. Before Thayer's own death he presented this egg, along with much else, to the Museum of Comparative Zoology.

Rowland Ward (1848-1912).

TUKE'S EGG

Egg no.60 (Grieve no. 25; Parkin no. XVI; Tomkinson no. 44)
Origin: Eldey
Colour: yellowish white, sparsely spotted with black and brown
Present location: Cambridge - Museum of Comparative Zoology, Harvard University, Cambridge, Massachusetts 02138, USA.

Hewitson, W. 1842-6 (2nd ed) *and* 1853-6 (3rd ed.). *Coloured Illustrations of the Eggs of British Birds*, p.413 (2nd ed.) *and* p.470 (3rd ed.). Roberts, A. 1861.

Tuke's Egg (natural size) - circa 1910.

Skins and Eggs of the Great Auk. *Zoologist*, p.7353. Champley, R. 1864. The Great Auk. *Annals and Magazine of Natural History*, p.236. Fatio, V. 1868. Liste des divers représentants de l'*Alca impennis* en Europe. *Bullétin de la Société Ornithologique Suisse*, tome II, pt.1, p.85. Blasius, W. 1884. Zur Geschichte der Ueberreste von *Alca impennis*. *Journal für Ornithologie*, p.157. Grieve, S. 1885. *The Great Auk or Garefowl*, p.87, p.103 *and* appendix p.28. Grieve, S. 1888 *and* 1896-7. Recent information about the Great Auk *and* Supplementary note on the Great Auk. *Transactions of the Edinburgh Field Naturalists' and Microscopical Society*, p.114 (1888) *and* p.260 (1896-7). Newton, A. 1905. *Ootheca Wolleyana*, p.376. Parkin, T. 1911. The Great Auk. *Hastings and East Sussex Naturalist*, vol.1, pt.6 (extra paper) pp.19-20. Thayer, J. 1912. Great Auk eggs in the Thayer Museum. *Auk*, p.209 *and* pl.12. Tomkinson, P. & J. 1966. *Eggs of the Great Auk*, p.112 *and* pl.44.

Tuke's Egg is an especially fascinating one because it can be matched with another (egg no.6). According to Alfred Newton (1905) there need be no doubt that both of these were laid by the same bird. The Cambridge professor noticed semispiral depressions at the pointed end of each and these he considered to be conclusive evidence of the relationship. Newton was, of course, a man of very great experience when it came to eggs.

What is known of the history of both specimens tallies neatly with this idea and carries the implication that the poor bird - one of the very last of her kind - lost her egg on two successive years. It seems likely that Newton's Egg (no.6) was taken on Eldey during 1841. As far as Tuke's is concerned, Newton (1905) wrote:

> Mr. Tuke bought his egg, in May 1841, of Reid of Doncaster, who had it from Friedrich Schulz, then of Dresden, to whom it had been sent by Brandt of Hamburg - who had it, of course, of Siemsen, of Reykjavik. It seems pretty safe to suppose it was taken on Eldey in 1840.

Mr. Tuke, a banker from Hitchen, Herts., owned the egg for more than 50 years but after his death his executors sold it. In April 1896 a Mr. Heatley Noble bought it at Stevens' Rooms on behalf of Mr. William Newall. He paid £168. By 1912 it was in the hands of the Rowland Ward company and the Piccadilly taxidermists sold it to J.E. Thayer of Lancaster, Mass. The egg eventually became part of the gift that Thayer gave to Harvard.

VICOMTE DE BARDE EGG NO.3 (POTTS'S EGG)

Egg no.61 (Grieve no. 46, Parkin no. XVII; Tomkinson no. 46)
Origin: unknown
Colour: yellowish white with blotches and streaks of grey, black and brown
Present location: Cambridge - Museum of Comparative Zoology, Harvard University, Cambridge, Massachusetts 02138, USA.

Potts, T. 1870. Notes on an Egg of *Alca impennis* in the Collection of the writer. *Transactions and Proceedings of the New Zealand Institute*, p.109. Blasius,

W. 1884. Zur Geschichte der Ueberreste von *Alca impennis*. *Journal für Ornithologie*, p.162. Grieve, S. 1885. *The Great Auk or Garefowl*, p.89 and appendix p.31. Anon. 1891 (April 24th). [Monthly summary] *New Zealand Herald*. Grieve, S. 1896-7. Supplementary note on the Great Auk. *Transactions of the Edinburgh Field Naturalists' and Microscopical Society*, pp.261-2. Butler, A. 1896-8. *British Birds with their Nests and Eggs*, vol.6, pl.23 (fig.464). Parkin, T. 1911. The Great Auk. *Hastings and East Sussex Naturalist*, vol.1, pt.6 (extra paper), p.6 *and* p.20. Thayer, J. 1912. Great Auk eggs in the Thayer Museum. *Auk*, p.209. Tomkinson, P. & J. 1966. *Eggs of the Great Auk*, p.113 *and* pl.46.

This seems to be the only Great Auk egg ever taken to the southern hemisphere. For a period of almost 40 years during the second half of the nineteenth century it was in New Zealand, the property of T.H. Potts

(Left). The Nunappleton Egg (no.62). (Right). Potts's Egg (no.61). Both natural size. Chromolithograph after a painting by F.W. Frohawk from A.G. Butler's *British Birds with their Nests and Eggs* (1896-8).

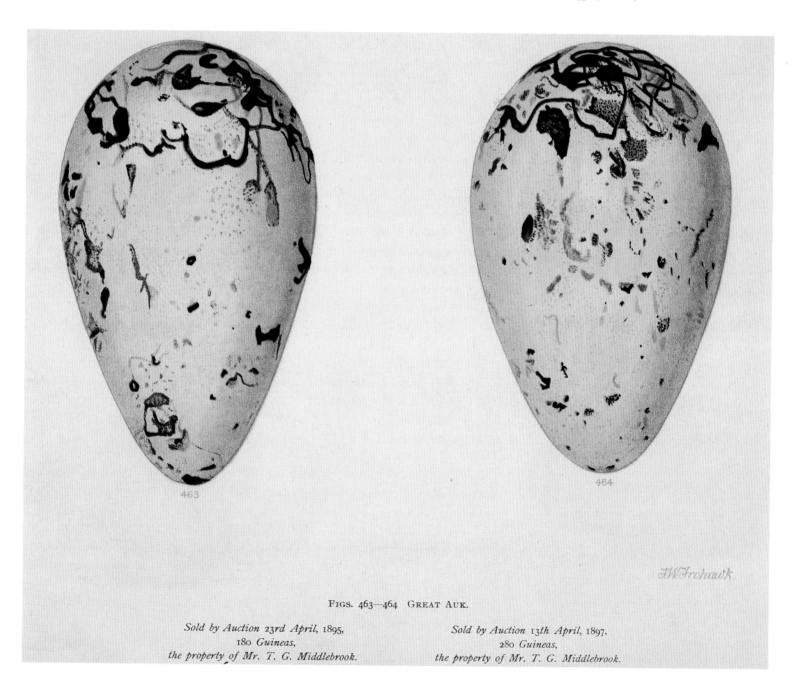

FIGS. 463—464 GREAT AUK.

Sold by Auction 23rd April, 1895,
180 *Guineas*,
the property of Mr. T. G. Middlebrook.

Sold by Auction 13th April, 1897,
280 *Guineas*,
the property of Mr. T. G. Middlebrook.

one of the pioneers of ornithological research in that outpost of the British Empire.

The egg was one of three owned by the Vicomte de Barde (*see also* eggs 19 *and* 40) that went, along with the rest of the Vicomte's collection, to the Boulougne Museum in 1825. De Barde is said to have had them for more than 30 years but nothing is known of their actual origin. In 1852 the Boulougne Museum exchanged the three for an Ostrich skin. The recipient was the London taxidermist James Gardiner and he sold them, almost immediately, to Potts. Within a few months of making his purchase Potts took the decision to emigrate to New Zealand and he sold two of his eggs. This one he chose to retain and he took it with him.

Once established in his new home at Ohinitahi on the South Island, Potts quickly formed a reputation as a naturalist. He made several important contributions to ornithological literature in the colony, most notably, perhaps, in his book, *Out in the Open: A Budget of Scraps of Natural History gathered in New Zealand* (1882). During 1888 he died- suddenly- and his egg passed to his wife. At about this time Symington Grieve- on a world tour- visited Mrs. Potts at her home in Christchurch and advised her on how to sell the egg but it was not until 1891 that she actually did so. Henry Forbes, Director of the Canterbury Museum, bought it, apparently for a friend in England.

This egg is next recorded in the possession of Leopold Field a gentleman who much interested himself in Great Auks and their eggs during the last decade or so of the nineteenth century. By 1897 he'd sold it to Rowland Ward who sent it to Stevens' Rooms where it was offered up at two thirty in the afternoon on the 13th of April. At around this period Mr. T.G. Middlebrook, proprietor of *The Edinburgh Castle*- a public house in Camden Town, London- was avidly acquiring Great Auk eggs for display at his licensed premises; he bought this one for £294.

Rowland Ward got the egg back again in 1906 and sold it to J.E. Thayer who donated it to Harvard at the end of 1931.

THE NUNAPPLETON EGG

Egg no.62 (Grieve no. 45; Parkin no. XIV; Tomkinson no. 31)
Origin: unknown
Colour: yellowish white with black, brown and grey blotches and streaks
Present location: Cambridge - Museum of Comparative Zoology, Harvard University, Cambridge, Massachusetts 02138, USA.

Roberts, A. 1861. Skins and Eggs of the Great Auk. *Zoologist*, p.7353. Champley, R. 1864. The Great Auk. *Annals and Magazine of Natural History*, p.236. Fatio, V. 1868. Liste des divers représentants de l'*Alca impennis* en Europe. *Bullétin de la Société Ornithologique Suisse*, tome II, pt.1, p.84. Blasius, W. 1884. Zur Geschichte der Ueberreste von *Alca impennis*. *Journal für Ornithologie*, p.162. Grieve, S. 1885. *The Great Auk or Garefowl*, p.88, p.104 *and* appendix p.31. Bidwell, E. 1894-5. [Minutes of meeting] *Bulletin of the British Ornithologists' Club*, p.32. Grieve, S. 1896-7. Supplementary note on the

THE EGGS

Great Auk. *Transactions of the Edinburgh Field Naturalists' and Microscopical Society*, p.261. Butler, A. 1896-8. *British Birds with their Nests and Eggs*, vol.6, pl.23 (fig.463). Middlebrook, T. (c.1900). *The Edinburgh Castle Free Museum-Illustrated Catalogue*, fig. on p.2. Newton, A. 1905. *Ootheca Wolleyana*, p.383. Parkin, T. 1911. The Great Auk. *Hastings and East Sussex Naturalist*, vol.1, pt.6 (extra paper), p.16 *and* p.18. Thayer, J. 1912. Great Auk eggs in the Thayer Museum. *Auk*, pp.208-9. Tomkinson, P. & J. 1966. *Eggs of the Great Auk*, p.108 *and* pl.31.

> While ... staying at Düsseldorf, in November 1847, I heard ... there was a Great Auk's egg to be had at Perrot's, an out-of-the-way shop down by the Seine in Paris. As I was returning to England I stopped in Paris ... and purchased the egg ... for 200 francs (about £8, 3s. 4d).
>
> Sir William Milner

Sir William Mordaunt Milner (1820-67).

It seems remarkable that the rumour of a Great Auk's egg in a curiosity shop in Paris could reach all the way to Düsseldorf during the far-off days of the 1840's when the speediest line of communication was by horse drawn carriage. Perhaps it is equally remarkable that Sir William Milner was sufficiently fired with enthusiasm to rush off and get it before returning to his home at Nunappleton, Tadcaster, Yorkshire. Here, the egg was soon joined by a stuffed bird (now in Edinburgh) and it remained in the possession of the Milner family until 1895 when Sir William's son, Frederick, decided to part with it. Like so many other eggs it was offered for sale at Stevens' where it was bought for £189 by T.G. Middlebrook, proprietor of *The Edinburgh Castle*. At this London public house it was displayed for several years but when Middlebrook died it was sent to the other celebrated disposer of Great Auk remains - Rowland Ward. In January 1906 Ward sold it to J.E. Thayer and it eventually passed to the Museum of Comparative Zoology.

In the late nineteenth century this egg was regarded as being especially remarkable due to the pitted nature of its surface.

THE GREEN-BLOTCHED ST. MALO EGG

Egg no.63 (Grieve no. 3,4 *or* 5; Parkin no. XV; Tomkinson no. 54)
Origin: unknown, but probably Iceland
Colour: cream, blotched and streaked with pale green
Present location: Cambridge - Museum of Comparative Zoology, Harvard University, Cambridge, Massachusetts 02138, USA.

Newton, A. 1870. On Existing Remains of the Gare-fowl. *Ibis*, p.261. Blasius, W. 1884. Zur Geschichte der Ueberreste von *Alca impennis*. *Journal für Ornithologie*, p.152. Grieve, S. 1885. *The Great Auk or Garefowl*, p.88, p.104 *and* appendix p.25. Grieve, S. 1888. Recent information about the Great Auk. *Transactions of the Edinburgh Field Naturalists' and Microscopical Society*, pp.116-7. d'Hamonville, L. 1888. Note sur les quatre Oeufs d'*Alca impennis* appartenant a notre collection oologique. *Mémoires de la Société Zoologique de France*, pp.226-7 *and* pl.6 (fig.C).

d'Hamonville, L. 1891. Addition à une note sur les quatres œufs du *Pingouin brachyptère*. *Bullétin de la Société Zoologique de la France*, p.34. Harting, J. (ed.) 1895. Sale of Great Auk's Egg. *Zoologist*, p.269. Parkin, T. 1911. The Great Auk. *Hastings and East Sussex Naturalist*, vol.1, pt.6 (extra paper), pp.12-4, pp.18-9 *and* pp.21-3. Thayer, J. 1912. Great Auk eggs in the Thayer Museum. *Auk*, p.209. Tomkinson, P. & J. 1966. *Eggs of the Great Auk*, p.116 *and* pl.54.

This egg, quite unlike any other, is marked with pale green blotches. Although there is no real proof for a story that it came from Iceland and was obtained during the early 1830's by a St. Malo ship owner, the tale is probably true. The gentleman from St. Malo is said to have taken the egg to France together with another (egg no.64) and to have bequeathed it to the Comte Raoul de Baracé of Angers. After the Comte's death his collection passed to Baron Louis d'Hamonville but in 1895 the Baron decided to part with this egg and it was sold at Stevens' for 165 guineas on 25th June. For a short while it was the property of Jay and Co., fur merchants of the Mourning Warehouse, Regent St., London but by June of 1897 it was back in the auction rooms, although this time it fetched five guineas less than on its previous outing. It was purchased by T.G. Middlebrook who was then determinedly acquiring eggs to show at his public house in Camden Town. Eight years later it was again offered for sale, this time by Rowland Ward. Ward managed to extract £200 from J.E. Thayer and it stayed in his private museum at Lancaster, Massechusetts until the early 1930's when he gave it to Harvard University.

The Green-blotched St. Malo Egg (natural size). Chromolithograph from *Mémoires de la Société Zoologique de France* (1888).

THE EDINBURGH CASTLE ST. MALO EGG

Egg no.64 (Grieve no. 3,4 *or* 5; Parkin no. XVIII, Tomkinson no. 55)
Origin: unknown, but probably Iceland
Colour: brownish yellow with blackish blotches
Present location: Cambridge - Museum of Comparative Zoology, Harvard University, Cambridge, Massachusetts 02138, USA.

All references are identical to those of egg no.63, except for:

d'Hamonville, L. 1888. Note sur les quatre Oeufs de l'*Alca impennis* appartenant a notre collection oologique. *Mémoires de la Société Zoologique de France*, pl.6 (fig.B). Parkin, T. 1911. The Great Auk. *Hastings and East Sussex Naturalist*, vol.1, pt.6 (extra paper), pp.12-4, pp.18-9, pp.21-3 *and* pp.33-4. Tomkinson, P. & J. 1966. *Eggs of the Great Auk*, pl.55.

The known history of this egg is virtually identical to that of the preceeding one despite the fact that they have twice been separated. Like its companion it was supposedly obtained in Iceland by a St. Malo ship owner and - via the collection of the Comte de Baracé - came into the possession of the Baron d'Hamonville. The French Baron kept this shell for rather longer than he kept its green fellow but four years after he'd sold the green egg he decided to part with this one and it was auctioned on July 19th 1899 at Stevens' Rooms. Here it was bought by T.G. Middlebrook for the spectacular price of £315 and became the fourth egg on display at his public house in Camden Town, *The Edinburgh Castle*. Middlebrook had previously bought the first St. Malo egg and so the two - temporarily parted - were now - temporarily - reunited. When, a few years later, Middlebrook died, his eggs were dispersed.

Curiously, this egg was not put up for sale at Stevens' but was offered instead at another Covent Garden auction house, Messrs. Debenham, Storr and Sons. The sale, on January 30th 1908, was something of a flop, the egg being bought by Rowland Ward for just £110, barely a third of its price just a few years previously. Ward was acting on behalf of J.E. Thayer and the egg soon crossed the Atlantic to join the rest of the Thayer collection. Here, for the second time, it was reunited with its original companion, which Thayer had already bought. Together ever since, they are now in the Museum of Comparative Zoology.

(Below, left). T.G. Middlebrook, proprietor of *The Edinburgh Castle*, inspecting The Edinburgh Castle St. Malo Egg. *Courtesy of Robert Ralph.*

(Below, right). The Edinburgh Castle St. Malo Egg (natural size). Chromolithograph from *Mémoires de la Société Zoologique de France* (1888).

BIDWELL'S EGG

Egg no.65 (Parkin no. XIII; Tomkinson no. 70)
Origin: unknown
Colour: brownish yellow, streaked and blotched with black, brown and grey
Present location: Cambridge - Museum of Comparative Zoology, Harvard University, Cambridge, Massachusetts 02138, USA.

Bidwell, E. 1894. [Minutes of meeting] *Ibis*, pp.422-3. Anon. 1894 (April 25th). A Good Investment. *Daily Graphic.* Grieve, S. 1896-7. Supplementary note on the Great Auk. *Transactions of the Edinburgh Field Naturalists' and Microscopical Society*, pp.241-2. Parkin, T. 1911. The Great Auk. *Hastings and East Sussex Naturalist*, vol.1, pt.6 (extra paper), pp.15-6, p.23 *and* p.33. Tomkinson, P. & J. 1966. *Eggs of the Great Auk*, p.121 *and* pl.70.

> I know very little of the Auks' eggs, sold at Little Hermitage. I am glad you did not write to my father as I hope he does not know of their sale as he is in poor health and it would only worry him... he never knew they were there or he would have told me ... and I have not asked him about them. He knew very little about what was in the house and took ... little interest in the curiosities it contained ... The only explanation I can offer why no-one knew of their existence is that ... my grandfather died when my father was 10 years old so that there was rather a break in the family tradition. If you mention these eggs ... I should prefer you to suppress our name.
>
> Cecil James Gladdish Hulkes

Cecil Hulkes included these remarks in a letter he wrote to egg fanatic Edward Bidwell on June 2nd 1894. Hulkes had recently sold off - by auction - the contents of his elderly father's house and overlooked the fact that a box of curiosities contained two Great Auk eggs. These passed through the sale virtually unnoticed and sold for a song (36 shillings) to the only person who suspected their true nature, a young man by the name of Wallace Hewett.

The discovery of hitherto unrecognised Great Auk eggs always caused a commotion in Victorian Britain and it quickly became common knowledge that poor Mr. Hulkes was the uninformed vendor and that he'd made the most enormous *faux pas*. The extent of his mistake became quickly apparent when, a month after making his lucky find, Wallace Hewett put his two little purchases into another auction sale, this one at a venue rather more suitable for them. The auction house he chose was, of course, Stevens' of Covent Garden where his first egg fetched 260 guineas and the second 175.

It is this second egg that is now in the Museum of Comparative Zoology; the other (egg no.70) has disappeared, its present whereabouts being unknown. Before reaching Cambridge, Mass. the 175 guinea egg passed through several hands. A certain Henry Munt bought it at Stevens' and promptly sold it to Edward Bidwell. Six years later, on June 20th 1900, it re-appeared at the Covent Garden saleroom and reached the slightly higher figure of 180 guineas. The purchaser was the well known dealer

---

The Great Auk, though long extinct, is beginning to bestir himself. Hearing that Mr. Stevens recently sold one of his—or perhaps we should say her—eggs for 300 guineas, he is beginning to put fresh specimens upon the market. Yesterday two more eggs were offered for sale at Stevens' Auction Rooms. But the fact is the Auk is rather overdoing it, for there is no doubt in the minds of collectors that this is not a good time for eggs. Eggs, they say in fact, will improve by keeping; and though this may not seem quite reasonable to those of us who are only familiar with the so many a shilling variety to be found on our breakfast tables, it is certain that yesterday's auction proved it to be true so far as the Great Auk is concerned.

Mr. Stevens rattled through the preliminary lots of the eggs of the grey shrike, the parsimonious cuckoo, the rose-coloured pastor, and such like, and at last came to the lot of the day—the first of two eggs of the Great Auk. The first of these, Mr. Stevens said, was a truly magnificent egg. "And when," added Mr. Stevens, "I say a magnificent egg, I mean that it is really (with conviction) a fine egg." Having thus defined his vocabulary, Mr. Stevens went on to give a little sketch of the career of the egg. Two months ago a young man who was interested in fossils and things of that kind went to a sale in the South of England where some boxes of specimens were to be sold. In turning these over he came upon what he thought might be a model of the Great Auk's egg. He put it back and (prudently) said nothing about it. On the day of the sale the boxes of "shells and fossils" were put up at 2s., and ran up to 8s. Just, however, as the prudent young man was about to become their possessor at this figure a lady appeared in the bidding, and with the obstinacy of her sex ran the lot up to 36s. At 36s., however, prudence (and perspicacity) prevailed, and the young man became the possessor of the hidden Auk's egg. In rummaging over the fossils he found another egg. "And here," said Mr. Stevens, amid deep silence," comes the most peculiar part of the story. He tied his eggs up in a silk handkerchief. And, gentlemen, he put them behind him on the saddle of his bicycle and rode home with them!" A sigh went round the room. It was clear that most of those present would have contemplated riding home on a can of nitroglycerine with less horror. The auctioneer added a few more particulars, such as that this brought the known number of eggs up to seventy, that this was one of the most beautiful eggs ever seen, and that it had everything about it which a well-constructed Auk's egg should have, markings, blotches, everything handsome about it, and then said he would now await further remarks from gentlemen anxious to purchase it.

An unexpected diffidence seemed to oppress bidders; and Mr. Stevens added that the egg had evidently been laid by a bird in the prime of life. "If it had tried for years," he observed appreciatingly, "it couldn't have laid anything finer."

"Fifty," said a bidder.

"Thank you; fifty guineas is the first bid."

"Sixty."

"Thank you. Sixty, seventy, eighty, 100, 110, 120, 140, 160! One hundred and sixty guineas only? 180, 200! An absolutely perfect egg. 210! Not going quite so fast as I could wish—220, 230, 240, 250. The last egg stopped for some time at this price. Well, I shall have to wait again while you make up your minds." And here Mr. Stevens beguiled a few minutes with particulars of the domestic life of Auks, and the history of the increasing value of their eggs. But nobody would move beyond 260gs., and at last the auctioneer came to the fatal words. "260gs. for the first time," he said in tones in which manly fortitude struggled with emotion. "260gs. for the second time. Gentlemen, I am terribly disappointed"—the bidder of the 260gs. endeavoured with indifferent success to look sympathetic—"260gs.—thrice —gone!" The purchaser was Mr. Herbert Massey.

After this the auction of the second egg, not quite so beautiful as the first, laid by an older bird, and slightly damaged, proceeded gloomily. It was knocked down eventually, after some painful scenes, for 175gs. Mr. Stevens said again how great his disappointment was—only a beggarly £456 for two eggs of the Great Auk. When one recalls the price recently paid for them, it is hard indeed to suppress the hope that the young man who rode home with them on a bicycle may not be similarly distressed.

James Gardiner who sold it almost immediately to Sir J.H. Greville Smythe of Ashton Court, Somerset, a serial Auk egg collector who managed a grand total of four. After Smythe's death the egg made its third visit to Stevens' but this time its price had fallen. On April 17th 1912 Rowland Ward paid 150 guineas for it on behalf of J.E. Thayer who owned it for almost twenty years before presenting it to Harvard.

ROYAL COLLEGE OF SURGEONS EGG NO.10 (THE CINCINNATI EGG)

Egg no.66 (Grieve no. 67; Parkin no. VI; Tomkinson no. 28)
Origin: unknown
Colour: whitish with brown and black spots
Present location: Cincinnati - Cincinnati Museum of Natural History, 1301 Western Av., Cincinnati, Ohio 45203, USA.

Champley, R. 1864. The Great Auk. *Annals and Magazine of Natural History*, p.236. Newton, A. 1865. The Gare-fowl and its Historians. *Natural History Review*, pp.483-4. Fatio, V. 1868. Liste des divers représentants de l'*Alca impennis* en Europe. *Bullétin de la Société Ornithologique Suisse*, tome II, pt.1, p.84. Blasius, W. 1884. Zur Geschichte der Ueberreste von *Alca impennis*. *Journal für Ornithologie*, p.168. Grieve, S. 1885. *The Great Auk or Garefowl*, p.88, pp.105-6 *and* appendix p.34. Sclater, P. & Saunders, H. (eds.) 1887. Sale of a Great Auk's Egg. *Ibis*, p.152. Grieve, S. 1888 *and* 1896-7. Recent information about the Great Auk *and* Supplementary note on the great Auk. *Transactions of the Edinburgh Field Naturalists' and Microscopical Society*, p.113 (1888) *and* p.260 (1896-7). Parkin, T. 1911. The Great Auk. *Hastings and East Sussex Naturalist*, vol.1, pt.6 (extra paper), p.7 *and* p.11. Tomkinson, P. & J. 1966. *Eggs of the Great Auk*, pp.106-7 *and* pl.28. Spink and Son Ltd. (c.1970). *Romantic Histories of Some Extinct Birds and Their Eggs*, pp.12-3 (specimen 9). Knaggs, N. 1976 (Sept.). The Great Auk Expedition. *Explorer's Journal*, fig. p.101. Bourne, W. 1993. The story of the Great Auk. *Archives of Natural History*, p.275.

This egg, one of those discovered in 1861 in the Hunterian Museum of the Royal College of Surgeons, was acquired by the Cincinnati Museum from the estate of Vivian Hewitt - via Spink and Son Ltd - during 1974. It was bought together with a stuffed bird and the two specimens cost $25,000.

In 1865, exactly 100 years before Hewitt's death, the egg was one of four selected by the Royal College as surplus to requirements and offered for sale at Stevens' Rooms. Here on 11th July it was sold as 'Lot 142' for 30 guineas to the Rev. H. Burney of Woburn, Bedfordshire. Twenty-two years later it was offered again at the same venue and this time was purchased by Leopold Field. Its price had escalated to £168. Field kept the egg for just two years but still made a handsome profit on it. Mr. Herbert Massey, something of a serial egg collector, bought it for £220. During 1939 his executors sold the egg to Vivian Hewitt for an undisclosed sum.

Bidwell's Egg (natural size). Watercolour by F.W. Frohawk. *Courtesy of Haslemere Museum.*

(Facing page). An extract from *The Daily Graphic* for April 25 1894.

PHILIP'S EGG

Philip's Egg (¾ natural size). *Courtesy of Peter Southon.*

Egg no.67 (Tomkinson no. 75)
Origin: unknown, but said to be Newfoundland
Colour: whitish, streaked, blotched and spotted black and brown
Present location: New York - American Museum of Natural History, Central Park West at 79th St., New York 10024-5192, USA.

Witherby, H., (ed.). 1918 (April 1st). An Unrecorded Egg of the Great Auk. *British Birds*, p.264. Seth-Smith, D., (ed.). 1918. [Exhibition of an egg on Jan. 9th 1918] *Bulletin of the British Ornithologists' Club*, XXXVII, p.40. Sheppard, T. 1922 (August-Sept.). Unrecorded egg of the Great Auk. *The Naturalist*, p.254. Jourdain, F. 1935 (March). The Sale of G.D. Rowley's Collections. *The Oologist's Record*, p.79. Tomkinson, P. & J. 1966. *Eggs of the Great Auk*, p.122 *and* pl.75.

The earliest surviving record of this egg comes in the form of a notice in the journal *British Birds* with the title, 'An Unrecorded Egg of the Great Auk.' As the journal's date is April 1st 1918 it might have been thought an April Fool's Day joke. It wasn't. The journal's editor announced:

> It is somewhat surprising to find that a hitherto unrecorded egg ... should still exist but at the meeting of the British Ornithologists' Club on January 9th 1918, Mr. E. Bidwell showed a handsome and well preserved specimen, free from flaws ... which is the property of Mr. F.R. Rowley.

No other details are given.

On May 25th 1922 a letter written from the Royal Albert Memorial Museum, Exeter arrived at the American Museum of Natural History in New York. The letter was from the Mr. F.R. Rowley mentioned in *British Birds*, a gentleman with no apparent connection to the well known Great Auk enthusiast George Dawson Rowley but who was the Curator of the Royal Albert. He had reached an agreement to sell his egg (which seems to have had nothing to do with his curatorial duties) to an American gentleman by the name of P.B. Philip and the letter was a request to the AMNH to act as broker in the deal:

> I am to send the egg to you by registered mail ... you will hand it to Mr. Philip and receive from him Exchange on London payable to me for £315. Will you be good enough to confirm this ... After hearing from you, I will pack and forward the egg ... I should add for the information of Mr. Philip that my specimen was bought with a miscellaneous lot of eggs at a sale in the country some years ago. A label attached to it (which I will send) gave the location as Newfoundland. The previous owner was dead and I could not discover how the egg came into his possession.

In due course the egg arrived safely in New York and went off to Mr. Philip while the money went in the opposite direction. During 1937 Mr. Philip's egg came back to New York when he bequeathed it to the American Museum of Natural History along with a very extensive collection of birds' eggs. The Newfoundland data may or may not be accurate.

DES MURS' PHILADELPHIA EGG

Egg no.68 (Grieve no. 53; Tomkinson no. 66)
Origin: unknown
Colour: brownish yellow with streaks and dashes of light and dark brown
Present location: Philadelphia- Academy of Natural Sciences, 1900 Benjamin Franklin Parkway, Philadelphia PA 1903-1195, USA.

Des Murs, O. 1860. *Traité Général d'Oologia*, p.468. Des Murs, O. 1863. Notice sur l'œuf de l'*Alca impennis*. *Revue et Magasin de Zoologie*, pp.4-5 *and* pl.2. Newton, A. 1870. On Existing Remains of the Gare-fowl. *Ibis*, p.261. Blasius, W. 1884. Zur Geschichte der Ueberreste von *Alca impennis*. *Journal für Ornithologie*, p.164. Grieve, S. 1885. *The Great Auk or Garefowl*, p.89, p.103, appendix p.20 *and* p.32. Grieve, S. 1896-7. Supplementary note on the Great Auk. *Transactions of the Edinburgh Field Naturalists' and Microscopical Society*, pp.263-4. Duchaussoy, H. 1897-8. Le Grand Pingouin du Musée d'Histoire naturelle d'Amiens. *Mémoires de la Société Linnéenne du Nord de la France*, p.101. Tomkinson, P. & J. 1966. *Eggs of the Great Auk*, p.119 *and* pl.66.

Philadelphia is a city with something of a reputation. 'I'd rather be a lamp post in Denver than Mayor of Philadelphia,' Charles 'Sonny' Liston is supposed to have muttered as he left the City of Brotherly Love for the last time. The legendary heavyweight champion wasn't the first - nor will he be the last - to declare his bad intentions. 'On the whole I'd rather be in Philadelphia,' reads the famous headstone of W.C. Fields, surely a particularly spectacular example of damning with faint praise. This city, then, is not for the delicate! For Great Auk egg enthusiasts it has a splendidly ugly mystique. Out of three eggs sent to Philadelphia from France in 1849, just one remains intact. Another vanished long ago and the third was rejected and sent to Washington, having first been smashed.

During the 1840's, Philadelphia's Academy of Natural Sciences made many important acquisitions. Its representatives scoured Europe in the hope of uncovering items or collections of particular interest. Among the trophies sent back to America were the Great Auk eggs that once belonged to Monsieur O. Des Murs. According to Des Murs there were three of these, although in a paper he published in 1863 (more than a decade after selling them) he illustrated just two. Curiously, there is no record of there ever being more than a pair in Philadelphia but this probably means little. Records at the Academy were not always well kept; sometimes they were not kept at all. Assuming that a third egg did reach America, it disappeared at an early date. Tomkinson and Tomkinson (1966) speculated that this supposed third egg was perhaps badly broken in transit, disposed of on arrival and never catalogued. The fate of the second egg is clearer. At an unknown date in the nineteenth century, someone broke it. The damaged remains were then presented to the Smithsonian Institution, presumably by shame-faced officials not wishing to regularly confront the evidence of their own carelessness. This leaves just a single egg surviving at the Academy and, apparently, it is still intact.

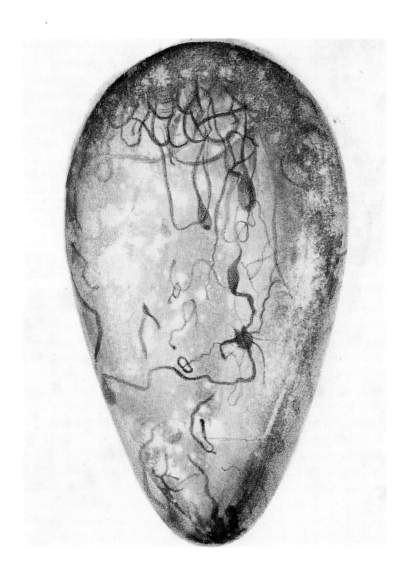 

(Above, left). Des Murs' Philadelphia Egg. (Above, right). Des Murs' Washington Egg. Both natural size. Hand-coloured lithographs from *Revue et Magasin de Zoologie* (1863).

Its origin is unknown. Both of the Des Murs eggs that remain were purchased from Paris dealers, one for five francs from Launoy on June 3rd 1830, the other from Bevalet on 10th May 1833 for three. It is not certain which is which but this, of course, is hardly of vital importance.

DES MURS' WASHINGTON EGG

Egg no.69 (Grieve no. 66; Tomkinson no. 67)
Origin: unknown
Colour: dirty brownish white with streaks and blotches of black and brown
Present location: Washington - National Museum of Natural History, Smithsonian Institution, Washington DC 20560 (cat. no.USNM 15141).

Des Murs, O. 1860. *Traité Général d'Oologia*, p.468. Des Murs, O. 1863. *Notice sur l'œuf de l'Alca impennis. Revue et Magasin de Zoologie*, pp.4-5 *and* pl.1. Newton, A. 1870. On Existing Remains of the Gare-fowl. *Ibis*, p.261. Blasius, W. 1884. Zur Geschichte der Ueberreste von *Alca impennis. Journal für Ornithologie*, pp.167-8. Grieve, S. 1885. *The Great Auk or Garefowl*, p.89, p.103, appendix p.20, p.32 *and* p.34. Grieve, S. 1896-7. Supplementary note on the Great Auk. *Transactions of the Edinburgh Field*

Naturalists' and Microscopical Society, pp.263-4. Duchaussoy, H. 1897-8. Le Grand Pingouin du Musée d'Histoire naturelle d'Amiens. *Mémoires de la Société Linnéenne du Nord de la France*, p.101. Tomkinson, P. & J. *Eggs of the Great Auk*, p.120 *and* pl.67.

This is the egg presented to the Smithsonian by the Academy of Natural Sciences, Philadelphia after it was broken. Like the egg retained by the Academy, it was purchased by Dr. Thomas Wilson from Monsieur O. Des Murs in 1849. Its origin is unknown, Des Murs having bought it from a Paris dealer in 1830 or, perhaps, 1833.

In addition to this damaged shell, the Smithsonian also has the preserved membrane of an egg, a specimen prepared towards the end of the last century by Dr. F. Lucas (catalogue no.USNM 346214). This membrane, preserved in fluid, has no connection with the egg but seems to be an item collected from the mud of Funk Island (*see*, Lucas, 1890, p.514).

WALLACE HEWETT'S EGG

Egg no.70 (Parkin no. XII; Tomkinson no. 69)
Origin: unknown
Colour: cream with streaks, blotches and spots of black, brown and grey
Present location: unknown.

Bidwell, E. 1894. [minutes of meeting] *Ibis*, pp.422-3. Anon. 1894. (April 25th). A Good Investment. *Daily Graphic*. Grieve, S. 1896-7. Supplementary note on the Great Auk. *Transactions of the Edinburgh Field Naturalists' and Microscopical Society*, pp.241-2. Parkin, T. 1911. The Great Auk. *Hastings and East Sussex Naturalist*, vol.1, pt.6 (extra paper), pp.15-6, p.23 *and* p.33. Tomkinson, P. & J. 1966. *Eggs of the Great Auk*, p.120 *and* pl.69. Spink and Sons Ltd. (c.1970). *Romantic Histories of Some Extinct Birds and Their Eggs*, pp.15-6 (specimen 18).

This is one of the thirteen eggs owned by Vivian Hewitt that were offered for sale by Spink and Son following Hewitt's death in 1965. It was one of nine that remained unsold in 1976 when Dr. Jack Gibson became aware of them (*see* eggs nos.15-22) and it stayed at Spink for several more years although it was certainly gone by 1993. There is no proper record of any transaction but a rumour - probably unreliable - suggests it was sold to a South American museum.

Its origin, like its present location, is unknown. It first came to notice when it was auctioned (along with egg no.65) at Rochester in Kent during 1894. The successful bidder, a Mr. Wallace Hewett, seems to have been the only person at the sale who recognised what these objects were and he bought them for 36 shillings. A month later they appeared at Stevens' Rooms where this one was sold to Herbert Massey of Didsbury in Cheshire for 260 guineas, an amount that showed Mr. Hewett a rather handsome profit. After Massey's death, more than 40 years later, the egg was acquired by Hewett's (near) namesake Captain Vivian Hewitt.

With the zeal characterising the endeavours of many nineteenth century Garefowl enthusiasts, Edward Bidwell set about trying to determine the egg's history soon after its existence became known. He uncovered nothing of any significance. Both this egg and its companion were submitted to the sale in Rochester as part of the effects of a Mr. Hulkes, a brewer who'd found them among family possessions that had probably come down to him from his grandfather. Nothing more could be discovered.

SHIRLEY'S EGG

Egg no.71 (Parkin no. XXIV; Tomkinson no. 68)
Origin: unknown
Colour: whitish, boldy streaked, spotted and blotched with black and brown
Present location: unknown.

Grieve, S. 1896-7. Supplementary note on the Great Auk. *Transactions of the Edinburgh Field Naturalists' and Microscopical Society*, p.241 *and* p.263. Parkin, T. 1911. The Great Auk. *Hastings and East Sussex Naturalist*, vol.1, pt.6 (extra paper), p.4, pp.28-9 *and* pl.1. Parkin, T. 1912. Sale of a Great Auk's Egg. *British Birds*, pp.256-7. Tomkinson, P. & J. 1966. *Eggs of the Great Auk*, p.120 *and* pl.68. Morris, P. 1991. Photographs of a missing egg of the Great Auk. *Archives of Natural History*, pp.249-50 *and* fig.1 (p.250).

At the start of the nineteenth century this beautifully marked egg was, apparently, in a collection of natural history specimens belonging to a Bristol resident by the name of Sheppard. About the year 1820 it was

The sale of a lost egg - Stevens' Rooms during the auctioning of Shirley's Egg (June 7 1910).

Two views of Shirley's Egg (approx. 2/3 natural size) - circa 1912. *Courtesy of Pat Morris.*

bought by a Mr. Shirley and, being labelled the egg of a Penguin, it attracted little attention for almost three quarters of a century. Eventually the grandson of Mr. Shirley, his curiosity aroused by a newspaper article on Great Auk eggs, began to suspect that the egg he'd inherited might be of greater interest than he'd hitherto supposed. The shell was shown to Alfred Newton who immediately confirmed that it was indeed that of a Great Auk.

On June 7th 1910 Mr. Evelyn Shirley offered his egg for sale at Stevens' Rooms and it was sold for 250 guineas to a Mr. E.L. Ambrecht. Two years later, on November 21st 1912, the egg was back in the rooms and on this occasion it was knocked down to the Rowland Ward company for a mere 220 guineas. Here, unfortunately, Mr. Shirley's egg passes out of knowledge. It is likely that Messrs. Rowland Ward sold it fairly quickly but there is no record of where it went to or who bought it.

MALCOLM'S EGG

Egg no.72 (Grieve no. 54; Tomkinson no. 30)
Origin: unknown
Colour: yellowish white with black and brown blotches, spots and streaks
Present location: unknown.

Blasius, W. 1884. Zur Geschichte der Ueberreste von *Alca impennis. Journal für Ornithologie*, p.151. Grieve, S. 1885. *The Great Auk or Garefowl*, p.88, p.94, p.103, appendix p.21 *and* p.32. Anon. 1905 (Sept.9th). The Great Auk or Garefowl. *Oban Times*. Anon. (c.1905). *Catalogue of the Ornithological Collection at Poltalloch*, p.15. Tomkinson, P. & J. 1966. *Eggs of the Great Auk*, p.108 *and* pl.30. Spink and Son Ltd. (c.1970). *Romantic Histories of Some Extinct Birds and Their Eggs*, p.13 (specimen 10). Hywel, W. 1973. *Modest Millionaire*, pp.161-2.

Round about 1840 Mr. John Malcolm paid what he considered the very high price of £1 for this egg. He obtained it, just as he obtained a Great Auk's skin, from Leadbeater the taxidermist of Golden Square, London. For many years the egg, together with the bird, was kept safely at Malcolm's family home - Poltalloch, Argyllshire, Scotland - and then, in November 1904, a disastrous fire occurred. Although the egg escaped destruction it was cracked in several places when water (being used to quench the flames) flooded into the drawer in which it was kept and drove it against the sides and top. In this damaged condition it remained at Poltalloch for almost half a century longer. Then, in July 1948, Mr. Peter Adolphe - inventor of the table football game *Subbuteo* - purchased both the egg and the stuffed bird (which had also survived the flames) on behalf of Captain Vivian Hewitt.

Hewitt kept them until his death in 1965 after which they became the responsibility of Spink and Sons Ltd. Legal difficulties and changing fashion resulted in Spink experiencing some difficulty in disposing of all of the Captain's thirteen eggs - after all, the golden age of egg collecting was long since over. Eventually all were sold, however, even though it took until 1993 for the last of them to go.

In 1888, when a Great Auk egg realised the then fantastically high price of £225, a correspondent of *The Pall Mall Gazette* - writing on Tuesday 13th March - speculated on how things might be 100 years hence:

> If any of these specimens ... should be sold again in 1988, what will they bring? Will it be thousands? or, like the tulips a century ago, be down again to "pence?"

The answer is rather unspectacular. Great Auk eggs are worth a good deal more than mere pence (the prices Spink asked in the early 1970's ranged between £250 and £1,250) but in real terms the monetary value of these objects has fallen dramatically; the £225 mentioned in *The Pall Mall Gazette* represented a fortune in nineteenth century Britain. Who knows what they might fetch 100 years from now?

Despite its cracks, Malcolm's Egg was a specimen that found a buyer comparatively quickly. It was sold at an unknown date before 1976 but there is no record of where it went to.

CAPTAIN COOK'S EGG

Egg no.73 (Grieve no. 10; Tomkinson no. 38)
Origin: unknown
Colour: white, strikingly spotted and streaked with brown and black and showing a large hole at the pointed end
Present location: unknown.

Newton, A. 1870. On Existing Remains of the Gare-fowl. *Ibis*, p.261. Blasius, W. 1884. Zur Geschichte der Ueberreste von *Alca impennis*. *Journal für Ornithologie*, p.153. Grieve, S. 1885. *The Great Auk or Garefowl*, p.87 and appendix p.26. Jourdain, F. 1934 (Dec.1st). The Sale of G.D. Rowley's Collections. *The Oologist's Record*, pp.75-9. Jourdain, F. 1935. Sale of Skins

THE EGGS

and Eggs of the Great Auk. *Ibis*, p.246. Jourdain, F. 1935. The skins and eggs of the Great Auk. *British Birds*, p.234. Carter, G. 1935 (March). The Eggs of the Great Auk. *The Oologist's Record*, pp.3-5. Tomkinson, P. & J. 1966. *Eggs of the Great Auk*, pp.110-1 *and* pl.38. Spink and Son Ltd. (c.1970). *Romantic Histories of Some Extinct Birds and Their Eggs*, pp.7-8, p.13 (specimen 12) *and* pl.8 (lower fig.). Tree, I. 1991. *The Ruling Passion of John Gould*, p.200.

The story of this lost egg is one of greed, lies, subterfuge and sheer opportunism, with the villain of the piece being none other than John Gould, celebrated producer of magnificent bird books.

It begins in London on an afternoon early in 1863 and the setting is a small indoor market once known as *The German Bazaar*, Regent Street. The tale, according to Gould, is nothing more than a story of a remarkable piece of good fortune. He had, apparently, gone to the house of Joseph Wolf - the famous wildlife painter - to collect a picture he'd commissioned (one, incidentally, that the roguish Gould would later attempt to pass off as his own work). This picture wasn't quite ready so Gould, knowing that Wolf worked for him only reluctantly, sat down to wait for it to be finished. With nothing to occupy him the eager entrepreneur quickly got bored and eventually left the house intending to while away the time by going for a haircut. He returned from his short trip to the barber in a state of some excitement and took from the handkerchief he held in his hand nothing less than the egg of a Great Auk. He'd bought it, he said, in *The German Bazaar* for the bargain price of £13, having assured the owner - a bird dealer by the name of Whitaker - that it was no more than a freak double-yoked egg of a less exciting species.

If this story doesn't exactly show the greedy Gould in a favourable light, the 'Garefowl Books' of Alfred Newton (lodged today at the University Library, Cambridge) tell a far more dishonourable tale. According to Newton's notes, a certain Mr. Arthur Crowley was strolling through *The German Bazaar* when he spotted what he took to be a Great Auk's egg. Lacking the confidence to buy it there and then, he made a huge tactical blunder. He decided to first seek the opinion of a respected ornithologist and the man he chose was John Gould! When the distinguished birdman heard that the egg was in the hands of the dealer Whitaker he loftily advised poor Mr. Crowley not to buy under any circumstances. The egg, he said, was very likely a forgery. Crowley returned empty-handed that night to his house in Croydon but by the next morning had reconsidered and decided, despite Gould's recommendation, to buy the egg. According to Newton:

> Directly after breakfast he went up to town. On arriving at the Bazaar he told Whitaker that he would take the egg, but was answered, "No you won't, sir, for it's sold." Further inquiry elicited the information that Mr. Gould called ... at two o'clock and bought the egg for £13. Mr. Crowley had only left Gould's house at a quarter to two.

Whether or not Crowley confronted Gould or exacted any revenge is not on record but nothing, it seems, could deter Gould from his shameless conduct over this egg. His next move was to 'discover' a provenance for his new possession and, perhaps predictably, this provenance was by no means

(Top). John Gould (1804-81).

(Bottom). Joseph Wolf (1820-99).

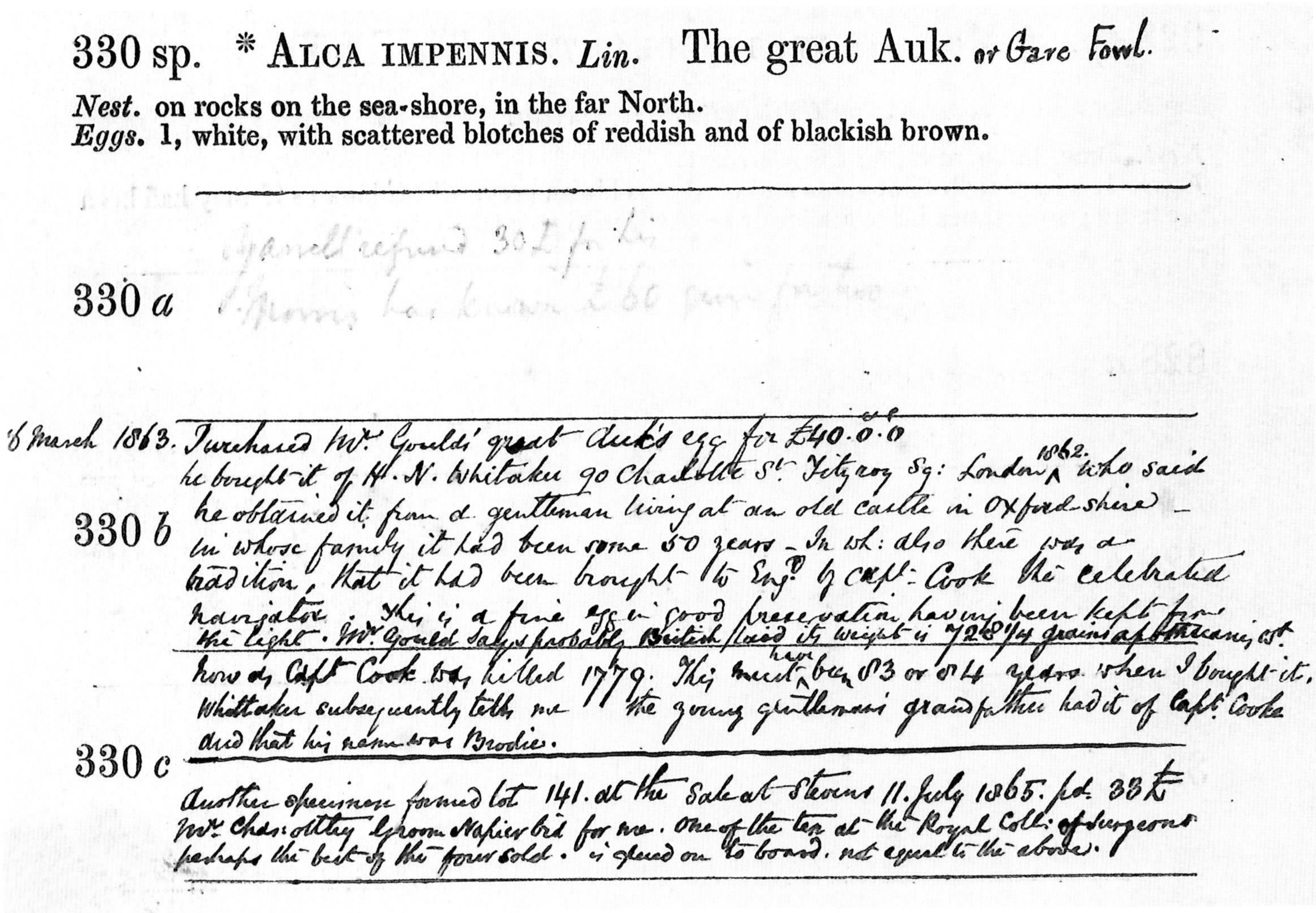

330 sp. * ALCA IMPENNIS. *Lin.* The great Auk. *or Gare Fowl.*

Nest. on rocks on the sea-shore, in the far North.
Eggs. 1, white, with scattered blotches of reddish and of blackish brown.

330 a *[handwritten notes]*

330 b *8 March 1863. Purchased Mr. Gould's great Auk's egg for £40.0.0 he bought it of H. N. Whitaker 90 Charlotte St. Fitzroy Sq: London 1862, who said he obtained it from a gentleman living at an old castle in Oxfordshire — in whose family it had been some 50 years — In wh: also there was a tradition, that it had been brought to Engd by Capt: Cook the celebrated navigator ... This is a fine egg in good preservation having been kept from the light. Mr. Gould says probably British. Laid its weight is 728 1/4 grains ... now as Capt Cook was killed 1779. This must be 83 or 84 years when I bought it. Whitaker subsequently tells me the young gentleman's grandfather had it of Capt: Cook and that his name was Brodie.*

330 c *Another specimen formed lot 141. at the Sale at Stevens 11. July 1865. pd 33£ ... Napier bid for me. One of the ten at the Royal Coll: of Surgeons perhaps the best of the four sold. is glued on to board. not equal to the above.*

an ordinary one. Gould maintained that the egg had once belonged to one of Britain's greatest seafaring heroes, Captain James Cook. Nor did it have an Icelandic or a Newfoundland origin like more ordinary examples. This egg, said Gould, came from a British Auk. Whether there was any basis at all for such opinions is not really known but it seems fair to assume that they originated in Gould's fertile imagination. Alfred Newton was completely sceptical about the considerable 'age' attributed to the egg:

> It has on the contrary a remarkably fresh look, and is I should fancy quite one of the late Icelandic ones, *et bien conservé*.

The Captain Cook connection was presumably invented as a device to extract the highest possible price for the egg and on March 16th 1863 Gould sold it for £40. The buyer was George Dawson Rowley author of the entertaining *Ornithological Miscellany* (1875-8) but before Rowley was given the privilege of paying the hefty sum asked, crafty Gould played another of his little tricks. The egg, he announced, could only be sold to a subscriber to his own current - and very expensive - book; Rowley promptly subscribed to *The Birds of Great Britain* (1862-73).

Although the story of a Cook connection is well enough known, the details - as told by Gould - have long been forgotten and are not in the published record. Fortunately, Rowley's own egg book - a catalogue to his collection - has recently been located and his scribbled notes reveal the full ingredients of Gould's story. According to this, the egg was shut up for more than 50 years in a house in Oxfordshire before it came to Whitaker. Whitaker himself rounded out the tale and told Rowley that the egg was obtained from Captain Cook by a man named Brodie and that it was from this gentleman's grandson that he'd got it. One can assume that Whitaker and Gould had come to some mutually beneficial agreement.

At the end of his life Rowley bequeathed the egg, along with his five others, to his son Fydell and after Fydell's death it was offered for sale at Stevens' on November 14th 1934. Much of Rowley's Great Auk material was purchased at this sale by Captain Vivian Hewitt but this particular egg escaped him and it was bought instead by Sir Bernard Eckstein of Uckfield, Sussex for 260 guineas. This loss to Eckstein was only temporary and during 1947, shortly before Sir Bernard died, Hewitt managed to buy it through the agency of Watkins and Doncaster, a well known naturalists' supply company.

Like the rest of Hewitt's eggs, this one eventually went to Spink and Son where it was sold at an unknown date sometime before 1976. There is no record of who purchased it and its present whereabouts are unknown.

Captain Cook's Egg ($\frac{3}{4}$ natural size) - circa 1970.

MÉEZEMAKER'S EGG

Egg no.74 (Grieve no. 6 *or* 7; Tomkinson no. 57)
Origin: unknown
Colour: creamish with streaks and large blotches of black and brown
Present location: unknown.

Olphe-Galliard, L. 1862. [Letter to the Editor] *Ibis*, p.302. Dubois, Ch.F. 1867. Note sur le *Plautus impennis. Archives Cosmologiques. Revue des Sciences Naturelles* (Bruxelles), no.2, pp.33-5 *and* pl.3 (upper fig.). Newton, A. 1868. Recent Ornithological Publications. *Ibis*, p.112. Fatio, V. 1868. Liste des divers représentants de l'*Alca impennis* En Europe. *Bullétin de la Société Ornithologique Suisse*, tome II, pt.1, p.84. Dubois, Ch.F. 1872. *Les Oiseaux de l'Europe et leurs œufs*, vol.2, pl.27 (upper fig.). Blasius, W. 1884. Zur Geschichte der Ueberreste von *Alca impennis. Journal für Ornithologie*, p.152. Grieve, S. 1885. *The Great Auk or Garefowl*, p.89 *and* appendix p.25. Duchaussoy, H. 1897-8. Le Grand Pingouin du Musée d'Histoire naturelle d'Amiens. *Mémoire de la Société Linnéenne du Nord de la France*, pp.103-6 *and* fig. p.105. Tomkinson, P. & J. 1966. *Eggs of the Great Auk*, p.117 *and* pl.57.

This egg, along with egg no.12, was allegedly brought to France by the captain of a whaling ship. It was given (perhaps sold) to a merchant at Bergues near Dunkirk who passed it on to a young egg collecting acquaintance. Years later this person's entire collection was obtained by Monsieur de Méezemaker, another Bergues collector.

The egg stayed in the Méezemaker family until 1924 when it was sold to Monsieur Heim de Balzac of 34 Rue Hamelin, Paris for approx-

(Facing page, above). A note in the egg book of George Dawson Rowley - the only record of the details of Gould's story. *Courtesy of Peter Blest.*

(Facing page, below). George Dawson Rowley (1822-78). *Courtesy of Peter Rowley.*

(Top). Méezemaker's Egg (no.74). (Bottom). Vivian Hewitt's Egg (no.12). Both natural size. Hand-coloured lithograph in *Archives Cosmologique. Revue des Sciences Naturelles* (1867). *Courtesy of Chris Smeenk and the Nationaal Natuurhistorisch Museum, Leiden.*

Archives cosmologiques, 1867.

Pl.3.

Oeufs du Plautus impennis.

imately £100. It was still in this gentleman's possession in the 1960's but its current whereabouts are unknown. Perhaps it is still owned by his descendants.

THE COMTE DE TRISTAN'S EGG

Egg no.75 (Tomkinson no. 73)
Origin: unknown
Colour: pale yellow with brown and black blotches and streaks
Present location: unknown.

de Tristan, Comte. 1913. Note sur un œuf d'*Alca impennis*. *Revue Française Ornithologique*, p.118. Tomkinson, P. & J. 1966. *Eggs of the Great Auk*, p.122 *and* pl.73.

In 1910 the Comte de Tristan opened a cupboard at his Château de l'Emerillon that had been untouched for 50 years. Inside he found a Great Auk's egg. It evidently belonged to the Comte's grandfather, a man who'd assembled many interesting and curious specimens but whose collections stayed largely undisturbed following his death in 1861. Evidence was found among the dead man's effects to suggest that his egg was brought back from Scotland before 1820; this information may or may not be reliable.

The egg was still owned by a descendant of the Comte when the Tomkinsons published the results of their survey (1966). The family perhaps still own it but it has not proved possible to determine this.

SCALES'S EGG

Destroyed egg no.1 (Grieve no. 42)
Origin: unknown
Colour: dirty white with blotches and spots of brown, black and grey
Destroyed by fire during 1872.

Newton, A. 1861. Abstract of Mr. J. Wolley's Researches in Iceland. *Ibis*, p.387. Champley, R. 1864. The Great Auk. *Annals and Magazine of Natural History*, p.236. Fatio, V. 1868. Liste des divers représentants de l'*Alca impennis* en Europe. *Bullétin de la Société Ornithologique Suisse*, tome II, pt.1, p.84. Blasius, W. 1884. Zur Geschichte der Ueberreste von *Alca impennis*. *Journal für Ornithologie*, p.161. Grieve, S. 1885. *The Great Auk or Garefowl*, p.88, p.107 *and* appendix p.30. Newton, A. 1905. *Ootheca Wolleyana*, pp.380-1 *and* pl.21. Tomkinson, P. & J. 1966. *Eggs of the Great Auk*, p.123.

The egg that once belonged to Mr. John Scales was destroyed by fire in Cork during 1872. Before this unfortunate event it was copied by the famous Newcastle taxidermist John Hancock. In 1858 he made four replicas: one for himself, one for Mr. Scales, one for John Wolley and one for Alfred Newton. This was certainly a very serious undertaking in Victorian Britain and an entry in John Wolley's egg book (*see* Newton, 1905), richly flavoured with the preoccupations of the time, reflects the importance attached to such activities:

Mr. Scales's egg now before me is wonderfully represented by the copies ... I went over them spot by spot, and I find each spot a map,

Ootheca Wolleyana Tab.XXI.

H. Grönvold, pinx.

Bale & Danielsson, L.ᵗᵈ lith.

Scales's Egg (natural size), painted from a replica in the Zoological Museum of Cambridge University. Lithograph by H. Grönvold from J. Wolley's and A. Newton's *Ootheca Wolleyana* (1864-1907).

and every maplet in its right place. The general effect is admirable ... I went over the casts individually... Each has its own merits... In one spot I have preferred mine, in another Newton's. In the general surface I have thought mine the best, and certainly in one place it has a very decided advantage, and that is at a place within the boundary of the lower third of the egg, in nearly the same longitude as the largest black spot. There, in Newton's cast, is a little too much drawing together of the picture laterally, and there are only three spots where there should be four; but this is over-criticism, it only shows that one can tell which is the real egg and which the copy.

There can be little doubt that the illustration of one of these copies by the outstanding egg artist Henrik Grönvold for Newton's *Ootheca Wolleyana* is a faithful representation of this lost egg.

Little is known of the real egg's history. It is reported to have come from the Orkneys but the grounds for this statement are unknown. Such an origin is, of course, highly unlikely. More certain is the information that Scales, who lived to be 90, bought it in Paris from Dufresne in either 1816 or 1817 and that he kept it until its destruction.

A photograph of Scales's Egg taken before its destruction in 1872.

THE LISBON EGG

Destroyed egg no.2 (Grieve no. 29; Tomkinson no. 64)
Origin: unknown
Colour: whitish with brown and grey streaks
Destroyed by fire in 1978.

Sclater, P. 1884. The National Museum of Lisbon. *Ibis*, p.122. Blasius, W. 1884. Zur Geschichte der Ueberreste von *Alca impennis*. *Journal für Ornithologie*, p.158. Grieve, S. 1885. *The Great Auk or Garefowl*, p.89 *and* appendix p.29. Grieve, S. 1888. Recent information about the Great Auk. *Transactions of the Edinburgh Field Naturalists' and Microscopical Society*, p.117. Tomkinson, P. & J. 1966. *Eggs of the Great Auk*, p.119 *and* pl.64.

Along with many other treasures, this egg was destroyed in the fire that ravaged the Museu Bocage, Lisbon on April 18th 1978.

It was discovered among the old contents of Lisbon's Museu Nacional during the early 1880's and nothing is known of its early history. There is a tradition that it was obtained in Italy by a Portuguese king; this tradition may or may not arise from confusion with the history of the stuffed bird that was for many years exhibited with the egg and was similarly doomed.

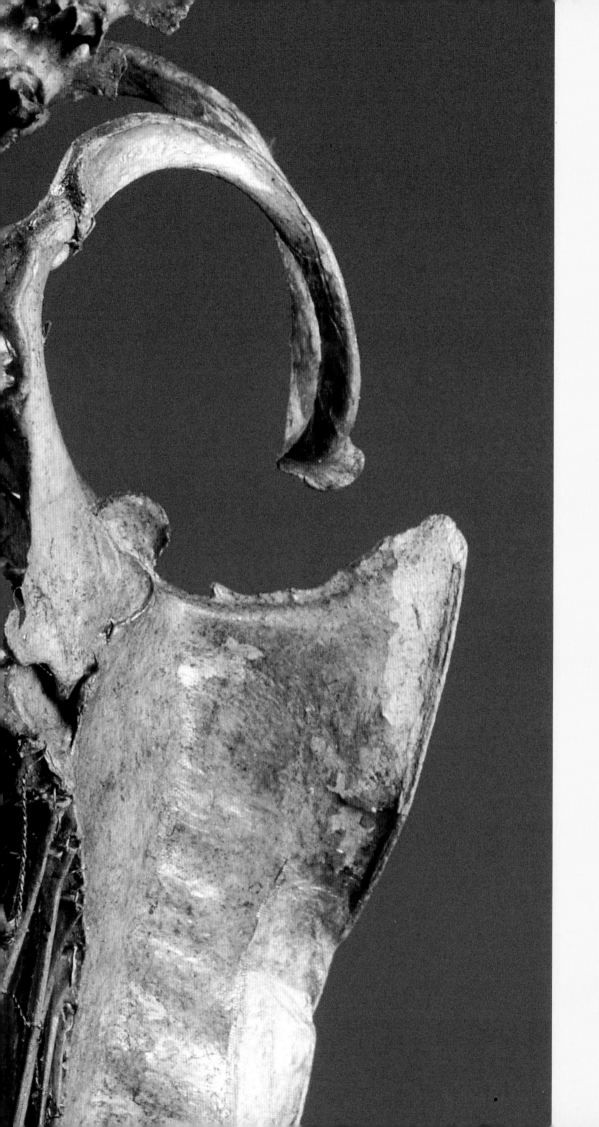

*SKELETONS,
MUMMIES,
BONES AND
PREHISTORY*

SKELETONS, MUMMIES, BONES AND PREHISTORY

Here ... the bones of myriads of Garefowl lie buried in the shallow
soil formed above their mouldered bodies, and here, in this vast
Alcine cemetery, are thickly scattered slabs of weathered granite,
like so many crumbling tombstones marking the resting place
of the departed Auks.

F.A. Lucas (1890)

Great Auk skeletons - complete or almost so - are by no means as rare as
might be expected. Many museums have them. This comparative lack of
scarcity is made additionally curious by the fact that almost all come from
just one source: Funk Island. The number of bones preserved in the cold
mud and soil of this bleak little piece of land has been great enough to
provide plenty of scope for skeleton preparators to exercise their talents
and construct example after example. Such skeletons are not necessarily
the remains of a single bird, however. Although there are some that may
be, the vast majority are composites made up from the parts of several
individuals.

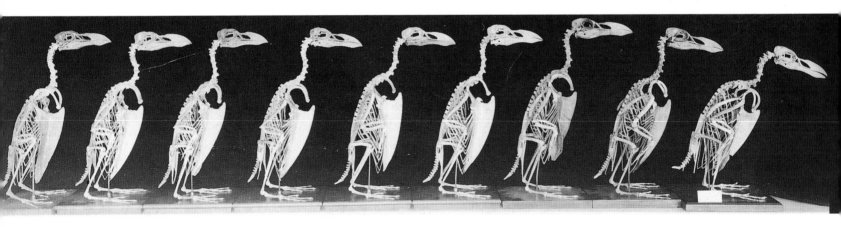

Auk skeletons. (Above). Nine skeletons in the
Museum of Comparative Zoology, Cambridge,
Mass. (circa 1920). (Facing page). A single speci-
men in the Dresden Museum. Photograph by H.
Höhler. *Courtesy of Siegfried Eck and the Staatliches
Museum für Tierkunde, Dresden.*

The nature of the Great Auk colony on Funk Island, and the ferocity
with which it was destroyed, particularly facilitated the preservation of
remains. Throughout the eighteenth century, and indeed in earlier years,
dead birds were simply crushed down into the topsoil where the climatic
conditions and the chemical properties of the earth favoured some degree
of preservation. During the latter half of the nineteenth century great quantities of preserved Auk material were taken away from the island, although
only a small percentage of this was suitable for use in the construction of
skeletons.

These skeletons did not - in general - fire the popular imagination, nor
for the most part did they catch the attention of private collectors. Attached

to them are hardly any of the curious tales that so characterise the histories of the stuffed birds and the eggs. These skeletons passed with little notice or comment into museum collections and any intriguing stories connected with them are few.

One of the more eventful tales concerns an example obtained by the Newton brothers, Alfred and Edward, for Cambridge University. This is a skeleton that was made up - nearly - from the bones of a single individual.

In the half-frozen mud of Funk Island, among all the more fragmentary pieces, were found a number of perfect, or near perfect, natural mummies. One of the first to be discovered was acquired, through the intervention of the Bishop of Newfoundland, by Alfred Newton. From it, Alfred and his brother Edward were able to extract enough material to make an almost

A skeleton preparator with his work - probably George Nelson of the Museum of Comparative Zoology, Cambridge, Mass (circa 1910).

344

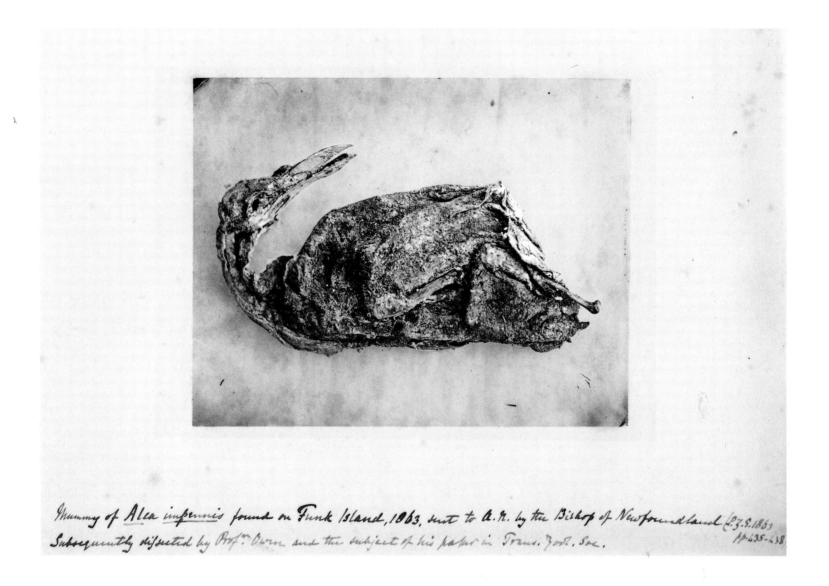

Mummy of Alca impennis found on Funk Island, 1863, sent to A.N. by the Bishop of Newfoundland (P.Z.S.1863 pp.635-638 Subsequently dissected by Prof.n Owen and the subject of his paper in Trans. Zool. Soc.

complete skeleton, only the bones of the extremities being missing. The Newcastle taxidermist John Hancock was able to lend to the Newtons the wanting parts- which he'd carefully removed from his own stuffed specimen. The now complete example was immediately sent on loan to Richard Owen, the famous comparative anatomist and coiner of the term 'dinosaur,' who soon published a minutely detailed technical description, together with a fine lithographic plate, in *The Transactions of the Zoological Society* (1865). On the skeleton's return to Cambridge Hancock's bones were sent back to Newcastle but the taxidermist's deftness of hand had given the Newtons an idea. They simply made good the new deficiency in their skeleton by supplying it with parts carefully withdrawn from the stuffed Garefowl (one belonging to the Cambridge University collection) in their care.

Perhaps the most important investigation at Funk was that conducted during 1887 by F.A. Lucas and his colleague Dr. Palmer. Under the direction of the Secretary of the Smithsonian Institution, Professor Spencer Baird, Lucas and Palmer sailed to the island aboard *The Grampus* and made as thorough a study as conditions would allow (*see* Lucas, 1890).

Although the vast majority of the existing skeletons undoubtedly came from material collected in the nineteenth century, there has been scope

Newton's Mummy. Found on Funk Island in 1863, by 1865 Newton had dissected it and reduced it to a skeleton (*see* Hancock's Auk). *Courtesy of the University Library, Cambridge.*

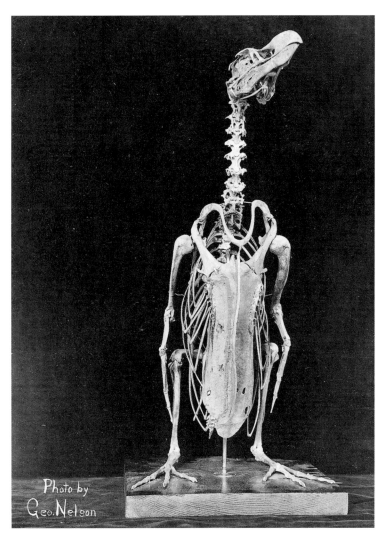

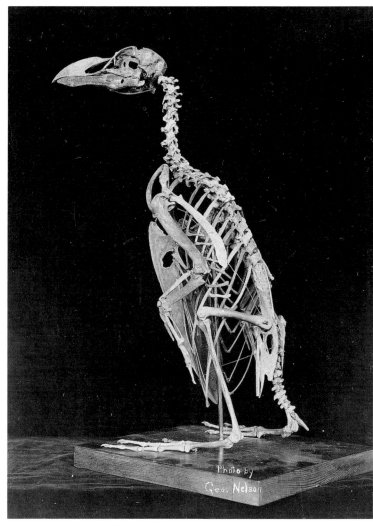

Two views of a skeleton.

for discovery on Funk during the twentieth. One attractive story concerns E. Thomas Gilliard who for many years was an employee at the American Museum of Natural History but who is best known to ornithologists for his outstanding - and, sadly, posthumously published - book, *Birds of Paradise and Bower Birds* (1969). As a young man Gilliard became fascinated by the story of the Great Auk and with a friend, Samuel K. George, he conceived the notion of exploring Funk in the hope of finding skeletal remains. His museum had little time for what was considered 'a boy's lark' and, with no financial or other help forthcoming, the two friends pooled their resources and came up with just enough money for the trip. During 1936 they set off for Newfoundland where they hired a local seaman to sail them through the treacherous waters around Funk. Once on the island they found- to the apparent surprise of many- just what they were looking for: sackfuls of bones. There were enough, both in quantity and variety, to enable the building of an almost complete skeleton on their return to New York. This object Gilliard displayed proudly on his desk as he rose through the museum ranks towards his eventual position as Curator of Birds. He died suddenly, at a comparatively early age, in 1965 and the skeleton - private property rather than the museum's - passed to his companion on the Newfoundland trip,

Samuel K. George. In the early 1990's, after Samuel George's own death, his widow presented his skeleton to the museum in which it had stood several decades previously.

Although the Funk Island deposits have yielded by far the most spectacular hauls of skeletal material, bones have been found at many, many other places around the coasts of the North Atlantic. In most cases these finds consist of just an isolated bone or two, sometimes no more than a fragment of a single piece is recovered. Preserved in cave deposits, bogs, kitchen middens (both prehistoric and more recent) and other kinds of archaeological and geological sites, their presence - whether found singly or in numbers - is clear proof that Great Auks were once familiar creatures to those who inhabited North Atlantic shores. The remains are widespread, being found on both sides of the ocean from far in the north down almost to the Tropic of Cancer.

Above all else, these bones demonstrate that *Alca impennis* was once a species distributed over vast areas and one that was regularly eaten by early (and not so early) man. They imply that the nesting sites and colonies known to us from the historical records of the last few centuries represent mere remnants of the Great Auk's once mighty empire.

That the species was in decline - or that its range was drastically changing- long before these historical records began to be kept cannot sensibly be doubted. Nor can it be doubted that it was the influence of man that pushed the Garefowl to final extinction. But while humans exterminated Great Auks whenever and wherever they found the means to get at them, other factors probably contributed to the apparent changes in range.

Records of bones from Florida, Gibraltar and southern Italy are clear indications that the species once enjoyed a much more southerly distribution than might be expected in a bird so firmly associated with the north. Major climatic changes occurred during recent prehistoric times and the influence these changes had on Great Auk populations can only be guessed at. Presumably, ice sheets and colder temperatures advancing from the north pressed the birds further south but attempts to finely correlate the distribution of bones with known changes in climate are largely unrealistic. Too often the data is insufficient to support helpful conclusions and there remain too many uncertain factors that cannot be meaningfully applied.

What the evidence shows unequivocally is that Great Auks have frequented - at one time or another - the coasts of Europe from the Murman Coast of the Barents Sea right down and round to Italy and that they inhabited the shores of North America from the Bay of Fundy to Florida and Bermuda. Whether the more extreme of these records (Florida, for instance) reflect relatively uncommon occurrences or whether they result from more regular prehistoric incidence is a matter for speculation; presumably, they are records of a rather different climate - one in which the Great Auk was able to flourish - to the climate that prevails today.

Another matter for speculation is the apparent rarity of Great Auk bones at many of the kitchen midden sites where they occur. Certainly they are often less plentiful than the remains of other birds. Whether this statistical rarity corresponds to a genuine rarity of individuals cannot be determined. Perhaps it does, but perhaps it simply shows that these Auks weren't

Bone measurements

Coracoid
    length   59mm - 61mm
    width   5mm - 7mm

Humerus
    length   100mm - 104mm
    width   10mm - 10.5mm

Ulna
    length   53mm - 58mm
    width   4.5mm - 5mm

Scapula
    width   15mm - 17mm

Tarsometatarsus
    length   50mm - 52mm

Femur
    length   71mm - 73mm

Great Auk bones. Illustration from J. Naumann's *Naturgeschichte der Vögel Mitteleuropas* (1896-1905); ¾ natural size.

available to human hunters in quite the same way as other species. Possibly their distribution was patchy or perhaps the fact that they were, as far as we know, only available for a month or so each year played a decisive rôle in any seeming rarity. Conversely, the regularity with which bones are obtained at so many sites might suggest that the living birds were relatively plentiful and easy to come by. We know nothing for certain about their movements when at sea; it may be unlikely but it is not entirely impossible that juveniles (or even adults for that matter) were inshore-birds which could be obtained at all seasons of the year.

Suggestions that perhaps the flesh wasn't good to eat and that therefore the species was not greatly sought after can probably be ruled out. The stomach of early man was not a delicate one, besides which historical evidence seems to indicate that Great Auks were tasty enough. Show a cook a picture of a Garefowl and the reaction is usually the same. 'What an enormous proportion of breast in relation to leg and wing,' he or she will say.

An additional difficulty lies in determining the age of bones. Sometimes this can be done with a degree of confidence, sometimes it can't. Records from one site may be reliably and carefully assembled. From another this may not be the case. Particularly in instances where finds were made many years ago, it may be impossible to judge the accuracy and merit of any conclusions reached. Only in the broadest sense is it possible to pull together some kind of meaningful pattern concerning the Great Auk's distribution in times long passed.

The species once lived over a vast watery area of the Northern Hemisphere but in late prehistoric and historic times its range altered. This is what the bones actually tell us. As far as the mechanics of this alteration are concerned we are left to speculate. Clearly the species responded to climatic changes; clearly it was affected by the persecution of man. Other than this, little can be said.

Any list of Garefowl bone sites is likely to be a long one. In North America bones are found in Indian middens from the Bay of Fundy to southern Massachusetts with more exceptional discoveries being made further south, most notably near Ormond, Florida. Among the best known localities are Nantucket Island, Calf Island, the shores of Frenchman's Bay, Gouldsborough, Mount Desert Island, Sorrento, Lamoine, islands in Casco Bay, Plum Island, Ipswich, Marblehead, East Wareham and Martha's Vineyard.

At Port au Choix on the northern coast of Newfoundland an altogether extraordinary discovery was made. A grave believed to be between three and four thousand years old was found to contain two hundred Great Auk beaks. These completely covered a human skeleton and it is speculated that they may be the decorative remains of a coat or cloak made from feathers.

In Greenland bones and bone fragments have been found in refuse heaps and middens at many Eskimo sites. Some of these indicate that, in addition to the deep southerly distribution already outlined, the species once ranged further to the north than is generally supposed. Bones dated around 2,000 years B.C. have been found in the Disko Bay area, a locality well within the Arctic Circle and one that was probably considerably

Watercolour by Maurice Wilson painted for the first edition of W. Swinton's *Fossil Birds* (1958).

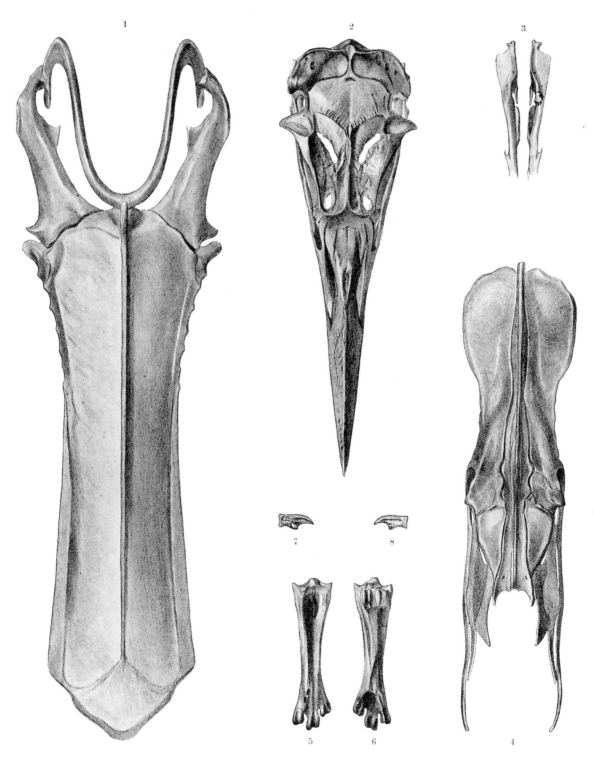

Alca impennis L. Riesenalk.

1 Brustbein mit Furcula und beiden Coracoidbeinen von vorn; 2 Schädel von oben; 3 Gaumenbeine von unten; 4 Becken von hinten; 5 Tarsometatarsus von vorn; 6 derselbe von hinten; 7 Krallenglied der rechten Mittelzehe von aussen; 8 dasselbe von innen. Nach T. C. Eyton 1875.

warmer then than now. Curiously, this is a locality associated with a much disputed - and usually rejected - record of a living bird early in the nineteenth century.

Another surprising place for Great Auk bones to occur is at an inland fjord site at Itivnera, some 60 miles (100 kilometres) from the open sea. Morten Meldgaard (1988) argues strongly that the find here implies that the birds were actually inhabiting the fjord rather than being bought inland by Inuit hunters after having been caught at the coast.

According to Meldgaard (1988), finds at older sites argue for a widespread distribution in West Greenland until around 1,500 B.C. after which the range appears to have shrunk southwards. This is probably due to climatic changes but perhaps also to the increasing penetration of man into the area. The same author also records that the vast majority (90%) of bones found in Greenland date from around 1,350 A.D. to 1,800 A.D. and are, therefore, comparatively recent. As far as Greenland's west coast is concerned these less ancient bones have been found at localities from Cape Farewell at the great island's southern tip to Søndre Strømfjord, virtually on the Arctic Circle.

As might be expected, bones have been found at sites in the British Isles, particularly in Scotland. Quite recently, good finds have been made in the Orkneys but the best known discoveries in these islands were made in a chambered cairn near Hullion, Rousay and in a broch close to the St. Marys area near Kirkwall. From the Island of Colonsay near Oronsay in the Firth of Lorne bones from several individuals are recorded; these relics are said to be more than 4,000 years old. The best known mainland records are from Caithness, near the villages of Elsay and Keiss. Here the

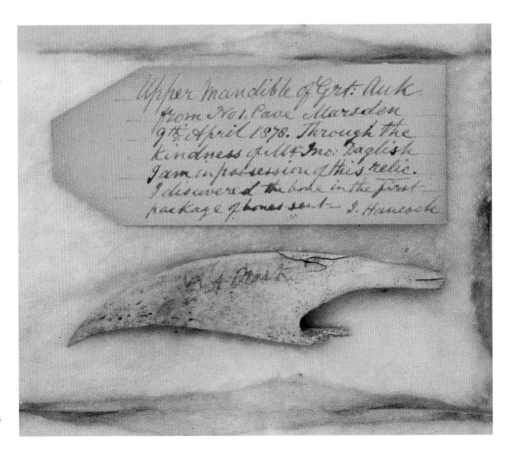

An upper mandible found near to South Shields in 1878. *Photograph courtesy of Keith Bowey.*

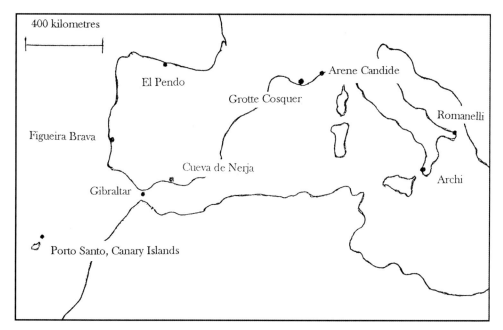

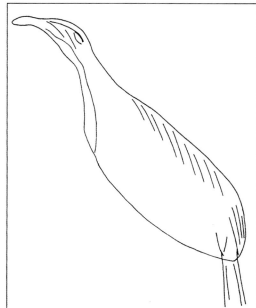

bones - again found in brochs - are supposed to be around 2,000 years old. Finds have recently been made at Dunbar.

An upper mandible discovered near to South Shields during April of 1878 is the best documented English find. This discovery was made amongst midden material in a cave between Marsden and Cleaden, approximately half a mile from the sea. The Auk was probably caught at the coast then taken the short distance inland. Another find from the north east of England was made by Deidre O'Sullivan of Leicester University's Archaeology Department when excavating a ninth century Saxon site at Greenshiel, Lindisfarne. From a mediaeval site at Castle Rushen came clear evidence of the former presence of Great Auks on the Isle of Man. The oldest known Garefowl bones come from a site at Boxgrove in southern England. Here, remnants some 500,000 years old have been identified by John Stewart an archaeologist with University College, London (*see* Harrison, C. and Stewart, J, *in press*).

In Ireland remains occur at a number of sites including Waterford, Sutton and White Park Bay. Interestingly, there is an eighteenth century record made by one Robert Gage (mentioned *in* Ussher and Warren, 1900) of Great Auks frequenting the Island of Rathlin. White Park Bay on the Antrim coast- the site of the first bone find in Ireland- is quite close by.

Scandinavia has many sites at which bones are present. From Donna and Leke, about half way up the Norwegian coast, right down to the shores of Denmark there exists good evidence of the species' former occurrence. Meijlgaard, Klinteso (where remains are particularly abundant), Sölager, Otterö Island, Ertebølle, Havelse, Serjrö, Rotekärrslid and Viste near Stavanger are among the better known sites. Some of the bones found in these places may be as much as 9,000 years old, others maybe no older than a millennium. The presence at Klinteso and Serjrö of bones supposed to come from young individuals suggests that breeding colonies once flourished nearby.

Sites around the Meditteranean at which bones and prehistoric paintings alleged to show Great Auks have been found

Copy of a painting in a cave at El Pendo that is alleged to show a Great Auk.

A scene outside Gorham's Cave, Gibraltar imagined by Maurice Wilson, well known for his watercolours of prehistoric subjects. A Great Auk lies dead behind the hunter and his family. *Courtesy of The Natural History Museum, London.*

Further north, bones from four adults were found at a late Neolithic site (circa 2,000 years B.C.) known as Mayak II. This place is situated on the Murman coast of the Barents Sea more than 250 miles (400 kilometres) east of the nearest known Great Auk site, one at Varanger Fjord in Scandinavia.

In the Netherlands, skeletal remains were found during excavations at a Roman site near Velsen. Bones have also occurred at a fossil locality known as de Maasvlakte.

Far more ancient than any of the Scandinavian or Dutch remains are bones found at La Cotte de St. Brelade on Jersey. These are estimated at between 70,000 and 90,000 years old. Supposed to be of similar date are the bone fragments found on Gibraltar in a cave near an old signal tower known as the Devil's Tower. Other pieces of bone have been recovered from Mousterian levels at Gorham's Cave. Recently, Great Auk remains were identified by Jo Cooper of the Natural History Museum, London from material gathered at Ibex Cave (*pers. comm.* but *see* Cooper, J. 1997). Other material from this site is thought to be 40,000 to 60,000 years old.

Bones taken from the Grotte Romanelli in Apulia, southern Italy are thought to be around 60,000 years old. Other Italian remains were discovered in Middle Palaeolithic beds at Archi in Calabria and at Arene Candide.

In Spain, a sternum was identified from Magdelanian beds at Cueva de Nerja, Malaga and in Portugal bones are reported from the Mousterian levels of Figuiera Brava. A fossiliferous deposit of unknown date has, apparently, revealed bones at Porto Santo on the Canaries and this represents the most southerly record.

Perhaps the most remarkable prehistoric relics are not connected with bones, however. At two different sites, one at El Pendo, Cantabria, Spain and the other at the Grotte Cosquer, Alpes Maritimes, France, there exist cave paintings that possibly show Great Auks.

The El Pendo painting may or may not be a depiction of an Auk. There are several kinds of bird with which it could be tentatively identified. The Grotte Cosquer pictures are altogether more convincing. Three creatures are shown and unless these are portraits of three plesiosaur-like water monsters, it is difficult to imagine what, if not Great Auks, they might be. It is estimated that they were painted around 20,000 years ago.

Other caves in southern Europe are alleged to contain Garefowl pictures but these images are even less convincing than the El Pendo ones. Similarly, there are in existence several artefacts found at archaeological sites - pins, batons, pendants, combs - supposed to be embellished with depictions of Great Auks. Judged impartially, however, none of these decorations are particularly convincing as representations of the species.

The Great Auks of the Grotte Cosquer. Photograph by A. Chêné, CNRS, Centre Camille Jullian. *Reproduced by kind permission of Jean Clottes and the Ministère de la Culture/Direction du Patrimoine.*

ISLANDS

ISLANDS

It need not be doubted that the Great Auk formerly bred on many islands and along many suitable stretches of mainland coast both on the west and the east sides of the North Atlantic. Of such long lost colonies little or nothing can be known. By the time written accounts began to be kept the species was restricted to a few islands and these islands were, typically, remote and inaccessible. Exactly how many were favoured by annual visits from the Garefowl is a mystery. Only a few of these haunts are remembered in the written records that survive, each of them last redoubts of a vanished empire.

THE GEIRFUGLASKER

Lying off Cape Reykjanes...is a small chain of volcanic islets, commonly known as the Fuglasker, between which and the shore, notwithstanding that the water is deep, there runs a Röst, nearly always violent, and under certain conditions of wind and tide such as no boat can live in.

Alfred Newton (1861)

The bird skerries just off the south-west tip of Iceland are probably the islands most powerfully associated with the Great Auk, simply because one of them - Eldey - has acquired such notoriety as the stage for the last scene in the history of the living Garefowl. But although Eldey was the setting for the single most dramatic act in the species' tale, another of the skerries was far more important to the bird during those years in which it flourished. This was the island known as the Geirfuglasker, an island some 25 miles (37 kilometres) from the mainland and one that for centuries provided a breeding station for Great Auks.

The whole area is one of intense volcanic activity and there are many records of eruptions during historical times. In the year 1210, then several times during the next thirty years, islands were thrust up from the sea only to disappear a few years later. In 1422, 1583 and again exactly 200 years later, more eruptions were recorded and the configuration of islands changed again.

The four main islands - those that seemed the most permanent - were Eldey (Fire Island), Eldeyjardrángr (Fire Island Rock), the Geirfuglasker (Garefowl Island) and the Geirfugladrángr (Garefowl Island Rock). Of these it was the Geirfuglasker that, as its name suggests, was most favoured by Great Auks. Whether or not the Geirfugladrángr was also used is not really on record.

As a home for the Great Auk the Geirfuglasker was possessed of two advantages over the other islands. First, it was comparatively easy for

Map of the bird skerries off Cape Reykjanes.

Great Auks in summer and winter plumages. Chromolithograph by J. G. Keulemans from J. Naumann's *Naturgeschichte der Vögel Mitteleuropas* (1896-1905). The island in the background is clearly meant to be Eldey.

Ordre 15.   Palmipèdes.

Werner del.   1/5 de nat.   Lith. de A.Belin.

Pingouin Brachiptère, plumage des noces (Alca Impennis, Linn)

Hand-coloured lithograph by J. Werner from J. Werner's and C. Temminck's *Atlas des Oiseaux d'Europe* (1828). This picture was almost certainly produced using a Geirfuglasker Auk as a model.

Garefowls to come ashore and then get off again; part of one side sloped relatively easily down to the sea. Second, the island was protected by treacherous currents. Except in the calmest of weather the Icelandic fishermen who lived on the adjacent mainland found it too dangerous to attempt a landing. When they did, they wreaked havoc and Garefowl numbers were probably seriously reduced. Through the course of many summers, however, there was no opportunity to land and during these periods the population had time to rebuild its numbers. Leaving aside the danger to Great Auks, the danger to fishermen is clearly illustrated in statistics given by Hjálmar Bárðarson (1986). He reports that twelve men were drowned in 1628. Eleven years later two large boats (from a fleet of four) failed to return from a raiding trip. In 1732, following an interval of 75 years during which there was no recorded trip to the skerry, two makeshift hovels were found together with three cudgels and the remains of human skeletons. Even during successful visits it was often necessary for a sailor to tie a rope around his waist before leaving his boat so that he could be dragged back to it through the surf when the time came to leave the island.

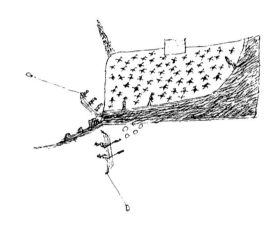

The Geirfuglasker. A translation of the Icelandic text accompanying this old drawing (circa 1760) reads, 'View from the north-east when approaching from the mainland, showing the two landing places and the East Reef.'

During their Icelandic journey of 1858 both Alfred Newton and John Wolley were shown a manuscript in the public library at Reykjavik that was written around the year 1760 and contained a description of the Great Auk and the Geirfuglasker. Newton had part of it translated:

> The space he [the Great Auk] occupies cannot be reckoned at more than a sixteenth part of the skerry ... and this only at the two landing places; further upwards he does not betake himself, on account of his flightlessness.

During 1830 there began the series of cataclysmic disturbances that resulted in the destruction of the Geirfuglasker. According to Newton (1861):

> Some ... miles ... out are the remains of the rock formerly known to Icelanders as the Geirfuglasker ... and to the Danes as Ladegaarden (Barn-building), in former times the most considerable of the chain, but which, after a series of submarine disturbances beginning on the 6th or 7th of March 1830, and continuing at intervals for about a twelvemonth, disappeared completely below the surface, so that now no part of it is visible, though it is said that its station is occasionally revealed by the breakers.

These are the circumstances under which the Great Auk lost its last true home. With its imperfect sanctuary on the Geirfuglasker gone, the remaining birds were forced to resort to a neighbouring island - Eldey.

ELDEY

Infamous for the events of June 3rd 1844, Eldey (Fire Island) lies some 10 miles (14 kilometres) off Cape Reykjanes, the south-west tip of Iceland, and 40 miles (65 kilometres) or so from Reykjavik. Approximately 220 yards (200 metres) long and half as wide, it is formed of volcanic tuff and rises precipitously from the sea to a height of almost 260 feet (80 metres).

Eldey's forbidding appearance is entirely appropriate for the place where the execution of a species was enacted; it looks like a gigantic, grey headsman's block. So sheer are its sides that there is no record of anyone reaching its flattish top until 1894, fifty years after the last Eldey Auks were killed.

The Island of Eldey. Photograph by James Fisher. *Reproduced by kind permission of Clemency Thorne Fisher.*

The Great Auk was, of course, quite incapable of reaching the top. Individuals of the species had to content themselves with landing on a small reef that stretches out from the island's northern end and they bred on those rocky ledges that were accessible and close by. Although this is the island most closely associated with Great Auks, Eldey was in reality quite unsuitable for the species. Indeed it was probably never used by Garefowls until after the volcanic activity of 1830-1 that wrecked the nearby Geirfuglasker. Probably, it was simply desperation that drove the last little band of individuals to it. Only for the brief period between 1830 and 1844 is there any record of Garefowls using Eldey and in that short span of years they were mercilessly hounded by Icelandic fisherfolk.

Most of the stuffed specimens that survive today are the remains of birds killed on Eldey, the remnant of the colony that had traditionally nested on the Geirfuglasker. Ironically, the island is now a bird sanctuary and cannot be visited without permission from the Icelandic Nature Conservation Council. Today it is home to an enormous colony of Gannets (*Sula bassana*). At around 16,000 breeding pairs, this is reputed to be the third largest gannetry in the world. Naturally, the Gannets can use the island's top to nest, a surface that - although seeming level from a distance - is remarkably uneven and usually ankle-deep in bird droppings.

Danish and English sailors used to give Eldey the name 'Meel-sækken' or 'the Meal Sack' on account of the island's generally lumpish shape and the droppings that stained its upper parts. Seen from a certain angle it resembles a half-filled bag of flour.

OTHER GEIRFUGLASKERS

Several islands off the coast of Iceland were given the name Geirfuglasker, a name obviously derived from 'Geirfugl' the Icelandic equivalent of 'Garefowl.' Clearly, all these islands were so named because they were once breeding stations for the Great Auk.

The most celebrated is the vanished Geirfuglasker off Cape Reykjanes, its greater fame being due to the late survival of the colony that frequented it. The other Geirfuglaskers probably boasted colonies of similar size but each had vanished before any proper records were made.

One of these Geirfuglaskers is situated some 30 miles (48 kilometres) from Iceland's east coast near to the island known as Papey. This Geirfuglasker was traditionally raided for eggs and birds around St John's Day (June 24th) but there are no records to indicate when the Auks that came here were finally wiped out. According to Newton (1861) this island was known to Danish seamen as 'Hvalsbak' (Whalesback).

Another Geirfuglasker is the most southerly of the islands that make up the Vestmannaeyjar Group. The birds that frequented this place were, apparently, culled regularly and although the colony lingered until late in the eighteenth century, it was probably gone by 1800.

A fourth Geirfuglasker perhaps existed and is mentioned by Preyer (1862). It is alleged that it was an island some 20 miles (32 kilometres) off the Breiðamerkursandr and it is marked on the small map that accompanies Japetus Steenstrup's paper of 1855. Newton (1861) thought - almost certainly correctly - that the island never existed and that the belief in it originated from confusion over the position of the Geirfuglasker near to Papey.

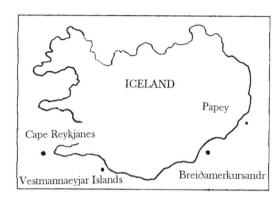

Map showing approximate positions of the Geirfuglaskers.

Olaf Worm's Auk.

THE FAROE ISLANDS

The Faroe Islands hold a small but significant place in the history of man's involvement with the Great Auk. It was from the Faroes in the mid-seventeenth century that Olaf Worm obtained a living Garefowl that he kept alive for a while in Copenhagen. A portrait of his bird was drawn and this much reproduced illustration is one of the earliest known representations of the species. A peculiarity is a narrow, white band around the throat - perhaps an embellishment of the artist, perhaps a collar by which the captive was led around.

Known in the Faroes as Gorfuglir, Garfogel or Gaarfugler, the birds were probably quite plentiful in the seventeenth century and even in the eighteenth some could, apparently be seen during most summers. There was a tradition that individuals, if taken alive, could be easily tamed but would not live for long away from the sea.

The species probably lingered on in the Faroes until early in the nineteenth century. An 81 year old man interviewed by H.W. Fielden (see *The*

Zoologist, 1872, p.3282) claimed to be the last living Faroese able to remember seeing one. His name was Jan Hansen and he recalled - rather precisely for an event that allegedly occurred some 60 years earlier - that on July 1st 1808 he visited an island called the Great Dimon in the company of a party of fowlers. On a ledge at the base of some cliffs they found a living bird and, as might be expected, they promptly killed it. Other stories, tales that appear to date from about the same time, indicate that individuals were still visiting Westmannshavn where one was allegedly killed with a stick while it sat on its egg. Another story told by an elderly man was heard by John Wolley. Wolley's informant stated that he'd seen a bird sitting on low cliffs during the early years of the nineteenth century. Such stories may have a basis of fact or they may not but tangible evidence of the Great Auk's occurrence in the Faroes was seen by Japetus Steenstrup when he visited the islands during the 1850's. He was shown a preserved head but what eventually became of it is unknown.

THE ORKNEY ISLANDS

That Great Auks visited the rather appropriately named Orkney Islands is beyond doubt. Whether or not they bred there during recent historical times is much less certain. There is simply a presumption in favour of the idea. Two birds, known locally as the 'King and Queen of the Auks,' visited the Orkneys for several years at the start of the nineteenth century and the skin of one of them still exists at the Natural History Museum's Sub-department of Ornithology at Tring, Herts. Perhaps these birds were the last representatives of a breeding group, perhaps they were just lost, disorientated creatures that belonged hundreds of miles away. As there is no earlier record of Garefowls in the area the issue remains unresolved.

Almost forty islands, plus a number of reefs and skerries, make up the group known as the Orkneys. They lie just off the northern coast of mainland Scotland and most of them are inhabited. It is one of the furthest flung islands of the group that is traditionally associated with the Great Auk. This is Papa Westray, virtually the most northerly of the islands, and it was from here that the jeweller turned museum proprietor William Bullock launched his well-documented hunt for a specimen in the summer of 1812. Here it was, some months afterwards, at a spot called Fowl's or Auk's Crag, that an individual - probably the bird Bullock chased - was killed and sent down to London. Although the precise circumstances of the death of this poor bird are recorded and even the spot where it rested was photographed, nineteenth century researchers (Buckley and Harvie-Brown, 1891 *and* Newton, 1898) ascertained that there was no really suitable spot on Papa Westray where a colony of Great Auks might have bred.

Newton (1898), after perusing a memorandum made by the late John Wolley some forty years earlier, came to the conclusion that any breeding colony might have frequented a small island just to the east of Papa Westray, an island known as the Holm of Papa Westray. Newton was so entranced by his closet deduction that he journeyed to the Orkneys on

The decorated title page of T. Buckley's and J. Harvie-Brown's *Vertebrate Fauna of the Orkney Islands* (1891). This book contains a remarkably full account of the Great Auk and its occurrence on Papa Westray. *Courtesy of Peter Blest.*

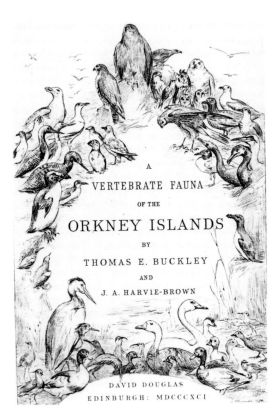

A

VERTEBRATE FAUNA

OF THE

ORKNEY ISLANDS

BY

THOMAS E. BUCKLEY

AND

J. A. HARVIE-BROWN

DAVID DOUGLAS
EDINBURGH: MDCCCXCI

at least four occasions to check its viability. Although he saw the Holm on each visit and thought the large slabs of sloping rock that flanked its westerly side seemed ideal, it was not until his fourth trip that weather and currents permitted a safe landing. On finally getting ashore, he found that the great slabs did indeed provide a perfect landing place for flightless, marine birds:

> Here would be room for a regiment of Auks to have landed at any state of the tide, and to have marched in line up the gentle ascent so far as they wished to go, even to the very turf-covered soil of the islet, while some three or four deep chasms running inward ... would serve ... to diminish the length of the land-journey to any aspiring Auk, or to facilitate the escape of one threatened with danger.

Despite his discovery Newton, discouraged by the lack of any historical tradition, entertained doubts as to whether Great Auks ever actually made use of this site.

ST KILDA

The Isle of Irte, which is agreed to be under the Circius and on the margine of the world beyond which there is found no land in these bounds.

John of Fordun in *The Scotichronicon* (circa 1380)

Lonely St Kilda - or Hirta as its inhabitants once called it - has a reputation as one of the most wildly beautiful spots on earth; certainly it is a very forlorn and spectacular place. Unrelenting cliffs that tower massively from the sea for over 1,000 feet (320 metres) and great wave-lashed stacks - the best known being Stac Lee and Stac-an-Armin - contribute to an unforgettably moody grandeur. Add to this the storms, mists and brooding

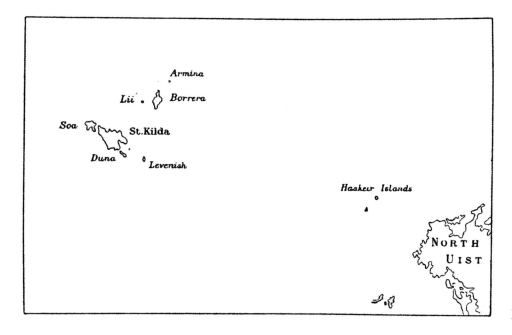

Map of the St Kilda group.

emptiness of the North Atlantic and St Kilda acquires a magical, timeless quality that few places can match and that few visitors fail to sense. Almost entirely cliff-girt, the only obvious place for a landing on the main island is a haven known as Village Bay, but even this access is virtually unusable when a storm blows up from the south east. Seen from this bay the rising hills - though bleak and treeless - look pleasantly rounded and lacking in menace. What makes St Kilda so dramatic is the fact that these hills have literally no back to them. They simply drop away sheer to the sea.

The most far-flung outpost of the British Isles, the small group of islands that make up St Kilda (Hirta, Boreray, Soay and Dun) are more than 50 miles (80 kilometres) west of the Outer Hebrides and 110 miles (175 kilometres) from the nearest approach of the Scottish mainland. A hundred miles further west is the stone known as Rockall that rears up some 70 feet (22 metres) above the waves. After this there is no other land till the New World is reached.

The aura of mystery in which St Kilda is steeped is heightened by the fact that for centuries a strange, tightly-knit little community inhabited it. These St Kildan people survived for generations until their remnant was evacuated to the mainland in 1930 due to straitened circumstances. In the isolation of their precarious, island-bound lives they'd developed their own customs and peculiarities and established a unique system of order and fair division of labour.

Their economy was largely based on harvesting the seabirds that lived on, or visited, St Kilda and the islanders were particularly adept at brinkmanship as they pursued their bird hunting livelihood on the crags. A typically curious tradition was connected with a rock that jutted out over the edge of a precipice and was known as the 'Mistress Stone.' In this place young men were required to prove their readiness for marriage with a display of balance and daring. Standing on the rock's most extreme edge and

Village Bay, St Kilda.

overlooking a 250 feet (75 metres) void the young cragsman was expected to balance on the heel of one foot, bend forward over the awesome drop and grasp his other foot in both hands. Then he was obliged to hold the position, all the time staring down at the jagged rocks and waves so far below, until his companions decided that courage could no longer be doubted. Needless to say, with such cavalier disregard for the perils of sheer heights, male St Kildans frequently ended their lives on the rocks at the bottom of the cliffs.

Right up to the time of their evacuation the St Kildans continued their exploits on the crags and their centuries old pursuit of birds. So dependent were they on the flesh of the Fulmar (*Fulmarus glacialis*) and the Puffin (*Fratercula arctica*) that they developed a distinctive smell thought to derive from the constant consumption of oils associated with these birds. The people's skin acquired a velvety texture, probably from the same cause.

In past centuries the St Kildans were certainly familiar with the Great Auk. They called it 'Gearrbhall' or 'Gernhell,' both words being obvious variants of 'Garefowl.' Several early accounts written by visitors to the island bear testimony to the Great Auk's association with St Kilda, the fullest account being that of Martin Martin (1698) in his book *A Late Voyage to St. Kilda*. The islands were used as a breeding station in the seventeenth century and this use continued, possibly, until the middle of the eighteenth after which time the species' visits became steadily more sporadic. These occasional visits certainly continued until 1821 (when a live bird was sent to Scotland) and probably went on till around 1840 at which approximate time the well known incident concerning the execution of a Garefowl for witchery supposedly occurred.

There is no genuine record of exactly which spots the Great Auks frequented. Much of the coastline is entirely unsuitable as a landing place for flightless birds due to the severity of the cliffs rising straight up from

The interior of a St Kilda cottage after the evacuation.

Chromolithograph by W. and A. Johnston (after Audubon) from G. Seton's *St Kilda - Past and Present* (1878). *Courtesy of Peter Blest.*

the sea. The bird caught in 1821 was on a ledge on the east side of the hill known as Oiseval and the individual from the 1840's is supposed to have wandered some way up Stac-an-Armin, which at 627 feet (200 metres) is the highest rock stack in Britain. J.N. Heathcote, in a book published in 1900 (*St Kilda*), claimed that the place where the Great Auk bred was called, 'The Rock of the Garefowl.' There seems to be no other record of this name, nor does the writer make any attempt to describe or identify the spot.

Finlay MacQueen catching birds in the traditional St Kilda manner.

THE ISLE OF MAN

Evidence of two quite different kinds suggests that Garefowls frequented the Isle of Man in historical times. The first category of evidence is osteological. Great Auk bones have been found on the island, most notably at a mediaeval site at Castle Rushen. The evidence of the second kind is more enigmatic. An old sepia drawing decorated with the words, 'These kind of birds are about the Isle of Man,' exists at the British Museum. It is one of a series of Manx views, all with a reliable provenance, and despite an extravagantly toucanesque development of the beak the image is unequivocally one of a Garefowl in breeding plumage. Drawn around 1652 by Daniel King it is perhaps the oldest of all known representations of the Great Auk; the prehistoric cave paintings at the Grotte Cosquer may, after all, be depictions of something entirely different and the pictures of Clusius (which pre-date the Daniel King drawing by about fifty years) are of uncertain value.

The earliest undoubted figure of a Great Auk. Drawing in sepia and watercolour by Daniel King (circa 1652). The picture's caption reads, 'These kind of birds are about the Isle of Man.'

K. Williamson (1939) pointed out that a considerable area of coastline on the south eastern side of the island is well supplied with low, flattish, shelving rocks that would afford easy access for flightless birds emerging from the water.

There is a curious record published by John Gawne in *The Peregrine* (the journal of the Manx Field Club) for March 1944 that may just possibly relate to Great Auks. Mr. Gawne says:

> This must be the bird Bill Corlett, a Port St. Mary fisherman, told me about in the summer of 1895. It was a flightless bird that used to come at certain times to the Flat Rocks south of the Point, and Bill ... pronounced the name 'Big Uig.' He had been told about it when a boy (perhaps 1840-50) by old men, the same as he told me when I was 13 years, and he 65. He was a very intelligent old man and died, aged 90 years, in 1920.

FUNK ISLAND

> Our two long boats were sent off to the island to procure some of the birds, whose numbers are so great as to be incredible, unless one has seen them; for although the island is about a league in circumference, it is so exceedingly full of birds that one would think that they had been stowed there.
>
> Jacques Cartier (1534)

Funk Island was home to by far the largest Garefowl colony of recent historical times. Estimates place the numbers at around 100,000 breeding pairs when the colony was at its strongest. Yet by 1800 all these birds were gone, exterminated by European sailors over a period of 300 years or so.

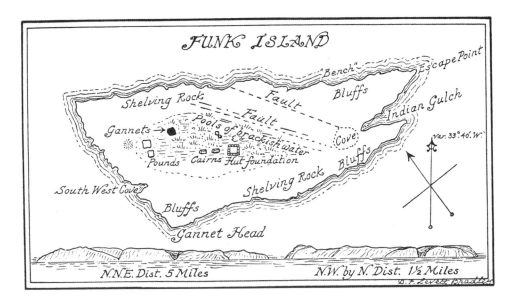

Map of Funk Island.

Although the Great Auk is long vanished, the island is still full of birds and supports one of the world's largest seabird colonies. The num-

Great Auks of Funk Island. Watercolour by Roger Tory Peterson. *Reproduced by kind permission of the late Roger Tory Peterson.*

bers are uncountable and in places the ground is absolutely carpeted with birds although evidence suggests that populations have declined in recent years. As might be expected from the name - and from an appreciation of the vast numbers of birds that live there - Funk Island is a rather foul smelling place covered in parts by vast accumulations of guano.

In size it is small, perhaps half a mile or so long (about a kilometre) by about a quarter mile (400 metres) wide and it is situated almost 40 miles (around 60 kilometres) off the north-eastern corner of Newfoundland. It takes the form of a rather low-lying rock of granite and on most sides mod-

Three views of Funk Island. *Photographs by Dick Wheeler; reproduced with his kind permission.*

An Auk's eye-view of Funk Island. *Photograph by Dick Wheeler; reproduced with his kind permission.*

estly sized, although sheer, cliffs face the ocean. Part of the island's north-eastern side slopes gradually to the sea and this is probably where the Garefowls came ashore.

Difficult of access and surrounded by treacherous currents and fogs, an easy landing cannot be taken for granted. Sometimes days may pass before the weather and tides are suitable. Quite apart from natural obstacles to a visit there are some unnatural ones too. Gilliard (1937) mentioned a superstitious taboo that left Newfoundland sailors and fishermen reluctant to land. The Island is supposed to be haunted. The spirit of 'The Madman of Funk' is said to outscream the gulls. Yet if Funk really is haunted it would surely be by Auks, for it was here that the birds were abused and tortured on a scale that is truly horrifying. Much of the island's topsoil is composed of rotted-down Garefowls and other seabirds, to such a degree that large quantities were commercially removed to provide fertiliser during the 1860's. Even today, bones and bone fragments can still be found preserved in the half frozen mould 200 years after the Great Auk's banishment from this side of the Atlantic.

As well as the bones, the island retains other reminders of the Great Auk's former residence. It is still possible to locate the remains of the walled pens into which the Garefowls were driven before being crudely despatched.

Even in prehistoric times the inhabitants of Newfoundland caught Garefowls and their remains have been found in several places. Probably there were then other colonies on other islands and suitable stretches of coast. There are several 'Penguin Islands' around Newfoundland and these together with a number of 'Bird Rocks' are likely former breeding stations for the species.

Funk Island itself was visited, at the appropriate season of the year, by Beothuk Indians almost up to the time when the colony vanished. These people laboriously paddled their canoes out to the island in the expectation of harvesting birds and eggs and their arrowheads and a paddle were found in a part of Funk known as 'Indian's Gulch.' According to Montevecchi and Tuck (1987) they dried yolks to make a sort of sausage. It was the coming of the European that spelled doom for the species, however, and from the sixteenth century onwards the birds were slaughtered with such ferocity that the outcome was inevitable. It is remarkable, perhaps, that they held out for almost 300 years.

MEN

MEN

Vivian Hewitt looking fondly at two of his Auks having just bought them at the Rowley Sale (November, 1934).

Many men have played a part in the tale of the Great Auk. The few remembered here are chosen for a variety of reasons. Some have made important contributions to Garefowl literature. Some have been more active players. Each displays some facet of man's involvement with the species. Many others could with equal justice have been included. The choice is, of course, a very personal one.

COPENHAGEN CURATORS - THE REINHARDTS AND JAPETUS STEENSTRUP

The establishment of the Royal Museum of Natural History, Copenhagen and the first development of its collections coincided with the years during which the only remaining breeding colony of Great Auks was that frequenting the islands and skerries off south-west Iceland.

In 1806 Johannes Hagemann Reinhardt (1776-1845) was the only candidate for the post of first Museum Inspector. The almost inevitable consequence was that he was given the job. He spent the next forty years lecturing, caring for the collection and trying desperately to increase it on only the meagerest of resources. One of the few advantages he enjoyed was the small supply of Garefowl specimens that trickled into Copenhagen from the far off Atlantic island. For reasons of geography and politics, Denmark held a virtual monopoly (challenged only by Hamburg) over trade between Europe and Iceland and Reinhardt was perfectly placed to secure any Auks that became available. These he used as small, but significant, bargaining chips in his mission to increase the scope and variety of his museum's collection. By means of sales, exchanges or even gifts to those he thought might benefit the museum he steadily improved his connections with other institutions and increased the influence and reputation of his own.

Bogged down by the demands of this work, he found little time for writing but one of the few papers he did manage to publish was a short treatise on the Great Auk. Although his literary contribution was small he played a crucial rôle in establishing the 'cult' of the Great Auk. It has often been implied that his part in promoting sales and generating interest brought about the species' extinction but this is a blame he scarcely deserves. The Icelandic fishermen who actually got the birds were almost as keen to kill them for food as to take them for specimens and the small price they received for each served only as an inducement to prevent the birds from reaching the table.

Surprisingly perhaps - in view of the importance of Reinhardt's museum position - there seems to be no surviving portrait of him.

An Icelandic Auk. Watercolour painted at Edinburgh on July 6 1839 by W. MacGillivray using Audubon's Auk as a model.

(Above, left). Japetus Steenstrup (1813-97).

Above, right). John Theodore Reinhardt (1816-82).

His son, John Theodore Reinhardt (1816-1882), followed in his father's footsteps becoming Inspector of Land Vertebrates at the Royal Museum. Although not associated so closely with the Great Auk (by his time the museum's duplicate specimens were almost all gone), he communicated much detailed information to Alfred Newton. His published works were sometimes crudely criticised but his perceptions were acute. He was, for instance, the first to recognise that the Dodo (*Raphus cucullatus*) was a giant, flightless pigeon. Before Reinhardt all kinds of fanciful affinities were suggested.

Although the Reinhardt's significance in the story of the Great Auk is undoubted, it was Japetus Steenstrup (1813-1897), one of their successors at the Royal Museum, who made the first serious attempt to define the species' history. His paper, 'Et Bidrag til Geirfuglens' (*Videnskabelige Meddelelser*, 1855), has provided a foundation for all later Garefowl research. Symington Grieve regarded Steenstrup as, 'The Father of Garefowl History.' Not only did Grieve find Steenstrup's paper invaluable, he found him to be an exceptionally courteous and helpful correspondent:

> He was kind and generous to a degree, affording information and writing long letters with quotations and references which must have required careful study. In failing health correspondence of this kind was written.

So touched was Grieve by Steenstrup's help that he dedicated *The Great Auk or Garefowl* to him. Several years later, when he heard of the older man's death, he wrote a short *In Memoriam* (see Grieve, 1896-7) to accompany one of his papers.

JOHN WOLLEY (1824-1859)

My Dear Newton,

> I felt sure your letter contained the fatal news as it had a black seal. The loss of poor Wolley is very great and I am most truly sorry for it.

George Dawson Rowley (Nov. 24th 1859)

John Wolley (1824-59).

John Wolley inspired affection and love in all who knew him and his death at the age of 36 deeply shocked his contemporaries. Just a year earlier, in seemingly perfect health, he'd accompanied his close friend Alfred Newton to Iceland where he conducted the research on which so much Garefowl lore is based. Shortly after their return he began to suffer occasional lapses of memory and uncharacteristic feelings of lassitude and by the summer of 1859 his brilliant mind showed alarming signs of giving way altogether. On November 17th, just a week before hearing the news of Woolly's death, George Dawson Rowley had written to Newton:

> The news concerning my poor friend ... caused me the greatest distress. I am indeed deeply shocked to hear of the wreck of so fine an intellect. A kinder hearted fellow does not exist ... We spent so many pleasant hours together in early times that I should feel ... one of the great landmarks of my life had been taken away should he be lost to us.

Yet lost he soon was. Incoherent and delirious, he rallied sufficiently, just before the end, to ask his closest relatives to ensure that his fabled egg collection be given to Newton. It is quite certain that the gift was not misplaced. Inspired by this collection, and by the circumstances under which he'd acquired it, Newton produced one of the great monuments of ornithological literature, the *Ootheca Wolleyana*. Begun in 1864 but not completed until 1905, this fine work is based on the meticulously kept notes of Wolley, augmented and supplemented by Newton's own rich knowledge and experience. Even to generations who find the whole notion of egg taking repellent, this masterly treatise is a profound and fascinating piece of work. W.H. Hudleston (see *The Ibis*, Jubilee Supplement, 1908) describes it as an 'enormous compendium of oological research.' Newton, who clearly felt honourbound to deliver this *magnum opus*, remarked that it was 'largely a record of ancient friendships.' By the time he published the last part more than 45 years had elapsed since Wolley's passing and Newton's own death was just a year or two off.

Although Wolley's contribution to the story of the Great Auk is little known, he was, in fact, one of the biggest players. Because he didn't live long enough to publish the fruits of his research his part in the tale is often overlooked, but one of his great obsessions - perhaps the greatest of all - was the Garefowl. During the last years of his life he formed the intention of publishing a monograph on the species to which end he made his excursion to Iceland. In 1858, the continuing existence of the Great Auk was something of a moot point; partly in the hope of resolving the issue Wolley and Newton travelled to the Reykjanes Peninsula and spent several weeks at the drab Icelandic fishing bay of Kyrkjuvogr. Their intention to visit Eldey and the neighbouring islands and skerries was thwarted by dismal weather that continued throughout the summer and made the short sea voyage impossible. Wolley made full use of the time, however, and spent the forlorn days of waiting in deep conversation with the Icelanders. He interviewed (interrogated is probably closer to the truth) every villager still living who'd had any kind of experience with the Great Auk and doggedly recorded the results of these interviews in notebooks now kept at the Cambridge University Library. After Wolley's death Alfred Newton inherited and then augmented these notebooks (eventually willing them to Cambridge as part of his own bequest) and he published an abstract of his friend's findings in *The Ibis* of 1861. Yet the considerably fuller record contained in the *Garefowl Books* (as Wolley himself labelled them) has been seen by very few and many Auk enthusiasts have remained largely unaware of them. They are certainly among the most important documentations of the life and death of the Great Auk. Without them, names such as Vilhjálmur Hákonarsson, Sigurðr Islefsson, Ketil Ketilsson, Jón Brandsson and Sigriður Thorláksdótter would have vanished into the mist of history. It is entirely due to Wolley that the circumstances surrounding the last raid on Eldey in June of 1844 are known. John Wolley is, then, one of the great unsung figures in the story of Garefowl research.

Born in 1824, educated at Eton and Cambridge, he devoted most of his short adult life to his passion for eggs. The collection he built up acquired almost legendary status at a time when cabinets full of eggs were something

So near yet so far! A view of Eldey (the tiny speck on the horizon) from the Reykjanes Peninsula. Wolley and Newton probably looked out on this view many times in their forlorn wait for the weather to change. The large rock in the foreground is called 'Karl' meaning 'Man' or 'Old Chap.' Photograph by Chris Smeenk. Reproduced with his kind permission.

of a commonplace. Yet Wolley was much more than an 'egger.' His contemporaries were united in their admiration for the power of his intellect and his meticulous attention to detail. Had he lived longer he would doubtless have added substantially to ornithological literature. In the event, it was left to Alfred Newton to ensure that at least some of his life's work was placed on published record.

Such was the affection he inspired that almost half a century after his death, Newton concluded a brief memoir of Wolley's life (see *The Ibis*, Jubilee Supplement, 1908) with the following words:

> His good qualities are treasured in the recollection of those who
> knew him, and especially of one to whom he gave the last token
> of his esteem, and who, having endeavoured (how imperfectly no
> one knows better than himself) to discharge a duty owing to the
> memory of a deeply lamented comrade, cannot conclude this
> sketch without an expression of gratitude at having been
> permitted to share so largely the intimacy of such an
> upright man.

ALFRED NEWTON (1829-1907)

I'm sometimes approached by people who tell me that they could write better books than mine. I usually reply, "Yes, you probably could. But you didn't, did you!"

Jackie Collins (1937-)

Alfred Newton- Cambridge scholar, erudite natural historian, man of letters- pursued a curious dream for most of his adult life, a dream that he was destined never to fulfil. With his fine mind and industrious nature he left behind at his death an extensive body of achievement, yet he never produced the one work that seemed closest to his heart. He never produced the definitive monograph on the Great Auk or, as he preferred to call it, the Garefowl. For thirty years or so (from the 1850's to the 1880's) he fussed and worried and niggled over it. In hot pursuit of his obsession he collected together every scrap of information he could uncover, he followed up every lead - no matter how unpromising - that might furnish him with more knowledge, he corresponded - almost daily - with colleagues, collectors and interested parties the length and breadth of Europe. He toured America and with his friend John Wolley he travelled to Iceland where, together, they conducted the research on which so much of our knowledge is based. When Wolley died, quite soon after their return and at the comparatively early age of 36, Newton, quite appropriately, inherited the fruits of their joint research.

Although he published a few fragments and several papers on the Garefowl, he still felt unable to produce his masterwork and eventually embarked on a second collaboration, this time with George Dawson Rowley, author of the beautiful and entertaining *Ornithological Miscellany* (1875-8). This partnership too was ended by death, for Rowley died suddenly in 1878. Once again Newton was left in sole possession of the most extensive collection of Garefowl lore ever assembled.

A drawing of Eldey made by Alfred Newton as- unable to land - he sailed regretfully by.

Eldey bearing S.W. by W.
27 July 1858

Alfred Newton (1829-1907).

With access to more information than anyone else could hope to gather, with a mind as subtle and sharp as any naturalist of his generation and full of an enormous fund of peripheral knowledge, it seemed no more than his natural destiny to produce a work that would match the scope of his life-long obsession. Certainly, no man seemed better equipped to do the job.

And then came an event that shattered the great plan. After thirty years of solid, perceptive labour, after he'd penned thousands and thousands of private words on his pet subject, someone else beat him to the punch. In 1885 a resident of Edinburgh, a man completely unknown to most of the naturalists of the day, did what Newton had consistently failed to do. Symington Grieve published *The Great Auk or Garefowl*. Rightly regarded as a classic of its kind and the cornerstone for all subsequent Garefowl research, Grieve's book came as a body blow to Newton.

Naturally enough - perhaps - he believed he could have done better. Much better. Probably he could have. The fact is, he didn't!

With an irony that seems almost merciless, it was to Newton that magazine and journal editors turned when they wanted Grieve's book reviewed. He was, of course, the leading authority on the subject. Symington Grieve himself takes up the story in a letter to Robert Champley dated August 18th 1885 and records both Newton's gentlemanly instincts and the difficulty he experienced in living up to them.

> I may tell you that Prof. Newton was asked to review my book and asked to be excused. He was told he might name any other gentleman he liked but none of those approached would take the work so he wrote and explained to me the position. I told him that as far as I was concerned it would give me pleasure if he would do what the Editor of *Nature* wanted. The next letter I got from the Professor told me he had written an article which was too egotistical and he would not call it a review ... so he was not sure if the Editor... would publish it. Last night I got the proof of the article from the Professor with a kind note saying that he would give effect to any alterations I cared to suggest if they did not too much alter his views. I have returned the article which I did not feel I could alter a word of.

The review was so brutal that it is easy to understand why Newton felt so squeamish and why Grieve's response was so disdainful. The Cambridge Professor may have begged to be relieved of the burdensome task of writing but he went ahead nevertheless, which perhaps suggests that the opportunity of taking a public swipe at Grieve's effort was irresistible

Newton was not an unfair man. Almost half of his review is taken up with an analysis of his own motives for penning such a poisonous description and he gives a carefully worded statement of his position. In short it is a declaration of interest:

> The force of circumstance has compelled me to set up a very high standard; and, when that standard has not been approached by any writer on the subject, it is almost impossible for me not to see his shortcomings, though many another might find... no fault at all. I therefore wish ... to record my opinion that in the present work the author has done the best that in him lies, and especially that this book, so far as it goes, is an honest book.

Such a style is fairly characteristic of Newton who liked nothing better than to outline the scope of his work or define his own intentions before actually beginning. He was the kind of man who was capable of writing an introduction quite as long as the main body of the work it fronted. Eventually, he came to the crux of his dilemma:

> If, after working at the subject for more than a quarter of a century, a man still finds himself unable, from one cause or another, to publish the results of his labour, it does not follow that he should be hard on anybody else who, with perhaps as many distractions, makes a praiseworthy attempt to set before the world what is known of a

Kenny Everett, taxidermist of Truro in Cornwall, holding Symington Grieve's personal copy of *The Great Auk*. Grieve continued to collect Garefowl lore long after publication of his book and this copy, which Mr. Everett currently owns, has many extra pages filled with the author's handwritten notes detailing information discovered too late for inclusion.

lost species, though he may not have devoted to the task a tenth of the time.

Having made this declaration, poor Newton found it impossible to live up to:
> There cannot be a dispute as to the great pains the author has taken with this work, but it would be inexpedient here to attempt any criticisms of its details, to an abundance of which exception may be taken.

He 'regrets to mention' that the book will not bear 'severe' criticism and, with undeniable truth, dismissively adds:
> The fact seems to be that up to a certain point the story of the Great Auk can be worked up and told by anyone willing to labour at it.

So that no-one could remain in any doubt about his position he continues:
> He [Grieve] has needlessly raised fresh difficulties for future investigators. Mistakes that have taken years of labour to correct ... are again set agoing ... suggestions are hazarded that have ... no firm ground ... Mr. Grieve has been unable to distinguish between good evidence and bad.

That Symington Grieve understood and sympathised with the human frailty so sharply exposed is clear from his letters and both men - remarkably perhaps - remained on relatively good terms with one another. Newton even went so far as to visit Grieve in Edinburgh. A letter from Grieve to Champley, this one dated September 30th 1885 indicates that at least some goodwill remained between them:
> Your letter arrived last night curiously enough when I was sitting having a chat with Professor Newton after dinner ... The Professor admits that there was nothing seriously wrong with my statements and that the errors he says he has found out are trivial.

So, Alfred Newton was thwarted from being the first to produce a major work on the Great Auk. Yet, an enigma remains. Having expressed such dissatisfaction with Grieve's work, it was almost incumbent on the Professor to give to the world the authoritative and measured tome of which only he was capable. But he never did.

He was to live for upwards of twenty more years during which time he never stopped gathering additional information and material, but he still never produced his book. In March of 1894 he wrote to J. Harting:
> I wish there were the slightest chance of my being able to finish a Great Auk book before I die - but it is impossible. Nevertheless, I go on collecting all the materials I can, and somebody who comes after me may make use of them.

At his death in 1907 he left to the Cambridge Museum of Zoology his enormous collection of books and correspondence as well as a wealth of manuscript notes and miscellaneous writings. Nor was this just his own material. All the notebooks and other natural history items that he'd inherited, so many years previously, from his friend John Wolley (including the

Lord Lilford's five eggs, four of which he gave to Alfred Newton. The fifth is now in The Natural History Museum, London.

> I give and bequeath to the ... University of Cambridge my Natural History Collections and Library together with the Cabinets, Cases and Apparatus thereto belonging including all my Copyrights, Books, Pictures, Prints, Drawings, Letters and Papers ... to be kept for the purposes of the said University and not for the purpose of sale ... I also give ... the sum of One Thousand Pounds ... to invest the same and to apply the annual income thereof to the keeping up and adding to the said Library.

It is still all there, including a Great Auk skeleton and seven eggs (Newton never actually owned a stuffed Garefowl - throughout his adult life he had daily access to the one belonging to the University Museum of Zoology), most of the letters and manuscript material now being housed in the University Library. The vast majority of the information he assembled on the Garefowl has never been published and is rarely looked at. It is the ornithological world's loss, as much perhaps as Newton's, that he never drove himself to finalise his Garefowl research.

Alfred Newton spent most of his life at Cambridge. According to Hans Gadow (1910) he was 'the life and soul' of the Department of Zoology. In many ways the embodiment of a thorough and cautious man of learning and science, he was much more than just that. He belonged to an age when such people seemed to be more profound and rounded men than is perhaps the case with zoological scientists today. There is no question that he had an elegant- even romantic and poetic- turn of mind. Long after most of his colleagues believed the Great Auk to be extinct, he persisted in the idea that somewhere in the chill waters of the Atlantic a few individuals survived. Never for a moment did he believe that the pair of birds killed on Eldey in 1844 were the last two. Only a man with a poetic sense could close the review of Symington Grieve's book in the way that Newton did:

> I most sincerely wish that I could accord higher praise to this work than I have been able to do, for Mr. Grieve's enthusiasm in the cause deserves greater success. It is seldom that anyone but a Fennimore Cooper or a Charles Kingsley feels the romance that clings around the history of an expiring race. Most men - men of science especially - nowadays believe in the survival of the fittest, and are content to let the dead bury their dead. The moral lesson I do not venture to draw, and in conclusion have only to ask pardon of the readers of *Nature* for putting myself so forward in this article.

What, one may wonder, did Symington Grieve make of it all? In his letter to Robert Champley of August 18th 1885, he wrote:

> To some minds the article will suggest a notice such as you sometimes see on lands protected for game - " Garefowl Preserve, Trespassers will be prosecuted."

SYMINGTON GRIEVE (1849-1932)

> In submitting these pages to the public the author has fears that they will not bear severe criticism; but he must plead as some excuse that they have been compiled during the relaxation of evenings that have followed the toils of active business life.
>
> Symington Grieve (Preface to *The Great Auk or Garefowl*, 1885)

The name of Symington (the *y* pronounced as the *i* in Simon, rather than as the same letter in simple) Grieve is, by virtue of its strangeness, quite unforgettable. By virtue of Mr. Grieve's celebrated monograph, no other man has a name more closely linked with the Great Auk. Yet when he published his book in 1885 he was hardly known to other natural historians and his work came as a complete surprise to them- as he wrote to Alfred Newton on August 15th 1885, "I am not a scientist, but a merchant." In a previous letter, written on September 19th 1884, he made clear his own view of his achievement:

> I can claim very little originality for my ... work, as it is prepared more with the idea of bringing wisdom within the reach of all ... I may mention that I am not publishing with any desire for profit. In fact I anticipate a heavy loss and possibly may allow my book to be sold under cost price.

Symington Grieve (1849-1932). *Courtesy of Christian Thin and John Ballantyne.*

Without doubt this beautiful book is a labour of love and notwithstanding the author's modest opinion of it, it forms the basis for any serious study of the species.

A rare surviving bottle of Uam Var whisky, blended by the firm of Innes and Grieve. *Courtesy of Barry Sinclair and Drambuie Ltd.*

The title page of Grieve's book.

THE GREAT AUK, OR GAREFOWL

(*Alca impennis,* LINN.)

Its History, Archæology, and Remains

BY

SYMINGTON GRIEVE
EDINBURGH

LONDON
THOMAS C. JACK, 45 LUDGATE HILL
EDINBURGH: GRANGE PUBLISHING WORKS
1885

Despite its unequivocal success and a reputation persisting to this day, Symington himself remains a remarkably elusive figure. For a man who produced important books and papers on a variety of subjects and who was the President of one learned society and an honoured member of several others, the story of his life has proved surprisingly difficult to piece together. There seems, for instance, to be no picture of him in any institutional collection in Scotland. Only through the intervention of Mr. John Ballantyne (of the Scottish Ornithologists' Club) who painstakingly tracked down Grieve's grandchildren (who, in turn, were kind enough to provide a photograph) has it proved possible to feature one here. It is thanks to the reminiscences of Symington's granddaughter - who still remembers his visits on Sundays when he would pick wild flowers with her or, after church, read to her from Rupert Bear books - that a few details of his life can be recorded.

One of the reasons for his descent into relative obscurity may be the diverse range of his interests. No sooner had he built a reputation in one field of activity than he moved on to another. A second reason may be that he lived to the ripe old age of 82, by which time most of those who'd known him in his prime were dead themselves and unable to contribute posthumous reminiscences. Obituaries are, of course, an easy way to determine the facts of a person's life but the few obituaries that exist for Grieve - fulsome though they may be - are little more than lists of his literary achievements. Yet another reason for his elusiveness must surely lie in his personal modesty and the disdain he showed for any act of aggrandisement. He tended to advance the cause and praise the achievements of others while disparaging his own. The letter written to Newton in 1884 makes this abundantly clear:

> When I think of the nature of your research and the knowledge you will be able to bring to bear upon the subject, I almost feel ashamed to make my compilation public.

His letter of the following year makes his generosity of spirit towards his Garefowl rival even more apparent:

> My wife and I are not grand folks ... however, if you will honour us with your company during the few days you are to be in Edinburgh, both of us will give you a true welcome and make our house your home for the time being. We have no family so you can have perfect quietness or see as many friends as you like.

What makes this kind invitation all the more extraordinary is the fact that it was written soon after Newton had penned his vitriolic, green-eyed review of Grieve's book. Symington was obviously a man of great compassion and tolerance for the failings and weaknesses of others. Perhaps understandably, Newton felt unable to take up the Grieves's offer and restricted himself to accepting an invitation to dinner.

Very much a 'gentleman amateur' Symington Grieve's scholarly achievements are evident even though he was the first to dismiss them. His interests were catholic. As well as his famous work on the Great Auk he produced a splendid two volume study of the Scottish islands of Colonsay and Oronsay called, appropriately enough, *The Book of Colonsay and Oronsay* (1923). This work represents the culmination of 45 years research into the history, archaeology and folklore of the islands. He published an important and influential paper in the *Transactions and Proceedings of the Botanical Society of Edinburgh* on 'The Floating Power of Seaweed' and he also produced work on corals. As far as birds are concerned his attraction to the rare and the curious is clear. In 1889 during a round the world trip that took in India, Sri Lanka (Ceylon), Australia, New Zealand, Hawaii and North America, he became acquainted with the New Zealand Kiwi and many years later produced an interesting memoir on the subject. On another journey, to Dominica in the West Indies, he undertook a study of a rare species of Petrel. This journey also inspired another important book, *Notes upon the Island of Dominica* (1906), a work described by R.A. Myers in the *World Bibliography Series-* Dominica (1987) as, 'the fullest general description of early twentieth century Dominica.'

Born in 1849, Symington lost his father when he was thirteen and from then on was obliged to make his own way in the world. More or less self-educated, at least in natural history matters, he was sent from his home in Edinburgh to London where he was apprenticed into the grocery trade. On his return to Scotland he joined forces with his brother Somerville and became a wine merchant and whisky blender. Together they established a blending plant in Edinburgh, their firm being known as Innes and Grieve. The brand of whisky they produced is, like the Great Auk, now long defunct but was sold under the name of *Uam Var*.

That Grieve was a man of enormous personal charm and one who showed every consideration to others is evident in the few writings of a personal nature that survive. Always these are characterised by their thoughtfulness and modesty. In a letter to Robert Champley dated January 26th 1885, he wrote:

> Many thanks ... for so kindly subscribing for three more copies but while I will put your name down for these I would not for anything that you should take them merely to encourage me as I have quite made up my mind for some loss and do not wish to make anyone else suffer unless they really have a use for the book ... I begin to fear it may not reach the point of excellence you desire though ... I trust it may meet a felt want in ornithological circles.

Thomas Parkin photographed while touring Australia during 1891. *Courtesy of Ralfe Whistler.*

Engraving from Thomas Parkin's pamphlet *The Great Auk or Garefowl* (1894).

THOMAS PARKIN (1845-1932)

Thomas Parkin, outstanding amateur cricketer (a 'tremendous slogger,' recalled one of his contemporaries), uninhibited grangerizer of books, resident of Hastings, East Sussex, keen traveller and avid collector was a man whose means were just sufficient to enable him to spend a lifetime indulging his varied interests. An obituarist noted that he, 'was of much too joyous a nature to devote himself to hard work,' and his long life was largely passed in pleasing himself. One of his abiding passions was the Great Auk and he contributed two papers to the literature, one of them, 'The Great Auk- A Record of Sales of Birds and Eggs by Public Auction'

A drawing of Thomas Parkin (1845-1932) by Cyril Davenport.

(1911), being well known to antiquarian natural history booksellers simply as 'The Parkin Pamphlet.' This is a detailed compilation of the history of each stuffed bird and egg known by Parkin to have passed through the saleroom.

Parkin was particularly intrigued by sales of Auk eggs and attended many of those that occurred during his lifetime. One of his favourite little jokes was to always start the bidding with a call of £50 then sit back and take no further part in the proceedings. Eventually he did summon up the financial courage to take the plunge, acquiring an egg to take back to his home in Hastings and paying an undisclosed sum in the process. He kept it in a secret cupboard in a room that Anthony Belt (1933) described as:

> Perhaps ... the most remarkable room in Hastings ... every inch of wall-space covered with treasured possessions, the beams of the low ceiling adorned with ancient pistols, guns and bayonets, the floor crowded with overladen tables, cabinets, book-cases and a great horse-hair sofa never used apparently but as a receptacle for the innumerable things that were always needing to be put down somewhere.

Here, according to Belt:

> Parkin would sit with his friends, for ever re-lighting a pipe always forgotten after a few puffs, constantly jumping up to fetch books to illustrate the subject of the moment, piling them into the visitors lap without giving time to look at what they had been brought for, ceaselessly talking, suddenly breaking off to inquire with a courteous inclination of the head, "And how is Madame? Well I hope."

A few months before his death Parkin took the decision to sell his Great Auk egg. It is easy to imagine the sadness with which he parted from what was, probably, an object he treasured more than any other. He was, said one of his obituarists:

> Not childish but charmingly child-like ... there was nothing small or mean about him; he never failed a friend or traduced an opponent... thinking no evil of anyone it was natural to him to credit others with good intentions; the soul of generosity, he was very sensitive of kindness and sought to repay it. In short- a Christian gentleman.

WILHELM BLASIUS (1845-1912)

Every bit the equal of Alfred Newton in terms of erudition, Wilhelm Blasius, a native of Brunswick in Germany, set very high standards in all his writings. Indeed, he was probably even more meticulous and thorough than his English counterpart. He wrote several papers on the Great Auk including a painstakingly researched memoir on eggs and mounted specimens that he generously allowed Symington Grieve to publish- in translated and somewhat reduced form- as an appendix to his book. His work culminated in the magnificent entry on *Alca impennis* that he wrote for the third edition of Naumann's *Naturgeschichte der Vögel Mitteleuropas* (1896-1905).

Sonderabdruck aus Naumann, Naturgeschichte der Vögel Mitteleuropas, Band XII.

Der Riesenalk, Alca impennis L.

Neu bearbeitet von
Geh. Hofrat Prof. Dr. Wilhelm Blasius in Braunschweig.

Gera-Untermhaus.
Lithographie, Druck und Verlag von Fr. Eugen Köhler.

This highly detailed account is without doubt one of the landmarks of Great Auk literature.

His brilliance was recognised early and in 1871, aged just 26, he was appointed Director of the Professorial Chair at the Carolo Wilhelmina University in Brunswick. At the same time he was made Director of the Herzogliches Naturhistorisches Museum (the Staatliches Naturhistorisches Museum, Brunswick), succeeding his father, Dr. Johann Heinrich Blasius, who had died during the previous year.

It is a curious fact that Blasius showed considerably more kindness to Symington Grieve in the run-up to the publication of Grieve's book than did most of the professional naturalists in Britain. This natural generosity of spirit was recognised in one of his obituaries (see *Auk*, 1912, p.571) where it is noted:

> He visited America during 1907 as a delegate to the International Zoological Congress in Boston and was remembered as a man of the most lovable disposition and sterling qualities...with a personality which attracted all with whom he came into contact.

(Above, left). Wilhelm Blasius (1845-1912). *Courtesy of the Staatliches Naturhistorisches Museum, Brunswick.*

(Above, right). The wrappers to Blasius's Great Auk entry for J. Naumann's *Naturgeschichte der Vögel Mitteleuropas* (1896-1905). This work is described by Sitwell, Buchanan and Fisher in *Fine Bird Books* (1953) as, 'The most scholarly and complete ornithological text book of its time (perhaps of any time).'

VIVIAN HEWITT (1888-1965)

Captain Vivian Hewitt, proud possessor of four stuffed Great Auks and thirteen eggs, was a twentieth century enigma. Fiercely rich, his fortune inherited from family brewing interests in Grimsby, his eccentric career lurched between racing fast cars, building steam engines, making pioneering flights in Blériot aeroplanes and collecting- on a variety of fronts- with a truly grand passion. This was a man who would, as a matter of course, answer any begging letter by popping a £5 note into an envelope and dispatching it immediately, who, despite the great wealth that he was born to, enrolled as an apprentice locomotive builder at the British Rail depot in Crewe and joyfully stuck at the self-inflicted task for four years, who employed ten men, week in, week out, year after year, to build an immense wall around the house he lived in - the exact purpose of which he never disclosed. This house was, in fact, not actually his. Dismayed by any kind of official document, he baulked at the idea of registering a house in his own name. Instead, he purchased the property in the name of his housekeeper thereby rendering himself simply a lodger in his own home. He was indeed a very spectacular character.

Vivian Hewitt- resident of Anglesey in Wales and Nassau in the Bahamas- was by virtue of his birth, a very lucky man and he used that luck to indulge whatever whim, passion or fancy seized him. Like Citizen Kane he used his money simply to 'buy things.' Perhaps his grandest gesture was to fire off a cheque for £50,000 to the Chancellor of the Exchequer, an interest free loan issued to the British Government in 1941 as part of Captain Hewitt's contribution to the war effort.

His collecting mania was legendary. Stamps, coins, guns, *objets d'art*, were all pursued with a white hot frenzy. Yet one of the Captain's chief delights appears to have been the careful unwrapping of any treasure he'd just bought then its meticulous re-packing - never to be re-opened again. At his death many of his possessions - just like Citizen Kane's - were found still packed in the boxes they'd been delivered in. His biographer William Hywel (1973) described the situation:

> So easy would it have been to build a special room with display cabinets, where he could have tabulated his specimens and exhibited them with pride. But no, they were hidden away in every conceivable

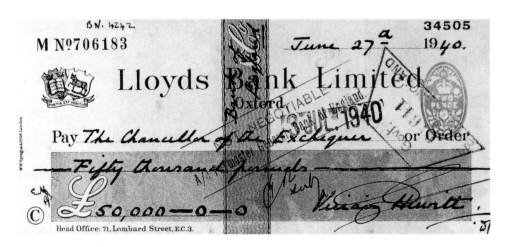

Vivian Hewitt's cheque for £50,000 - his personal contribution to the war effort.

The Animated Picture of VIVIAN HEWITT'S FLIGHT TO IRELAND is the sole copyright of Mr. A. Cheetham, who Cinematographed the Flight, and this Picture can only be seen at the MARKET STREET Cinema, Rhyl.

Photo by JNO. WILLIAMS   RHYL

Vivian Hewitt - aviator. *Courtesy of Christopher Frost.*

and unlikely corner ... His bed-sitting room, as well as the passages, were stacked high with countless boxes and all available drawers were equally crammed. He himself had an uncanny knack of laying his hand on a required box, but he never appeared to have the inclination of showing off his wares to anyone else.

Of all his collecting passions none was greater than his passion for birds and their eggs. Not that he himself was an active nest raider; his preference was for raiding those who had actually performed the robbery. Indeed, so avid a protectionist did he become that he would not allow shooting on his property. Thus he hunted out old collections, always buying the choicest and rarest examples. By the end of his life it is reckoned that he'd amassed nearly a million eggs.

The Great Auks were, of course, his pride and joy and whenever an example- either in stuffed or egg form- came onto the market, he did his best to acquire it. One by one he added eggs to his collection until he reached the grand total of thirteen - almost 20% of those known to exist. The stuffed birds proved a little more difficult but he still managed to obtain four of them. In this century the only institution to equal that feat is the Natural History Museum of London. No private individual has come close.

In November 1934, he purchased the two Great Auks that once belonged to George Dawson Rowley together with two eggs. The bill came to £1,659, a vast sum of money by the standard of the day. William Hywel (1973) described the scene when the birds arrived at the Captain's home in Anglesey:

> The Captain, unable to suppress his joy at their safe delivery,
> summoned all the workmen ... The occasion was worthy of an
> audience and the building of the wall assumed secondary
> importance for the time being. He launched into an oration
> on the Great Auk, its habits and history, finally concluding
> with the price he had paid for these much sought-after additions
> to his collection. The men were stunned, almost unbelieving.
> It was yet another of the imponderables which surrounded
> their master and shrouded his actions in mystery. When one
> considers that at this time the average wage for a farm labourer
> was £1 a week it is not surprising ... At the current rate of wages,
> their employer had paid ... what it would have taken a farm
> labourer thirty-two years to earn! One of the men could not
> contain himself.
>
> "Pay me that much," he said to the Captain, "And I'll go
> in a ruddy cage myself and sing to you."

Hewitt remained unmarried and the break-up of his estate was beset with complications. The chaotic nature of his collecting habits and his complete failure to catalogue his holdings caused considerable difficulty after his death in 1965. As far as his egg collection was concerned, there was an added problem. Legislation aimed at protecting birds had come into force during his lifetime and the eggs of wild birds could no longer be bought or sold. The person who inherited the collection of eggs, in despair at not being able to legitimately dispose of them (and having no wish to house such a vast quantity), threatened to throw them all off a Welsh cliff-top. Hewitt himself would, naturally, have found such a situation dreadfully distressing. In a letter he wrote to C.R.S. Pitman written on March 18th 1947, he wrote:

> I wouldn't leave eggs of mine to any museum ... None of them seem
> to take any interest ... and I cannot see a valuable and historical
> collection such as mine treated in this way.

Anyone who has spent much time in museum basements and seen the cavalier treatment often handed out to collections of many kinds might sympathise with the Captain. Yet his failure to make any proper provision for his birds and eggs almost brought them to an end far worse than any he might have imagined. The dreadful scene on the cliff-top was averted at, it seems, the eleventh hour when the new owner was persuaded to donate all to the *British Trust of Ornithology*. Rather ironically, this organisation has since been allowed to sell them and the majority were bought by the crazed billionaire John Dupont (currently- 1997- having just been tried for murder) for the Delaware Museum of Natural History.

Hewitt's Auk material was never part of this fiasco. Exempt from any protective legislation because, of course, there was no living creature left to protect, it was all eventually sold by Spink and Son Ltd. of London. His amazing collections of stamps, coins and other material were similarly dispersed.

Perhaps Vivian Hewitt is best summed up by his biographer and friend William Hywel in a vivid description of the day when the Captain first showed him a stuffed Great Auk, perched precariously on top of a chest of drawers, and asked Hywel to guess how much it had cost:

> In my ignorance ... I remained dumb but curious ... Having failed
> to entice me to drop a monumental brick, he casually disclosed
> the figure and left me flabbergasted and wondering what use I
> might have made of an equivalent amount of money. He then
> opened a drawer underneath the ornament, for as such I regarded
> it, and there, wrapped up in various articles that comprised his
> underclothes, he finally unearthed a cardboard box. Inside,
> nestling in layers of cotton wool, were two large eggs.

They were, of course, Great Auk eggs!

THE PENGUIN. —

—Alca impennis.

NAMES

NAMES

Mergus Americanus Clusius, 1605, *Exoticorum Decem Libri*, vol.5, p.103.
Alca impennis Linnaeus, 1758, *Systema Naturae*, 10th ed., vol.1, p.130.
Plautus impennis Brünnich, 1772, *Zoologiae Fundementa*, p.78.
Pinguinus impennis Bonnaterre, 1790, *Tableau encyclopédique et méthodique-Ornith.*, vol.1, p.28.
Pingouin impennis Buffon, 1817, *Histoire Naturelle* (Lacépède edition), vol.14, p.313.
Alca borealis Forster, 1817, *Synoptical Catalogue of British Birds*, p.29.
Chenalopex impennis Vieillot, 1818, *Nouveau Dictionnaire d'Histoire Naturelle*, vol.24, p.132.
Alca major Boie, 1822, Ornithologische Beiträge, *Isis*, p.872.
Mataeoptera impennis Gloger, 1842, *Hand-und Hilfsbuch der Naturgeschichte*, p.475.

Penguin, Garefowl, Apponath, Wobble, Esarokitsok, Riesenalk- these are just some of the names that have been applied to *Alca impennis*. Great Auk, surely the most widely recognised, is one of the most recent. It appeared, quite suddenly, around the middle of the eighteenth century but the coiner's name is unknown. The derivation is quite obvious, however, and a similar expression is used in German - *Riesenalk*, which translates literally as *Giant Auk*. The word *Auk* appears to come from the old Norse *Alca* which is probably a phonetic rendering of the cry of several Alcid species.

Some of the older names are more mysterious, perhaps none more so than the name *Penguin*. The application of this word to the Great Auk and also to the familiar birds of southern oceans has caused endless confusion, but it is not widely realised that the Garefowl and its relative the Razorbill were actually the original Penguins. For how long the term was in use and the exact nature of its derivation are matters of some uncertainty but, clearly, it was once very popular among sailors. There is little doubt that it was sailors who first transferred it to the birds they saw - and thought were similar - in southern seas. As they ventured more frequently way below the Equator so the name was applied with increasing regularity to the birds now regarded, almost universally, as Penguins. The decline of the Northern Penguin doubtless hastened this trend as generations of seafaring men grew up who were never likely to encounter the species. With the rise in popularity of the name *Great Auk*, the term *Penguin* steadily fell into disuse for the northern bird - at least in the English-speaking world. In France it persisted until much more recent times. Here, perversely, the Penguins of the southern hemisphere were often called by the name *Gorfue*, an obvious corruption of the old northern word *Garefowl*, while *Pingouin* was reserved for the Razorbill and *Grand Pingouin* for the Great Auk.

The Penguin. One of a series of prints produced by *The Society for the Promotion of Christian Knowledge* (circa 1860). Unable to decide whether he was painting a Penguin or a Great Auk, the artist has created a charming hybrid. *Courtesy of Mrs Samuel K. George.*

THE PENGUIN.——*Alca impennis.*

This bird has such small and short wings, that it would in vain try to fly. But these wings, though small, are of great use to it when it seeks its food. The Penguin is fond of fish, and moves with amazing swiftness, by help of its wings, under water, in search of its food. It passes the chief part of its life on or in the sea; and, being usually very fat, it does not suffer from remaining a long time in a wet and cold state. When on land, flocks of these birds may be seen walking upright in a formal, stately manner, holding their heads high. They look, from a little distance, like a company of soldiers. As the feathers on the breasts of some of the species are beautifully white, with a line of black running across, they have sometimes been compared, when seen afar off, to a number of children with white aprons tied round their waists with black strings. The Penguin loves a cold climate. It sleeps very sound, and is extremely tenacious of life. The female lays a single egg. She makes a slight hollow in the earth, just large enough to prevent her egg from rolling out. The manner in which the Penguins and Albatrosses, with a few other species of sea birds, lay out together a piece of ground, (four or five acres,) for their nests, and superintend their charge, is given in Bishop Stanley's work on Birds, vol. ii. p. 276, &c. Edit. 1840. Some Penguins in the South Sea Islands are called Hopping Penguins, and Jumping Jacks, from their habit of leaping quite out of the water, sometimes to the height of three or four feet, on meeting with any check in their course through the sea.

(Right). The complication over the names 'Penguin' and 'Garefowl' is shown in the captions to this coloured plate produced for H.R. Schinz's *Naturgeschichte und Abbildungen der Vögel* (1833).

(Below). The plainly derivative image of the Great Auk in Schinz's first edition was substituted by another (equally derivative) in his second (1846-53).

This confusion is reflected in many older books where pictures of the Great Auk are sometimes captioned *Penguin* and pictures of Penguins are captioned *Great Auk*. Very occasionally, illustrators produced rather bizarre hybrids incorporating characteristics of both creatures. Similarly, older text descriptions are sometimes very unclear and it is not always possible to determine which kind of bird is referred to.

As far as the origin of the name *Penguin* is concerned, there are several schools of thought. The most likely idea is that it is Welsh or Breton in origin. In the third volume of *Hakluyt's Voyages*, in the middle of a section describing a sixteenth century voyage to Newfoundland, is the following curious passage:

ALC.1 impennis. Der nordische Pingouin. I Geirfugl.

Another picture with a caption showing the possibility for confusion over names. Hand-coloured engraving by J. Walter from his *Nordisk Ornithologie* (1828-41).

Madock ap Owen Gwyneth [a twelfth century Prince of North Wales] ... gave to certaine islands, beastes and foules, sundry Welsh names, as the Island of Pengwin ... there is likewise a foule in the saide countreys called by the same name at this day, and is as much to say in English, as Whitehead, and in trueth the saide foules have white heads.

Another possibility is that the name comes from the Latin *pinguis*, meaning fat; yet another is that it is derived from pinion-winged.

Among the peoples of the far north - Scandinavians, Icelanders and Gaelic speakers - the Great Auk was known by the name of *Geirfugl* and its variants - *Garefowl, Gejfuglen, Goirfuglir, Gearbhul* etc. The etymology of this word and the course of its development through the north is a much debated subject. In Icelandic the word *geyr* means *spear* and it is likely that this meaning lies at the root of the name. Probably it has something to do with the weapon-like nature of the beak but perhaps it relates to the spear-like movement of the bird through the water. In this case it might be compared to *Gyrfalcon*, the name of a species with spear-like movement through the air. There are many variations on the word but most are instantly recognisable. *Fugl*, of course, is the word from which *fowl* derives and is the same word as *vögel*, the German word for bird.

One name that may, or may not, have been applied to the Great Auk is *Apponath*. The evidence for its application is rather sketchy and it may well have indicated another species of Auk - perhaps the Razorbill. Various explanations are given for the word, a term that first entered the literature in accounts of the voyage of Jacques Cartier. This famous expedition took place in 1534 when the Frenchman took two ships from St. Malo to Newfoundland. Some philologists have maintained that *Apponath* - spelled variously as *Apponatz, Aporath,* or *Apponar* - is clearly of Eskimo or North American Indian origin. Others, including Steenstrup (1855), believed it might be a French corruption of the English word *harpooner*. The spelling

Hand-coloured lithograph by C.F. and A. J. Dubois from their *Oiseaux de l'Europe et de leurs Oeufs* (1861-72).

Haute brachiptère

apponar is, perhaps, particularly suggestive of English spoken with a heavy French accent.

The name *Wobble* has been associated with the Great Auk with just about as much certainty as *Apponath*. Its derivation, however, appears much easier to resolve. Presumably it refers to the species' rather uncertain means of progression when out of the water.

Even more certain is the meaning of the word *Esarokitsok*. Used by the Inuit of Greenland it means stump or small-winged. Similarly clear is the meaning of *Brillenalk* which can be best translated as *Spectacled Auk*.

The technical name for the Great Auk is *Alca impennis* and this, with a few aberrations is how the species has been listed since the mid-eighteenth century when Linnaeus devised the modern nomenclatural system. Quite recently it has become fashionable in 'scientific' ornithological circles to call the species *Pinguinus impennis*. While the name is undeniably attractive, its use is based on the wish to generically separate the Great Auk from the Razorbill (*Alca torda*). Bearing this in mind, it is difficult to appreciate the reasons for it having taken such root.

There are, of course, guidelines or rules as to whether one species should be included in a genus with others, but inevitably all such decisions become matters of common sense applied to zoological knowledge. Closeness of relationship between Razorbill and Great Auk seems so immediately apparent that recognition in their respective technical names is entirely appropriate. Although there are clear differences in external appearance and skeletal structure, common sense suggests that the Great Auk is neither more nor less than a modified Razorbill. In the absence of strong physical evidence to the contrary, it seems self-evident that the relationship between these two species is far closer than their relationship to other Alcids and if the generic unit is to retain real meaning the breaking up of such a genus makes little sense. Giving little clarity to the nature of any supposed relationships within the family, use of the name *Pinguinus* may be an indulgence based on emotion rather than reason.

Two engravings by Clusius (1605) that may or may not show Great Auks. (Top). The bird Clusius calls *Anser magellanicus*. (Bottom). Clusius's *Anser americanus*.

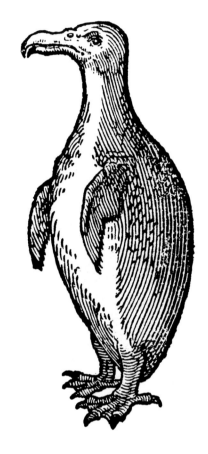

*LATE RECORDS,
ANOMALOUS
SIGHTINGS AND
CRYPTOZOOLOGY*

LATE RECORDS, ANOMALOUS SIGHTINGS AND CRYPTOZOOLOGY

Now, is the Great Auk really extinct? Some few naturalists fondly hope to hear of its discovery ... Newspaper paragraphs sometimes allude to it as they do to the fabulous sea serpent and the extinct moa; but it would appear that, although looked for, it is not to be found.

J. Milne (1875)

The idea that a creature has survived its imagined date of extinction is always a tempting one and there is some evidence to suggest that Great Auks were encountered after the year 1844. Even without any consideration of such evidence, the supposition that the birds killed on Eldey in June of 1844 were actually the very last of their kind is hardly a realistic one. Despite the wild romance and climactic drama of this tale the notion that a few other birds lingered here and there is altogether more likely. Probably, a few scattered individuals were still wandering northern seas, far from their breeding colony. Others, perhaps, were swimming in the waters off Eldey at the time of the raid. The men of the landing party were only ashore for a few minutes and there is no evidence that they returned that summer or, indeed, in subsequent summers. Like desert flowers, one or two birds might easily have wasted their sweetness and come ashore unnoticed. The incident of June,1844 represents nothing more nor less than man's last definite encounter with living Great Auks. Later records - no matter how convincing they might seem - are all open to doubt.

Occasional stories of living Garefowls have continued right up to recent times and each can be judged according to its merits. The most outrageously cryptic tale has nothing to do with a late record, however. It concerns a freshwater pond near Marlow, Buckinghamshire - not two miles from the River Thames - on the estate of Sir William Clayton. Professor J. Fleming (1824) recorded, apparently on the authority of William Bullock, that a Great Auk was taken alive from this pond in the early years of the nineteenth century. What seems to be a variant of this tale is noted by J. Milne (1875), but in this account the action takes place in an Edinburgh pond and the time is set less approximately, with the year 1810 being suggested. Exactly what kind of creature really figures in these stories remains a mystery, but it seems sure that Great Auks do not lie at the bottom of them. Another extravagant tale - this one of a Great Auk killed near Southwold in Suffolk - was allegedly founded on the testimony of the celebrated botanist Sir William Hooker, a man whose word could hardly be

doubted. Unfortunately for the credibility of the story Hooker denied knowing anything about it and generously assumed that someone had completely - but genuinely - misunderstood something that he'd said.

Similarly enigmatic are several stories of Great Auks being killed or found dead on both sides of the English Channel. These stories date from the early years of the nineteenth century and concern individuals reported at Hastings, Cherbourg and Dieppe. They are all records that were generally discounted. Whether any of them has any substance cannot be determined but there is no reason why such visits couldn't have occurred. Individuals were probably blown off course and isolated from others of their kind fairly regularly and such unlucky birds would almost inevitably turn up in unexpected places.

There may also be truth in the often recounted tradition of Auk visits to Lundy in the Bristol Channel. There is little room to doubt the suitability of Lundy as a breeding station for Great Auks but the lateness of some alleged records makes them rather problematical. A letter (*see* Mathew, 1866, *The Zoologist*, p.100) from the appropriately named Reverend H.G. Heaven (the Heaven family purchased the island in 1836) succinctly outlines the story:

The camera never lies - a living Great Auk about to be netted on an Orkney beach during 'The Great Auk Hunt' of the 1980's. Against all odds the Auk escaped and was never seen again! Perhaps it was just a hoax. Photograph by Mark Ellidge for *The Sunday Times*.

There is strong presumptive evidence ... that the Great Auk has been seen <u>alive</u> on the island within the last 30 years ... In the year 1838 or 1839 ... one of our men in the egging season brought us an enormous egg, which we took for an abnormal specimen of the Guillemot's egg ... This, however, the man strenuously denied, saying it was the egg of the 'King and Queen Murr,' and that it was very rare to get them, as there were only two or three 'King and Queen Murrs' ever on the island ... He said they were not like the 'Pickle-bills' [Guillemots] but like the Razor-billed Murrs; that they were much larger than either of them; and that he did not think they could fly, as he never saw them on the wing nor high up on the cliffs ... and that they ...'scuttled' into the water, tumbling among the boulders, the egg being only a little way above high water. He thought they had deserted the island, as he had not seen them or an egg for...15 years till the one he brought to us; but that they (i.e. the people of the island) sometimes saw nothing of them for four or five years but he accounted for this by supposing the birds had fixed on a spot inaccessible to the eggers ... The shell of the egg we kept for some years but, unfortunately, it at last got broken. It was precisely the Guillemot's egg in shape, nearly, if not quite, twice the size, with white ground and black and brown spots and blotches ... The man spoke of the birds in such a way that one felt convinced of their existence, and ... he himself had seen them; but he evidently knew no other name for them than 'King and Queen Murrs,' which he said the islanders called them 'because they were so big and stood up so bold-like'... Nobody, he said, had ever succeeded in catching or destroying a bird, as far as he knew, because they were so close to the water, and scuttled in so fast. The existence of these birds had been traditional on the island when he came to it, and even the older inhabitants agreed that there were never more than two or three examples. He himself never knew of more than one couple at a time.

Lundy is by no means alone in providing surprisingly late records. The year 1844, so portentous in Great Auk history, brought a story from Ireland.

During the month of February, after a heavy storm, a Great Auk was supposedly picked up on the long strand of Castle Freke, County Cork. What became of it is unrecorded. The following year a pair of birds were reported in Belfast Bay. W. Thompson, compiler of *The Birds of Ireland* (1849-56) stated that this sighting was made by a fowler, a Mr. H. Bell, in whom he had every confidence. Fifty years later, in 1894, Thomas Parkin - the dedicated Auk enthusiast - remained unconvinced:

I have spent part of a winter myself in the North of Ireland wild-fowl shooting, and know what *flights* of imagination the average Irish duck shooter is capable of.

Thompson would, of course, have argued that his man Bell was not average.

At about this same time the Island of Skye was, perhaps, favoured with a visit, although the record relating to it is not particularly convincing. Some 35 years after the alleged event a Mr. Mackenzie, factor for St Kilda, described how he'd once strolled along a Skye shore with one Malcolm Macleod. Macleod, it seems, was taking the odd pot-shot at

Hand-coloured lithograph by H. Meyer from his *Illustrations of British Birds* (1835-41). Meyer was one of those who continued to believe in the survival of the Great Auk long after 1844.

seabirds when out on the water the men saw a creature that, from its size and shape, they took to be a Great Northern Diver (*Gavia immer*). Macleod managed to shoot it but, as the poor bird was some distance out, it proved necessary to get a boat to retrieve the body. On close examination this body was found to be quite unlike anything either man had seen before but, unfortunately for the establishment of any credibility, it was only after 35 years of hindsight that Mr. Mackenzie decided to publicly identify the bird as a Great Auk. In a conversation with R. Scot Skirving (a man whose name occasionally crops up in Garefowl literature) that took place during 1880, Mackenzie finally made his identification. Although he probably believed all that he was saying, none of the features he described are particularly suggestive of *Alca impennis*. He failed to mention the size of the beak, nor did he notice anything unusual about the small wings. Instead he directed attention to large claws, a peculiarity that struck him so forcibly that he'd asked Macleod if he might have the feet. As is usually the case with such stories, the feet were lost by the time he met Scot Skirving.

Skirving was sufficiently intrigued, nevertheless, to track down Malcolm Macleod who was able to remember the incident. He recalled shooting a second bird of similar kind the day after the first but had never seen their like either before or since. He'd kept a head from one of the birds as a trophy but - again - it was long gone by the time he told his story. Just like Mackenzie, he remembered nothing unusual about the beak but thought the birds were dark above, white below and had small wings. Clearly, these two men had seen something outside their experience but whether or not Great Auks were involved is doubtful.

A story that gained some degree of acceptance was one of an individual shot in 1848 by a Herr Brodtkorb near Vardö, Varanger Fjord, Norway. Most nineteenth century researchers believed this tale to be based on a mis-identification but a few - most notably Robert Collett (1884) - thought it completely genuine. Collett knew Brodtkorb personally and found him well informed and entirely trustworthy. During the summer of 1883 Brodtkorb showed Collett the little strait - an arm of the sea separating Vardö from the islets of Hornö and Renö - where he'd shot the bird 35 years previously. A translation of a letter from Brodtkorb to Collett outlines the circumstances of the incident:

> In April 1848 I shot ... a strange bird the like of which I never saw before nor since ... I was rowing ... with some companions ... when we espied ... four large birds ... One of my companions ... asked me to fire at them ... in order ... to learn exactly what sort of birds these could be which, instead of flying, only paddled upon the water with their wings. I fired and one fell ... Its back was black, and ... so far as I can remember, its whole head and neck were of that colour ... it was in shape like an Auk [by which is meant Razorbill or Guillemot]. I remember particularly that we observed a white spot at the eye on the side of the head. On the other side the ball, which had gone through the head, had torn away a piece of the white spot and shattered the beak, so that as regards the form of the beak I can tell nothing. The wings were so small that we were all agreed that this circumstance was the reason why the bird only paddled.

In another letter Brodtkorb gave what may be the last description of the behaviour of living Great Auks:

> In swimming they used both wings and feet, and also dived, but did not stay long under the water. It almost seemed as if they only went through the tops of the waves. The birds kept together and did not seem afraid. We ... heard a cry... they emitted when they drew more closely together. It resembled a cackling, as if they wished to call one another ... When the shot went off all the four birds disappeared; but shortly after I saw the remaining three paddling on farther until they disappeared behind the surging waves.

To answer allegations that these birds were nothing more unusual than Divers, Collett made it clear that Brodtkorb was an eager sportsman who'd been familiar with all the local seabirds from boyhood.

The major problem with this story is not so much the lateness of the record but rather the place in which it occurred. Vardö is much farther north than any known recent locality for the Great Auk (although bones have been found nearby in prehistoric middens) and this factor weighed heavily in the negative assessment that men such as Steenstrup and Newton made. Brodtkorb mentioned that at the time of the incident a storm was blowing from the south. Could the effect of this storm, acting on birds that were perhaps already displaced and disorientated, be enough to account for their appearance in such an unexpected place? It will always remain a matter of opinion.

Alfred Newton, who actually interviewed Brodtkorb, was unconvinced. His friend John Wolley, who spoke to the Norwegian at the same time, believed the story. As for Brodtkorb, he could never provide the evidence that would have settled the issue. He'd thrown away the broken body on the beach when his boat landed. The next day, by which time he'd thought better of his hasty action, he found the tide had washed the corpse away.

In 1852 a report of a living Great Auk was circulated that is hard to reject. Colonel Henry Maurice Drummond-Hay, outstanding field naturalist, talented artist and first President of the British Ornithologists' Union was a man of impeccable credentials. He was also a man who claimed to

(Below, left). Ruthven Deane (circa 1900).

(Below, right). Colonel H. M. Drummond-Hay (1814-96), first President of the British Ornithologists' Union.

be the last to see a living Garefowl. After serving with his regiment in Bermuda and at Halifax, Nova Scotia he was returning to Europe during December of 1852 when, on the edge of the Newfoundland Banks, he spied a creature he confidently believed to be a Great Auk. It passed within 30 yards or so (27 metres) of the steamer he was aboard and, as might be expected with such a man, Colonel Drummond-Hay was well equipped with field glasses. He distinctly noted the bill and the white patches before the eye and felt sure that he could not be mistaken.

The following year a dead bird was allegedly found on the shore of Trinity Bay on the eastern side of Newfoundland. Three years later another was supposedly caught on the island's western shore.

Another rather anomalous record from the Newfoundland Banks - although coming a decade or two earlier - is given by John James Audubon (1831-9). On a passage from New York to England Henry Havell, the brother of Audubon's engraver:

> Hooked a Great Auk ... in extremely boisterous weather. On being hauled on board it was left at liberty on the deck; it walked very awkwardly, often tumbling over, bit everyone within reach of its powerful bill, and refused food of all kinds. After continuing several days on board it was restored to its proper element.

Various stories were circulated around the year 1860. A bird was, perhaps, shot in Greenland between Fortuna Bay and Engelman's Harbour- but, then again, perhaps it wasn't. Reports of the finding of eggs on an Icelandic skerry were shown by Alfred Newton (1861) to be entirely false. On March 10th 1862 George Dawson Rowley wrote to Newton concerning another tale:

> In 1859 a skipper who went to Rockall to fish for cod saw this species [Great Auk] diving several times. It was, I believe, near the island [Rockall lies approximately 100 miles (160 kilometres) west of St. Kilda] but being extremely rough at the time he did not continue to pursue it. He is *positive in his statement*, and knows the bird well from being at the British Museum and was struck with its large size in the water.

A tale that related to the early 1870's was recorded by Symington Grieve in 1888. An Icelander, described as an 'official,' claimed he'd seen a Great Auk near Mevenklint, a skerry some 40 miles (65 kilometres) north of Grimsey. This is another record placed too far to the north to be easily credible, Grimsey lying off the north coast of Iceland well within the Arctic Circle.

More realistic, although probably equally lacking in substance, is a rumour repeated by F.A. Lucas in *The Auk* for July 1888 that Garefowls could still be found on the Penguin Islands at the mouth of Gros Water Bay, Newfoundland.

A particularly peculiar saga that turned out to be an elaborate - but not very sophisticated - hoax, was launched in the pages of *American Naturalist* (vol.6, p.369) during 1872. Ruthven Deane, a gentleman who greatly interested himself in Garefowl matters, enthusiastically became its publicist. He'd heard a strange tale and communicated his information to the journal, anxious to discover it there was any real substance to it:

While at Montreal, in August 1871, Mr. Alfred Lechevallier, a naturalist who has collected largely in Labrador, informed me of a specimen of this supposed to be extinct species. It was found dead in the vicinity of St. Augustin, Labrador coast, in November 1870, by some Indians, from whom Mr. Lechevallier obtained it ... It was a male, and, although in a very bad state, he preserved it, and has recently (1872) sold it to a naturalist in France, who is to send it to Austria. Although it was a very poor specimen he realised 200 dollars.

(Above, left). Illustration by Charles Whymper from C. Dixon's *Lost and Vanishing Birds* (1888).

(Above, right). Watercolour by Ernest Griset (circa 1880). *Reproduced by kind permission of Barbara and Richard Mearns.*

Eleven years later this notice came to the attention of J.E. Harting who re-published it in *The Zoologist* (1883, p.470) together with the following request for more information:

We should be glad to know whether any of our readers have seen this specimen, and can add anything further to its history; also whether they can inform us where it is now deposited.

After the passage of some months Harting was able to update readers of *The Zoologist* (1884, p.141) with the information received. He'd discovered that

Ruthven Deane had corresponded with Lechevallier and been given the name of the naturalist in France to whom the preserved bird was allegedly sold. It was a man well known to Harting and one whose name is connected with the histories of at least two Great Auk eggs, Monsieur Fairmaire of Paris. As might be expected, Harting wrote to Fairmaire and the French dealer made a very definite response. He'd never heard of Lechevallier or his Great Auk. Harting ended the story with a few disgruntled words:

> Should this meet the eye of M. Lechevallier perhaps he will be
> good enough to explain, for the satisfaction of ornithologists, the
> discrepancy which appears to exist between the statements in
> his own letters and in that of M. Fairmaire.

Another tale circulating at around the same time is related in *The Land of Desolation* by I.J. Hayes (1871). During 1867 a Great Auk was supposedly killed and eaten by a Greenland Eskimo. Although the tale lacks substance another came from Greenland during the 1890's when a Garefowl was allegedly seen near the island of Ingmikertok in the Angmagaslik Fjord.

The mystery of the 'Arran Auk' will remain unsolved. Probably, it was just another Great Northern Diver but perhaps it was something altogether more curious. *The Field* for March 27th 1880 carried an article by 'G.C.G.' (in fact a vicar from South Devon, the Reverend G.C. Green) that focused on the possibility of Great Auks still occurring off the west coast of Scotland. The evidence presented, although intriguing, is hardly convincing.

As the undoubted encounters with living Garefowls passed deeper into history, so the allegations of new sightings became more infrequent. Eventually, the trickle of dubious claims all but dried up. There were few tales worthy of attention during the first decades of the twentieth century but then, quite suddenly, a spate of stories emerged from Norway. In the late 1930's Great Auks were, apparently, alive and well around the Lofoten Islands. There is little doubt that the sightings themselves were genuine.

The Arran Auk. An extract from *The Field* for March 27 1880.

During the voyage I had had a good deal of conversation about birds with our pilot, who seemed to be a most intelligent old man, and to have had many opportunities for observation during previous voyages with ornithological yachtsmen, and to have made the best use of those opportunities. He told me exactly at what point of our trip we should fall in with the Manx shearwater, and I found that he was quite right. After much talk about birds, many and various, that he had seen in his travels, he exclaimed, "But of all birds, sir, I wish I could show you an Arran auk." I thought at first that he was speaking of some kind of hawk, and had dropped the "h" in his pronunciation; but he soon showed me that he meant something quite different. He said the bird in question was a sea bird, as large as a goose or a turkey—much larger than a cormorant; that it was black above and white underneath; that it had a large sharp bill, but that, in spite of its being so large, it only had wings of the size of a young chicken, and never flew.

There was only one bird that I could imagine which would answer to that description, but I could not believe it possible that that could be the bird. I brought "Yarrell's Birds" to him and showed him the pictures of all the larger birds of that class—cormorants, divers, &c.; but he said he had seen all, or most of these, but they did not answer to his description: but directly he saw the picture of the great auk, he pronounced that to be the bird. He declared that it was to be found on the coast of Arran, but only in the month of March, and assured me that he himself had shot it there several times.

We had many conversations on the subject during the voyage, and he always maintained the same opinion so stoutly, and I found him so correct in his information about other birds where I had the opportunity of testing it, that I did not know what to think. The excitement about the Arran auk became quite a joke to my companions. However, he undertook, if he was alive the following March, to procure me a specimen. Alas! I never had the opportunity of properly testing the correctness of his statements. In the October after we left him the old man suddenly dropped down dead of heart disease one Sunday morning on his way home from church. He had not forgotten his promise to me, however, and had left instructions to his sons to try and procure me a specimen of the bird described; and accordingly, one day last March, a hamper arrived from Lamlash, containing a very fine specimen of the great northern diver, with a letter from his son saying that, according to his father's wishes, he had sent me a specimen of the auk found on those coasts.

Of course a great northern diver was not a great auk, but I do not feel at all sure that if the old man had lived this bird would have been sent. He himself had several times pointed out to me the difference between the bill of the great auk and that of the diver, besides the very conspicuous fact of its having such very tiny wings. I cannot help thinking that at some time or other he had shot the great auk in that neighbourhood; and having, in subsequent seasons, seen large diving birds about the same place without the opportunity of examining them very closely, he had taken it for granted that they were of the same species. His son, not knowing the difference, sent the bird which seemed nearest to answer to his father's description. And, after all, it was a handsome addition to my collection, and will always be especially valuable to me as recalling many a pleasant friendly chat with poor old James Ferguson of Lamlash.

Last Sighting. Wood engraving produced in 1993 by Colin See-Paynton. *Reproduced by kind permission of the artist.*

Unfortunately, they were simply cases of mistaken identity. The truth is that observers were deceived by King Penguins (*Aptenodytes patagonicus*), nine of which were released off northern Norway in August 1936. At least two survived until 1944, the last disappearing around the end of May - almost 100 years after the final encounter with genuine Great Auks on Eldey.

The disappearance of this Penguin, so far from its natural home, might have signalled the end of all reports of living Great Auks. And so it did. But only for a while. In 1971 J. Wentworth Day told another story from the Lofotens, this one from the island of Bö (pronouced boo). Although told long after the event, this tale curiously predates the introduction of the King Penguins. Wentworth Day heard it in 1927 from a boyhood hero of his - Edward Valpy, a man he decribes as, 'an explorer, a first rate naturalist and a man of unimpeachable truth.' Valpy, apparently, rented 50,000 acres of shooting on the island and on returning to London announced that he'd seen the last of the Great Auks. According to Day (1971) he said:

> It slipped off a rock near the boat-builder's shed by the quay. I saw it perfectly plainly. I've seen too many stuffed specimens to be mistaken. I called to the boat-builder and told him what I'd seen... He said that the Geirfugl had been about there for quite a long time. His sons often saw it. He seemed to regard it more or less as a fixture about the place. It had dived under his boat quite close to him one day.

It is impossible to say what lies at the bottom of this tale. Ill health prevented Valpy from ever returning to Bö and he died in 1929.

Years were to pass with no further suggestion that Garefowls were anything but extinct. Then, during the mid 1980's, there was a flurry of activity. Journalists from all over Britain descended on the Orkney Islands in the wake of rumours that living Great Auks still frequented the island of Papa Westray. 'The Great Auk Hunt,' as it came to be called, turned out to be nothing more than an elaborate hoax dreamed up to promote sales of a brand of whiskey.

The Great Auk is, alas, extinct. Were it to be re-discovered, the event would be as remarkable as any in the history of zoology.

APPENDICES

APPENDIX 1: THE MISSING BIRDS OF 1844

One of the more intriguing mysteries in ornithological history surrounds the fate of the skins of the last two Great Auks. What happened to them? Why did they vanish? Where are they?

The only reason for the fourteen man voyage to Eldey that resulted in the birds' destruction was the procurement of specimens. Yet Carl Siemsen, the merchant who actually commissioned the raid, never got his birds. It is an ironic fact that the deaths of these two unfortunate creatures were, from the perspective of the man who ordered them, utterly pointless.

Another curious aspect to the mystery is the fact that the specimens didn't vanish entirely. The whereabouts of the internal organs of both birds, preserved in spirits, are well enough known. They are, and have been since soon after the birds' deaths, in the Zoologisk Museum, Copenhagen (formerly the Royal Museum). The process by which these remains reached the museum - and stayed there - whereas the skins did not, is uncertain, however.

The compiling of Great Auk specimen lists had such enduring fascination for nineteenth century Garefowl specialists, that the absolute disappearance of these two high profile skins is rather difficult to understand. Why was it that men like Alfred Newton, Victor Fatio or Wilhelm Blasius were unable to discover what had happened to them? Certainly, they made considerable efforts to find out. Although the skins may be neither more nor less scientifically important than any others there can be no doubt that they are the ones with the most romantic associations and none of these men were blind to such sentiments. Yet one factor probably holds the key to why all researchers - even the most diligent - drew blanks. At the time of their deaths these birds seemed no more significant than any of the others; no-one knew that man might never encounter their kind again. There was neither need nor reason to specially mark them and by the time the full implication of their demise was appreciated, the trail of their remains had gone cold. Perhaps the one man who could have told their story was long since dead.

What is known of the sad pair of plump little corpses is briefly as follows. On the day following the fishermen's return to their home port the leader of the raiding party, Vilhjálmur Hákanarsson, set out to deliver the birds to Reykjavik where, apparently, Carl Siemsen was waiting for news. Along the way he met up with a certain Christian Hansen and Hansen somehow managed to persuade Hákanarsson to sell the birds on the spot. Perhaps the price offered - 80 rigsbank dollars (around £9) - was greater than the amount promised by Siemsen; perhaps the deal simply saved Vilhjálmur the inconvenience of a trip to Reykjavik. Hansen took the birds on to the Icelandic capital where he sold them to an apothecary called Möller. It was Möller, apparently, who skinned them and gave the skins a very rudimentary stuffing. Soon afterwards he sent the bodies and the skins to Denmark. At this point the skins themselves pass out of certain knowledge.

There are, however, one or two records and fragments that relate to them. In the Alfred Newton archive at Cambridge University, there are several hand-written pages detailing the recollections of Professor Reinhardt (the younger) concerning Great Auk specimens that passed through Copenhagen. The Reinhardts and their colleagues at the Royal Museum handled many skins, of course, so how clearly and accurately they could recall each is uncertain. Dated August 9th 1861, the recollections were penned seventeen years after 1844 and the disappearance of the skins.

Reinhardt believed that the two bodies and the two skins were purchased by Professor Eschricht of the Royal Museum. Although the bodies were deposited in the museum, Eschricht, for reasons unknown, took the skins to the Congress of German Naturalists held in the autumn of 1844 at Bremen where they were sold, fetching a disappointing price that barely covered the Professor's expenses.

The Evening of June 2nd 1844. Oil painting by Errol Fuller (with acknowledgements to J.M.W. Turner).

It is impossible to estimate how much reliance can be placed on Reinhardt's memory of Eschricht's actions but presumably it was on the basis of this recollection that a suspicion grew up at the Royal Museum (mentioned by Steenstrup *in* Grieve, 1885, appendix p.7) concerning the stuffed specimen then - and now - kept at the Bremen Museum.

The Copenhagen curators supposed that this bird, a specimen purchased at the time of the 1844 Bremen Congress of Naturalists, was indeed one of the last pair. And so it may have been. Yet it was bought, not from Eschricht but from a Hamburg dealer named Salmin and this opens up other possibilities. Although Salmin might have been acting as a middle man between Eschricht and the Bremen Museum, it is just as likely that he was operating on his own account. There are very good reasons to suppose he had received three specimens taken during a raid on Eldey that occurred three or four years previously and it is quite possible that he was still in possession of one of these. He may even have still had a bird from an earlier raid; during the 1830's he bought and sold several Great Auks including an entire body preserved in pickle (which vanished long ago). Doubtless he took advantage of the Congress to unload much of his old stock but whether the Great Auk was part of this old stock or whether it was something he'd just bought remains a mystery. It cannot necessarily be assumed, therefore, that this Salmin specimen was one of the birds taken to Bremen by Eschricht - if indeed Eschricht did take the last two Great Auks there and this, in itself, is by no means certain.

Although the Bremen Museum's specimen has at least some claim to be regarded as one of the birds killed in 1844, another example proposed by nineteenth century researchers can be discounted. This is the specimen now in Oldenburg, a bird with a history that can be traced back further than the year 1844. Yet another specimen proposed, that now in Kiel, was probably purchased for the town's university museum during 1844 and could conceivably be one of the last two birds.

The only other clue to solving the mystery lies in the rather shadowy figure known as 'Israel of Copenhagen.' Today his name is little more than an entry that sometimes occurs on the yellowing pages of dusty museum accession registers. From the nature of the specimens that these entries record, some conclusions can be reached. It is evident that he forged good trading links with Iceland and also with Greenland. Clearly, he was a dealer in items of natural history and one of his specialities was the Great Auk. He seems also to have dealt in tobacco and spent his winters in Copenhagen and the summers in Amsterdam. According to a letter from the dealer Frank of Amsterdam (copied by Newton into one of the notebooks now kept at Cambridge) Israel died in 1848, a date that may be of some significance.

His name is of interest here simply because he had two Great Auks in Copenhagen in 1845. Where he got them from is unknown but their presence in the Danish capital at this particular time is highly suggestive. After all, Israel was a dealer noted for obtaining Great Auks straight from Iceland- and the skins of the last two birds were certainly sent by Apothecary Möller from Reykjavik to Denmark. Yet there remains the rumour about Eschricht. Perhaps the Professor did get the birds and perhaps he did take them to the Bremen Congress during the autumn of 1844. But even if this is so, such factors would by no means eliminate Israel from consideration. It is quite likely that Israel visited the Bremen Congress (only a dealer lacking entirely in enterprise would have failed to do so) perhaps *en route* from Amsterdam to Copenhagen. If, as reported, the price was low, Israel may well have been the purchaser and the shrewd dealer would, of course, have taken the Auks with him for his winter sojourn in Denmark. Whether Israel bought his Auks direct from Iceland or whether he got them from Eschricht, it seems probable that the birds he had in stock in Copenhagen during 1845 were the birds killed the previous summer. The specimen in Bremen Museum, bought from Salmin, may well be one that the Hamburg dealer had had in stock for several years.

Alfred Newton often wondered why the Copenhagen curators failed to shed a more revealing light on the whole matter but the reason is probably quite simple. If the Israel hypothesis is correct, those at the museum would almost certainly have known little or nothing. Being a dealer, Israel was probably highly secretive and was under no obligation to advertise his arrangements.

The most important factors are, of course, the changes wrought by the passing of time. In the mid 1840's these two birds seemed of no more significance than any others and it was a decade or more before their true status was recognised. By that time the only man who knew the truth was dead and the story, it seems, had died with him.

Despite this, the subsequent history of Israel's specimens can be traced. In 1845, presumably on his way back to Amsterdam during the spring, he sold his birds to Lintz, a Hamburg merchant. Later in the year Lintz sold them to Amsterdam Frank, perhaps the greatest of all Garefowl dealers. A tantalisingly brief note in the Newton archive at Cambridge suggests that Frank did indeed regard these birds as the 'last' two Great Auks. Frank sold one to Viscount Bernard Du Bus Ghisignies, Director of the Brussels Museum and this bird has remained in Brussels ever since. The second he sold to Count Westerholt-Glikenberg and one facet of this transaction may also point to this being one of the birds of 1844. Frank maintained that when the Count first received the specimen he had it stuffed - and badly - by his own taxidermist. Does this imply that the bird was relatively fresh when received by the Count and stuffed only in a very rudimentary and temporary manner? The answer is - not necessarily, but perhaps. When Frank bought the bird back from the Count's heirs (some twenty years later) it caused him great trouble to have the bird re-stuffed and made, to his eye, perfect. Perhaps then, this is one more argument that can be used to slightly advance the Israel hypothesis. Since it left Frank's hands for the last time, the bird has had several owners and it is now in the Museum of Los Angeles County.

Perhaps it will soon be possible to construct tests whereby the tissue from suspected skins can be matched against samples from the preserved organs of the last of the Great Auks. Since many of the surviving specimens and eggs come from last dwindling band of Great Auks, DNA tests might reveal many interesting things about their relationships.

APPENDIX 2: EDWARD BIDWELL'S PHOTOGRAPHS

During the closing decades of the nineteenth century Edward Bidwell (1845-1929), an East Anglian gentleman of some means, formed a determination to list the whereabouts of every surviving Great Auk egg and then to photograph each of them. Although he didn't quite succeed in his self-appointed task, it is due to his dogged determination that a virtually complete photographic record can be provided today.

Bidwell was clearly an inveterate collector. In addition to an obsession with birds' eggs, he formed collections of antique lamps and candlesticks, candle-snuffers and a variety of curious fire-making implements. His collection of egg photographs has proved his most lasting achievement, however.

He was able to include 71 Great Auk eggs on the final version of his list and of these he managed to photograph 69. He seems to have supplied copies of these photographs, carefully mounted on pieces of card, to specimen owners in exchange for access to their eggs. Whether partial or complete sets were also offered for sale is not known. A number of very incomplete collections exist but full sets seem to be very rare. Fortunately, one of these was acquired during the 1930's by a Mr. Gerald Tomkinson. At his death it passed to his son J. W. Tomkinson who, together with his French wife, decided to publish the photographs in a catalogue of Great Auk eggs which included a potted history of each. To Bidwell's original collection they were able to add photos of the two he'd failed to obtain (those in Philadelphia and Washington, no's 68 and 69 respectively) and four other eggs (no's 2, 52, 67 and 75) which had come to light since the time Bidwell made his list. During 1966 the Tomkinson's published their invaluable catalogue in association with The Natural History Museum (London). Two photographs of each egg are included (one taken from each side) although in the case of egg no.42 (Tomkinson no.49) the photos are virtually identical and for egg no. 75 (Tomkinson no.73) only one picture was available. The last egg but one is labelled *d2* (representing *destroyed egg no.2*) and is the Lisbon Egg, destroyed by fire during 1978. The Erlangen-Nürnberg Egg (no.50) was not known to Bidwell or to the Tomkinsons.

Each egg is shown a little under half the natural size and its number corresponds with the numbering system followed in this book rather than that used by Bidwell and the Tomkinsons.

It has proved possible to trace the whereabouts of 69 surviving eggs. Another six probably still exist but their present locations remain unknown. Two others (now destroyed) are known from good pictorial representations and egg fragments have occasionally come to light testifying to the former existence of several more.

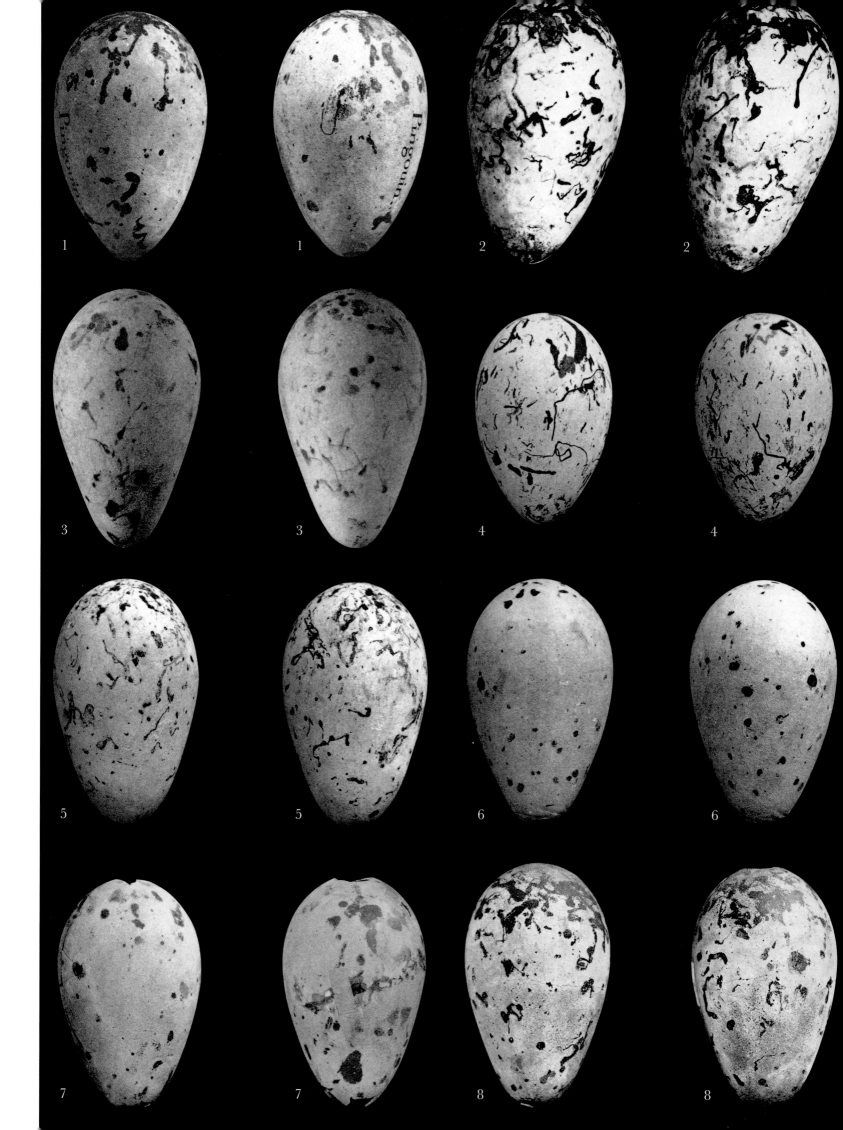

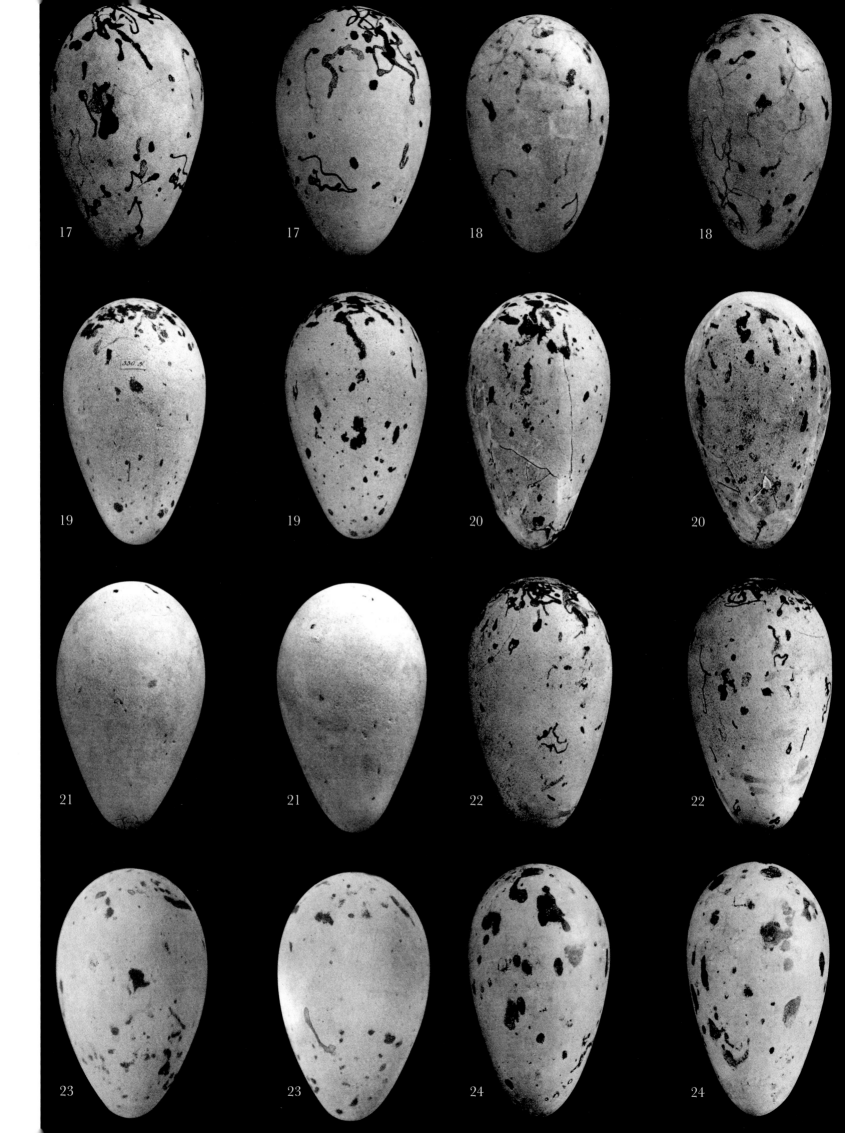

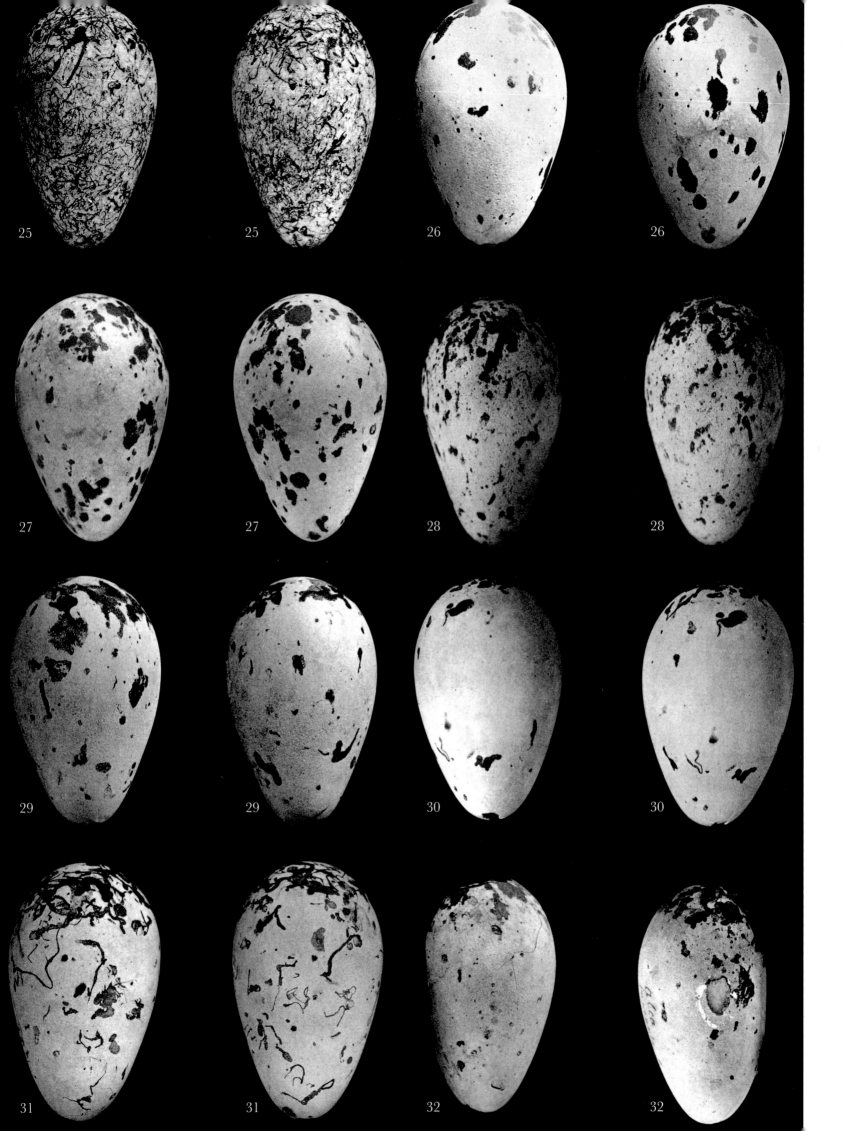

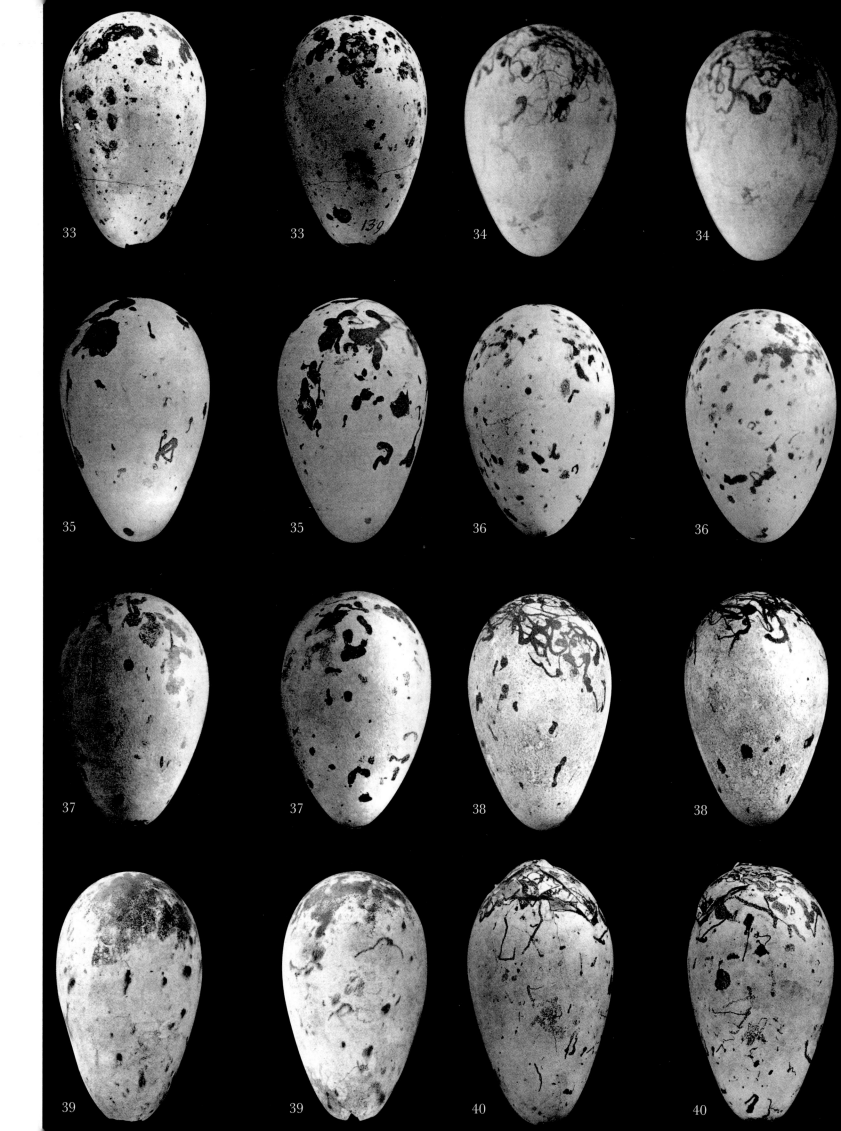

33   33   34   34

35   35   36   36

37   37   38   38

39   39   40   40

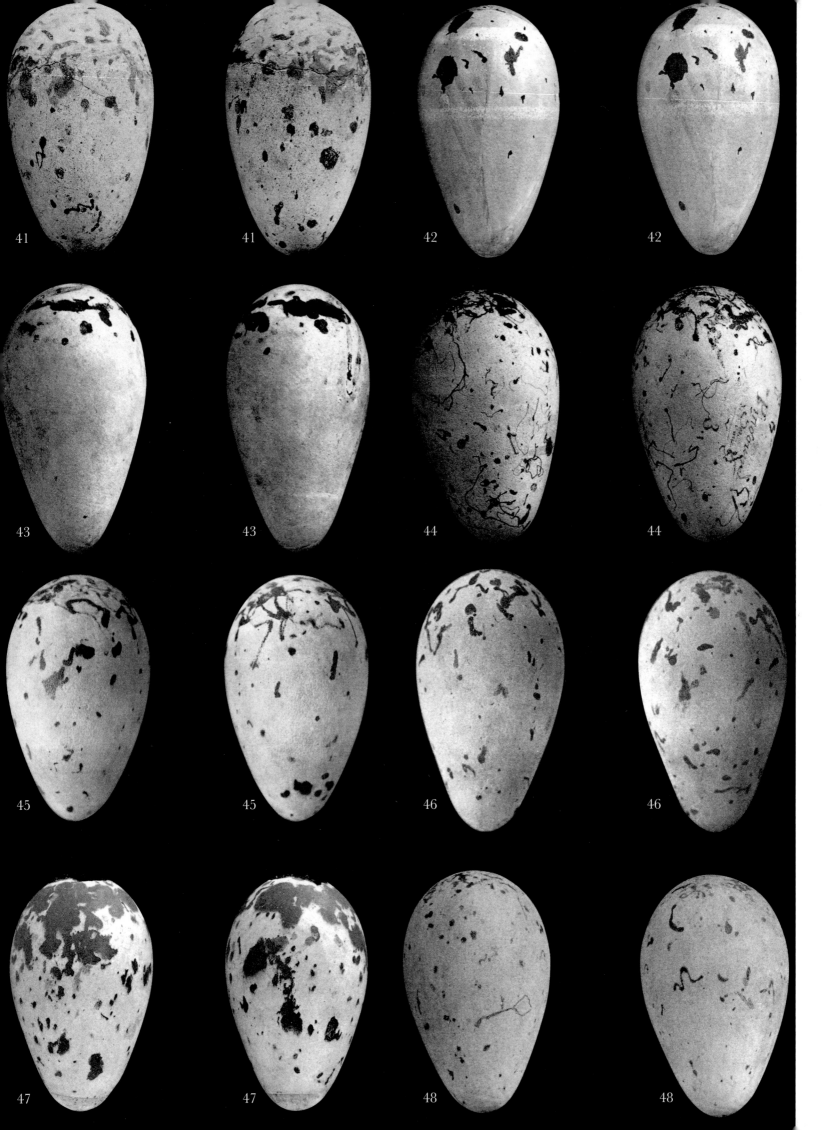

41   41   42   42

43   43   44   44

45   45   46   46

47   47   48   48

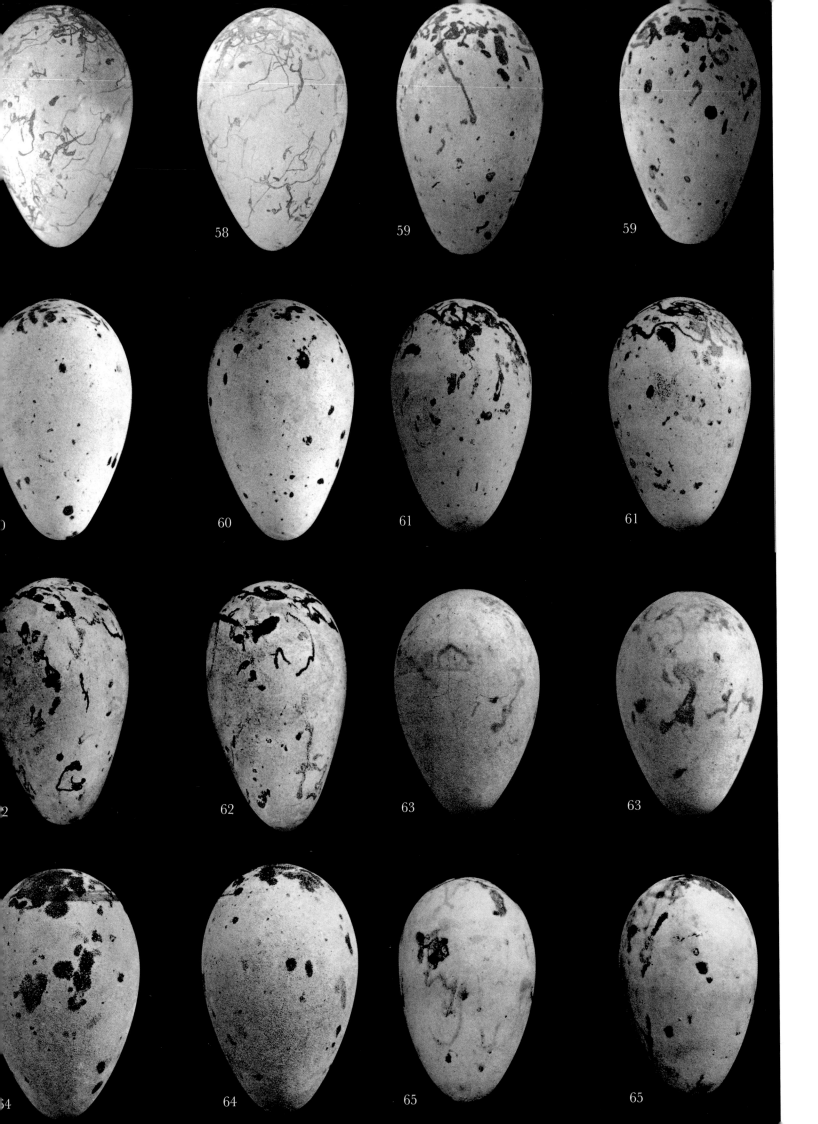

74 74 d.2 d.2

75

ACKNOWLEDGEMENTS

Many, many people - too many to list - have helped in the making of this book. Some have sent photographs, some have given permission to reproduce pictures to which they own rights. Others have given their time to answer complex questions, translate obscure pieces of text or comb museum archives for information. To all these I offer thanks.

There are several people who must be thanked particularly. My friends Brenda Ball, Peter Blest, Clemency Thorne Fisher and Peter Southon have each contributed several pictures and hunted down old and obscure items. This book would have been much poorer without the contribution that each has made. Dr. Jack Gibson, David Wilson, Dick Wheeler and John Ballantyne have also given enormous help in a variety of ways.

Lastly, I owe thanks to my friend Ray Harris Ching who painted the wonderful picture of a swimming bird specially for this book - a magical image that has inspired me, through countless difficulties and setbacks, to see this volume through to publication.

LIST OF SUBSCRIBERS

The publication of this book was helped by the support of those listed below who subscribed to a special edition.

Shell U.K. Exploration and Production, Aberdeen
Artis Library, University of Amsterdam
Belfast Public Library
Zoologisches Forschungsinstitut und Museum Alexander Koenig, Bonn
Staatliches Naturhistorisches Museum, Braunschweig
Balfour and Newton Library, Cambridge
Cambridge University Library
The International Owl Society, Chelmsford
Staatliches Museum für Tierkunde, Dresden
Glasgow Museum and Art Gallery
Trinity College, Hartford, Connecticut
Richard L. D. and Marjorie J. Morse Department of Special Collections, Hale Library, Kansas State University
The Scottish Natural History Library, Kilbarchan
Musée de Zoologie, Lausanne
Naturalis Library, Nationaal Natuurhistorisch Museum, Leiden
Alexander Library, Oxford
National and University Library of Iceland, Reykjavik (2 copies)
Queen Elizabeth II Library, Memorial University of Newfoundland, St. John's
Zentralbibliothek, Zurich

John Alexander, England
Pablo Almeida, Cuba
T. Andersen, Norway
Warren Anderson, Australia
B. A. Aston, Scotland
Charles Thomas Ball, England
Brenda Ball, England
John Ballantyne, Scotland
Aldwyth Bates, England
T. C. Bean, England
Peter Blest (10 copies), England
Gerard Brookes, England
B. N. Brown, England
Michael Bursill, England
Ann and John Burton, England
Buteo Books, U.S.A.
Sarah Chater, England
Rob Chinery, England
Ray Harris Ching, New Zealand
David Clugston, Scotland
Eugene Coqueral, Wales
Peter Cole, England
Sam Coleman, England

Rosemary Crane, England
Rev. J. H. K. Dagger, England
Michael Davies, England
Dr. P. G. Davies, England
George Dick, Scotland
Tony D'Souza, England
D. A. Elkins, France
Kenny Everett (2 copies), England
Mark Fairman, England
Russell A. Fink, U.S.A.
Dr. Clemency Thorne Fisher, England
Adam Fogerty, England
Linda Foord, England
Lawrence Foster, England
Cliff and Dawn Frith, Australia
Jill Fuller, England
R. Fuller, England
Jeremy Gaskell, England
Alan Gibbard, England
Dr. J. A. Gibson, Scotland
Henry and Linda Giller, England
M. B. Gray, England
Allen M. Hale, U.S.A.

Professor W. G. Hale, England
Celia Hammond, England
Peter Hansard, New Zealand
F. K. Hargreaves, England
Emma Hawkins, Scotland
J. B. Hawkins, Australia
Kenneth Haydek, U.S.A.
Dr. Horst Herrlick, Germany
Mick Hill, England
Bryan Holden, England
Larry Holmes, U.S.A.
Ewold Horn, The Netherlands
Stan Howe, Scotland
Da-Shih Hu, U.S.A.
Anthony Hull, England
Mervyn James, Scotland
Les Jessop, England
Harry Johnson, England
Robert Johnson, U.S.A.
Valerie Jones, Spain
Hew Kennedy, England
Gaël Lagadec, France
David M. Lank, Canada
Lord Lamont of Lerwick
Mike Latter, England
Danny Lewis, Ireland
Magnus Linklater, Scotland
T. A. Lording, England
Richard Losasso, England
John Metcalf, England
Irene Metcalf, England
Pat Morris, England
David N. Nettleship, Ph.D. (zool.), Canada

Mike Norris, England
H. Nuijen, The Netherlands
Peter A. O'Connor, England
T. Ormande, Ireland
A. N. Other (2 copies), England
Roger Page, England
Peregrine Books, England
Peter Petrou, England
Mary Philcox, England
Michael Rich, England
Tony Rowe, England
St. Ann's Books (10 copies), England
Bernard Sayers, England
Professor William J. L. Sladen, U.S.A.
Carol Sinclair Smith, England
Chris Smeenk, The Netherlands
Peter Southon, England
S. A. H. Statham, England
Patricia Steer, England
William Todd III, U.S.A.
Richard Wallace, New Zealand
David Waring, England
Hermione Waterfield, England
Richard Wheeler, U.S.A.
Sandra Wheeler, U.S.A.
Wheldon and Wesley Ltd., England
Ralfe Whistler, England
Jim Whitaker, England
Margaret Whitworth, England
Barry Williams, England
David Wilson, England
Simon Wilson, England
Keith Zabell, England

BIBLIOGRAPHY

Note: items relating to just a single stuffed bird or egg may be listed only in the references for that specimen.

In addition to the items listed here, there are two important collections of material relating to the Great Auk. The first of these forms part of the Alfred Newton Archive at the University Library, Cambridge. The second - a two volume scrapbook compiled by George Dawson Rowley and his wife Caroline - is the property of Peter Rowley and can be seen, with his permission, at Morcott Hall, Morcott, Rutland, England. The volumes are titled Alciana: A Monograph on the Gare Fowl, or Great Auk *and are dated 1866 although, clearly, they have been added to after this date.*

Agassiz, L. (1864). A perfect specimen of a mummified *Alca impennis. Report of the Museum of Comparative Zoology,* pp.16-22.

Allen, J. (1876). The Extinction of the Great Auk at the Funk Islands. *American Naturalist,* p.48.

Allingham, E. (1924). *A Romance of the Rostrum.* H.F. & G. Witherby, London.

Altum, B. (1863). Der Brillenalk. *Natur und Offenbarung,* 9: 15-23.

Andrews, C. (1920). Remains of the Great Auk and Ptarmigan in the Channel Islands. *Annals and Magazine of Natural History,* p.166.

Anon. (1986). The Great Auk Search Begins. *The Orcadian* (8th May).

Armstrong, E. 1939. *Birds of the Grey Wind.* Oxford University Press, London.

Ascenzi, A & Segre, A. (1971). A new Neanderthal child mandible from an Upper Pleistocene site in Southern Italy. *Nature,* 233: 280-3.

Audubon, J. (1827-38). *The Birds of America.* Published by the author, London.

Audubon, J. (1831-9). *Ornithological Biography.* Published by the author, Edinburgh.

Baedeker, F. (1855-63). *Die Eier der Europäischen Vögel.* Leipzig.

Baird, S. (1866). [Note on old works containing references to *Alca impennis.*] *Ibis,* pp.223-4.

Baird, S. (1866, May). Specimens Extant of the Great Auk. *Harper's New Monthly,* 41: 308.

Baird, S., Brewer, T. & Ridgway, R. 1884. The Water Birds of North America. *Memoirs of the Museum of Comparative Zoology,* vol.XIII. Little, Brown and Co., Boston.

Baldwin, S. (1873). The Great Auk. *Yorkshire Nature Recorder,* pp.165-6.

Bannerman, D. & Lodge, G. (1953-63). *The Birds of the British Isles.* Oliver and Boyd, London.

Bárðarson, H. (1986). *Birds of Iceland.* Published by the author, Reykjavik.

Barrett-Hamilton, G. (1896). The Great Auk as an Irish Bird. *Irish Naturalist,* pp.121-2.

Bate, D. (1928). Excavations of a Mousterian rock-shelter at Devil's Tower, Gibraltar. *Journal of the Royal Anthropological Institute,* pp.107-8.

Belt, A. (1933). Thomas Parkin [obituary]. *Hastings and East Sussex Naturalist,* 4 (6): 120-30.

Bengston, S. (1984) Breeding ecology and extinction of the Great Auk. *Auk,* pp.1-12.

Benicken, F. (1824). Beiträge zur nordischen ornithologie. *Isis,* pp.877-91.

Bent, A. (1919). Life Histories of North American Diving Birds. *U.S. National Museum Bulletin,* 107: 1-239.

Berlioz, J. (1985). Notice sur les spécimens naturalisés d'oiseaux éteintes existant dans les collections du Muséum. *Archives du Muséum d'Histoire Naturelle,* volume du tricentaire, ser.6, tome XII: 485-95.

Bewick, T. (1805). *A History of British Birds.* Published by the author, Newcastle.

Biggar, H. (1924). *The Voyages of Jacques Cartier.* Publications of the Public Archives of Canada no.11.

Birkhead, T. (1993). *Great Auk Islands.* T. & A.D. Poyser, London.

Blasius, W. (1881-3). Ueber die letzten Vorkommnisse des Riesen-Alks. *Jahresberichte des Vereins für Naturwissenschaft zu Braunschweig für die Vereinsjahre 1881-2 und 1882-3,* pp.89-115.

Blasius, W. (1884). Zur Geschichte der Ueberreste von *Alca impennis. Journal für Ornithologie,* pp.58-176.

Blasius, W. (1900). Der Riesenalk in der ornithologischen Literatur der letzten 15 Jahre. *Ornithologische Monatsschrift,* pp.434-46.

Blasius, W. (1896-1905). *See* Naumann.

Blyth, E. (1837). On the osteology of the Great Auk. *Proceedings of the Zoological Society of London*, pp.122-3.

Boie, F. (1822). Ornithologische Beiträge. *Isis*, pp.871-86.

Bolle, C. (1862). Notiz *Alca impennis* betreffend. *Journal für Ornithologie*, pp.208-9.

Bones, M. (1993). The Garefowl or Great Auk. *Hebridean Naturalist*, 11: 15-24.

Bourne, W. (1993). The story of the Great Auk. *Archives of Natural History*, 20 (2): 257-78.

Brandicourt, V. (1897). Un oiseaux rare. La Grand Pingouin. *Cosmos* (Paris), 46: 777-80.

Brandt, J. (1837). Rapport sur une Monographie de la famille Alcadées. *Bulletin Scientifique l'Académie Impériale des Sciences de Saint-Pétersbourg*, pp.170-190.

Breuil, H. (1911). *Les Cavernes de la région Cantabriques*. Monaco.

Brisson, M. (1760). *Ornithologie*. Paris.

Brodkorb, P. (1960). Great Auk and Common Murre from a Florida midden. *Auk*, 77: 342-3.

Buckley, T. & Harvie-Brown, J. (1891). *A Vertebrate Fauna of the Orkney Islands*. David Douglas, Edinburgh.

Burness, G. & Montevecchi, W. (1992). Oceanographic-related variation in bone sizes of extinct Great Auks. *Polar Biology*, 11: 545-57.

Burton, M. (1970). Killed by greed and folly. *Illustrated London News* (Jan. 10th).

Carruccio, A. (1902). Sovra un palmipede rarissimo e di gran valore *Plautus impennis* donata da S. M. Il Re Vittorio Emanuele III al Museo Zoologico della R. Università di Roma. *Bolletin Society Zoologico Italiane*, pp.1-15.

Carter, G. (1935). The Eggs of the Great Auk. *The Oologist's Record*, pp.3-5 (March).

Cartier, J. (ed. Michelant, M., 1865). *Le Voyage de Jacques Cartier au Canada en 1534*. Paris- *see also* Biggar, H.P.

Cartwright, G. (1792). *Journal of Transactions and Events, during a residence of Nearly Sixteen Years on the Coast of Labrador*. Allin and Ridge, Newark (England).

Champley, R. (1864). The Great Auk. *Annals and Magazine of Natural History*, pp.235-6.

Charlton, E. (1860). On the Great Auk. *Zoologist*, pp.6883-8.

Chigi, F. (1936). La morte delle specie animali. *Rassegna Faunistica*, pp.3-49.

Christiani, A. (1917). Om fund af Gejrfugle knogler paa Vardø. *Dansk Ornithologisk Forenings Tidsschrift*, pp.1-4.

Christiani, A. (1929). Om mogle jordfundne knogler fra Vardø. *Dansk Ornithologisk Forenings Tidsschrift*, pp.79-83.

Christy, M. (1894). On an early notice and figure of the Great Auk. *Zoologist*, pp.142-5.

Clark, G. (1948). Fowling in prehistoric Europe. *Antiquity*, 22: 116-30.

Clintock, F. (1860). The Great Auk. *Zoologist*, p.6981.

Clottes, J., Beltrán, A., Courtin, J. & Cosquer, H. (1992). The Cosquer Cave on Cape Morgiou, Marseilles. *Antiquity*, 66: 583-98.

Clusius, Caroli Clussi Atrebatis (1605). *Exoticorum libri decem: Quibus Animalium, Plantarum, Aromatum historiae describunter*. Ex Officina Plantiniana Raphelengii.

Collet, R. (1866). Briefliches über *Alca impennis* in Norwegen. *Journal für Ornithologie*, pp.70-1.

Collet, R. (1884). Ueber *Alca impennis* in Norwegen. *Mittheilungen des Ornithologischen Vereines in Wien*, 5: 65-9 *and* 87-9 (May).

Cooper, J. (1997). [The Great Auk in Pleistocene Gibraltar]. *Bulletin of the British Ornithologists' Club*, vol.117 (3), p.159.

Coues, E. (1868). Monograph of the Alcidae. *Proceedings of the Academy of Natural Sciences Philadelphia*, 20.

Coues, E. (1871). The Great Auk. *American Naturalist*, 4: 57.

Coulon, M. (1878). Prix actuels des peaux et des oeufs de l'*Alca impennis*. *Bulletin de la Société Neuchâtel*, 16: 294-5.

Dale, D. (1886). *The Great Auk's Eggs*. The Religious Tract Society, London.

Day, J. Wentworth (1971). The Tradgedy of the Great Auk. *Blackwood's Magazine* (April), pp.354-61.

Deane, R. (1872). The Great Auk. *American Naturalist*, pp.368-9.

Denys, N. (1672). *The Description and Natural History of the Coast of North America* (translated by W.F. Ganong, 1908). The Champlain Society, Toronto.

Des Murs, O. (1863). Notice sur l'oeuf de l'*Alca impennis*. *Revue et Magasin de Zoologie*, 15: 3-5.

Didier, R. (1934). Le Grand Pingouin. *La Terre et La Vie*, 4: 13-20 (January).

Donovan, E. (1794-1819). *The Natural History of British Birds*. London.

Dresser, H. (1871-96). *A History of the Birds of Europe*. London.

Dresser, H. (1910). *Eggs of the Birds of Europe*. London.

Dubois, C. (1867). Sur le *Plautus impennis*. *Archives Cosmologiques. Revue des Sciences Naturelles* (Bruxelles), 2: 30-5.

Dubois, C. & Dubois, A. (1861-72). *Les Oiseaux de l'Europe et de leurs Oeufs*. Brussels.

Duchaussoy, H. (1897-8). Le Grand Pingouin du Musée d'Histoire naturelle d'Amiens *and* Le Grand Pingouin du Musée d'Histoire naturelle d'Amiens- Notes additionnelles. *Mémoires de la Société Linnéenne du Nord de la France*, pp.88-129 *and* pp.241-51.

Duncan, J. (1900). *Birds of the British Isles*. London.

Dunning, J. (1872). Great Auk. *Zoologist*, p.2946.

Eastham, A. (1968). The avifauna of Gorham's Cave, Gibraltar. *Bulletin of the Institute of Archaeology*, 7: 32-42.

Eckert, A. (1964). *The Last Great Auk*. W. Collins, London.

Edwards, G. (1743-64). *A Natural History of Uncommon Birds* and *Gleanings of Natural History*. London.

Egede, H. (1741). *Det Gamle Grønlands nye Perlustration*. Copenhagen.

d'Errico, F. (1994). Birds of the Grotte Cosquer: the Great Auk and Palaeolithic prehistory. *Antiquity*, 68: 39-47.

d'Errico, F. (1994). Birds of Cosquer Cave. The Great Auk and its significance during the Upper Palaeolithic. *Rock Art Research*, 11 (1): 45-57.

Faber, F. (1826). *Über das Leben der hochnordischen Vögel*. Ernst Fleischer, Leipzig.

Faber, F. (1827). Geirfugl. *Isis*, 20: 679-85.

Fabricius, O. (1780). *Fauna Groenlandica*. Johannes Gottlob, Copenhagen and Leipzig.

Fatio, V. (1868). Quelques mots sur les exemplaires de l'*Alca impennis*, oiseaux et œufs qui se trouvent en Suisse. *Bullétin de la Société Ornithologique Suisse*, tome II, pt.1: 73-9.

Fatio, V. (1868). Liste des divers représentants de l'*Alca impennis* en Europe. Oiseaux, squelettes et œufs. *Bullétin de la Société Ornithologique Suisse*, tome II, pt.1: 80-5.

Fatio, V. (1870). Supplément a la liste des divers représentants de l'*Alca impennis* en Europe. *Bullétin de la Société Ornithologique Suisse*, tome II, pt.2: 147-57.

Fielden, H. (1869). [letter on eggs of *Alca impennis*]. *Ibis*, pp.358-60.

Fielden, H. (1872). Birds of the Faeroe Islands. *Zoologist*, pp.3277-94.

Fisher, J. (1945). Alfred Newton and the Auk. *Bird Notes*, 21: 75-7.

Fisher, J. (1954). The Great Auk. *Country Fair* (January), pp.32-6.

Fisher, J. and Lockley, R. (1954). *Seabirds*. W. Collins, London.

Fitzinger, F. (1862-4). *Wissenschaftlich-Populäre Naturgeschichte der Vögel*. Vienna and Leipzig.

Fleming, J. (1824). Gleanings of natural history during a voyage along the coast of Scotland in 1821. *Edinburgh Philosophical Journal*, 10: 95-101.

Fleming, J. (1828). *History of British Animals*. Edinburgh.

Foelix, R. (1996). Seltsame Vögel im Aargauischen Naturmuseum. *Aarauer Neujahsblätter, 1996*, pp.52-9.

Fordun, John of (c.1380). *See* Skene, W.

Forrest, H. (1899). *The Fauna of Shropshire*. Shrewsbury.

Fraipoint, J. (1910). *Collections Zoologiques du Baron Edm. de Sélys Longchamps. Catalogue Systématique et Descriptif- Oiseaux*. Hayez, Brussels.

France, A. (1907). *L'Ile des Pingouins*. Calmen-Levy, Paris.

Fritsch, A. (1863). Notiz über *Alca impennis*. *Journal für Ornithologie*, pp.295-7.

Fritsch, A. (1871). *Vögel Europa's*. Prague.

Fuller, E. (1987). *Extinct Birds*. Viking Rainbird, London.

Gadeau de Kerville, H. (1890-2). *Faune de la Normandie: Oiseaux*. Paris.

Gadow, H. (1910). The Ornitholgical Collections of the University of Cambridge. *Ibis*, pp.47-53.

Garðarsson, A. (1984). [Seabird cliffs of the Reykjanes Peninsula]. *Árbok Ferðafélags Íslands*, pp.126-60.

Gawn, J. (1944). A memory of the Great Auk. *The Peregrine- A Publication of the Manx Field Club*, 1 (March): 4.

Gibson, J. (1882-3). On a hitherto Unrecorded Specimen of the Great Auk in the Collection of the Duke of Roxburghe. *Proceedings of the Royal Physical Society*, pp.335-8.

Gilliard, E. (1937- Sept.). Bird Men Courageous. *Natural History* (New York), pp.480-90.

Gilliard, E. (1961). The Bony Treasure of Funk Island *in* Geres, J. (ed.), *Discovery: Great Moments in the lives of Outstanding Naturalists*. Lippinscott, Philadelphia.

Glegg, W. (1949). Great Auks reported from Lofoten Islands. Explanation: Introduction of King Penguins. *Bulletin of the British Ornitholgists' Club*, pp.120-1.

Glegg, W. (1949). The history of a Great Auk's egg presented to the British Museum by Lord Lilford, of the Great Auk's eggs bequeathed to the nation, and of the remians of a recently discovered egg. *Bulletin of the British Ornitholigsts' Club*, pp.77-80.

Gloger, C. (1860). Die frühere ausserordeutliche Häufigkeit der grossen oder Schwimm-Alke. *Journal für Ornithologie*, pp.60-3.

Godfrey, W. (1959). Notes on the Great Auk in Nova Scotia. *Canadian Field-Naturalist*, 73: 175.

Gore, J., ed. (1938). *Mary Duchess of Bedford, 1865-1937*. John Murray, London (printed for private circulation).

Gould, J. (1832-7). *The Birds of Europe*. Published by the author, London.

Gould, J. (1862-73). *The Birds of Great Britain*. Published by the author, London.

Gray, J. (1864). Notice of the skeleton of a Great Auk found in guano near Newfoundland. *Annals and Magazine of Natural History*, p.319.

Gray, R. (1871). *Birds of the West of Scotland including the Outer Hebrides*. Murray and Son, Glasgow.

Gray, R. (1880). On two unrecorded Eggs of the Great Auk discovered in an Edinburgh Collection; with remarks on the former existence of the bird in Newfoundland. *Proceedings of the Royal Society of Edinburgh*, pp.668-82.

Greenway, J. (1958). *Extinct and Vanishing Birds of the World*. American Committee for Wildlife Protection, New York.

Greenwell, J. (1987). Raiders of the Lost Auk. *International Society of Cryptozoology*, Newsletter 6 (Spring), pp.5-7.

Grieve, S. (1883). Notice of the discovery of remains of the Great Auk on the Island of Oronsay. *Journal of the Linnaean Society, London*, pp.479-87.

Grieve, S. (1885). *The Great Auk or Garefowl*. Thomas C. Jack, Edinburgh.

Grieve, S. (1888). Recent information about the Great Auk. *Transactions of the Edinburgh Field Naturalists' and Microscopical Society*, 2: 92-119.

Grieve, S. (1896-8). Supplementary note on the Great Auk or Garefowl *and* Additional notes on the Great Auk or Garefowl. *Transactions of the Edinburgh Field Naturalists' and Microscopical Society*, 3: 237-73 *and* 327-40.

Grieve, S. (1923). *The Book of Colonsay and Oronsay*. Edinburgh.

Grosvenor, G. & Wetmore, A. (1937). *The Book of Birds*. National Geographic Society, Washington.

Grote, H. (1914) *Alca impennis* im Jahre 1848 in Norwegen erbeutet? *Ornithologische Monatsberichte*, pp.5-6.

Guinness Book of Records, 19th edition (1972). Guinness Publications, London.

Gurney, J. (1868). The Great Auk. *Zoologist*, pp.1442-53.

Gurney, J. (1869). Notes on the Great Auk. *Zoologist*, pp.1639-43.

Gurney, J. (1872). Great Auk at Disco. *Zoologist*, pp.3064-5.

Gurney, J. (1913). *The Gannet*. Witherby, London.

Hachisuka, M. 1927. *A Handbook to the Birds of Iceland*. Taylor and Francis, London.

Hahn, P. (1963). *Where is that Vanished Bird?* Royal Ontario Museum, University of Toronto.

Hakluyt, R. (1552-1616). *The priciple navigations, voyages, traffiques and discoveries of the English Nation*. London.

d'Hamonville, L. le Baron. (1888). Note sur les quatres œufs *d'Alca impennis* appartenant à notre collection zoologiques. *Memoires de la Société Zoologique Francais*, pp.101-4.

Hancock, J. (1874). *A Catalogue of the Birds of Northumberland and Durham*. Newcastle-upon-Tyne.

BIBLIOGRAPHY

Hardy, F. (1888). Testimony of some early voyagers on the Great Auk. *Auk*, pp.380-4.

Hardy, J. (1841). *Catalogue des Oiseaux observés dans le département de la Seine-Inférieure.* Le Roy, Caen.

Harrison, C. & Stewart, J. (*In press*). The bird remains. *In*: Roberts, M. & Parfitt, S (eds.). The Middle Pleistocene Site at ARC Eartham Quarry, Boxgrove, West Sussex, U.K. *English Heritage Monograph Series, no.16*, London.

Harting, J. (1883). The last Great Auk. *Zoologist*, p.470.

Harting, J. (1884). The last Great Auk. *Zoologist*, pp.141-2.

Harting, J. (1901). *Handbook of British Birds.* John Nimmo, London.

Harvie-Brown, J. & Buckley, T. (1889). *Vertebrate Fauna of the Outer Hebrides.* David Douglas, Edinburgh.

Harvey, M. (1876). The Great Auk. *Forest and Stream*, 6 (July 20th), p.386.

Hay, O. (1902). On the finding of the bones of the Great Auk in Florida. *Auk*, pp.255-8.

Hayes, I. (1871). *The Land of Desolation, Being a Personal Narrative of Adventure in Greenland.* Sampson Low, Marston, Low and Searle, London (*see* pp.291-2).

Heathcote, J. (1900). *St Kilda.* London.

Heim de Balzac, H. (1929). Un nouvel œuf d'*Alca impennis. Alauda*, pp.366-8 (*see also* Rapine, J., 1930).

Hellman, A. (1860). Notizen über *Alca impennis. Journal für Ornithologie*, pp.206-7.

Hewitson, W. (1831-8). *British Oology.* Newcastle and London.

Hewitson, W. (1842-6). *Coloured Illustrations of the Eggs of British Birds.* London.

Hewitson, W. (1853-6). *Coloured Illustrations of the Eggs of British Birds.* London.

Hobson, K. & Montevecchi, W. (1991). Stable isotopic determination of trophic relationships of Great Auks. *Oecologia*, 87: 528-31.

Homeyer, A. (1862). Notiz zu *Alca impennis. Journal für Ornithologie*, p.461.

Horrebow, N. (1758). *The Natural History of Iceland.* London.

Hunt, J. (1815-22). *British Ornithology.* Norwich.

Hywel, W. (1973). *Modest Millionaire: the biography of Captain Vivian Hewitt.* Gwasg Gee, Denbigh.

Jardine, W. (1833-44). *The Naturalist's Library - British Birds*, pt.4 (vol. XXVII). Edinburgh.

Jones, J. (1870). The Great Auk from Funk Island. *Zoologist*, pp.2182-3.

Jones, T. (1869-73). *Cassell's Book of Birds.* London.

Jørgensen, B. (1973). *Gejrfuglen.* Brøndum, Copenhagen.

Jourdain, Rev. F. (1906). *The Eggs of European Birds.* R.H. Porter, London.

Jourdain, Rev. F. (1934). The Sale of G.D. Rowley's Collections. *The Oologist's Record* (Dec. 1st), pp.75-9.

Jourdain, Rev. F. (1934). The skins and eggs of the Great Auk. *British Birds*, pp.233-4.

Jourdain, Rev. F. (1934). Sale of Skins and Eggs of the Great Auk. *Ibis*, pp.245-6.

Kartaschew, N. (1960). *Die Alkenvögel des Nordatlantiks.* Ziemsen Verlag, Wittenberg.

Kingsley, C. (1863 - and later editions). *The Water Babies: A Fairy Tale for a Land Baby.* London.

Kjærbölling, N. (1851-6). *Danmarks Fugle (Ornithologia Danica: Icones Ornithologiæ Scandinavicæ).* Copenhagen.

Knowles, W. (1895). Remains of the Great Auk from White Park Bay. *Proceedings of the Irish Academy*, 3: 650-63.

Knox, A. & Walters, M. (1994). Extinct and Endangered Birds in the collection of the Natural History Museum. *British Ornithologists' Club*, occasional publications, no.1. Tring.

Koenig, A. (1931-2). *Katalog de Nido-Oologischen Sammlung im Museum Aklexander Koenig.* Bonn.

Lambrecht, K. (1964). *Handbuch der Palaeornithologie.* Asher and Co., Amsterdam.

Latham, J. (1781-5). *A General Synopsis of Birds.* London.

Lewin, W. (1789-94). *The Birds of Great Britain.* Published by the author, London.

Lewin, W. (1795-1801). *The Birds of Great Britain.* Published by the author, London.

Ley, W. (1935). The Great Auk. *Natural History*, 36: 351-6.

Ley, W. (1938). Great Auk Saga. *Zoo and Animal Magazine* (March), 2 (10): 30-1.

Lilford, Lord (1885-98). *Coloured Figures of the Birds of the British Isles.* London.

Livezey, B. (1988). Morphometrics of flightlessness in the Alcidae. *Auk*, 105: 681-98.

Løvenskiold, H. (1964). Avifauna Svalbardensis with a discussion upon the history and anatomy of the Great Auk. *Norsk Polarinstitutt Skrifter*, 129: 1-460.

Lucas, F. (1880). Great Auk notes. *Auk*, pp.278-83.

Lucas, F. (1888). Coastal variations in birds. *Auk*, pp.195-6.

Lucas, F. (1890). The Expedition to the Funk Island, with observations upon the history and anatomy of the Great Auk. *Report of the U.S. National Museum*, 1887-8, pp.493-529.

Lucas, F. (1890). The Great Auk in the United States of America. *Auk*, pp.203-4.

Lucas, F. (1891). Animals Recently Extinct or Threatened with Extermination, as represented in the Collections of the U.S. National Museum. *Report of the U.S. National Museum*, 1888, pp.609-49

Lucas, F. (1891). Explorations in Newfoundland and Labrador in 1887. *Report of the U.S. National Museum*, 1888-9, pp.709-28.

Lucas, F. (1894). Der Riesenalk. *Zeitschrift für Ornithologie*, 4: 25-6.

Luther, D. (1986). *Die Ausgestorbenen Vögel der Welt*. Ziemsen Verlag, Wittenberg.

Lyngs, P. (1994). Gejrfuglen- Et 150 års minde. *Ornitologisk Forenings Tidsskrift*, 88: pp.49-72.

Lysaght, A. (1971). *Joseph Banks in Newfoundland and Labrador*. London.

Macaulay, K. (1764). *The History of St. Kilda*. London.

Macgillivray, J. (1840). An account of the island of St Kilda. *Edinburgh New Philosophical Journal*, 32: 47-70.

Macgillivray, W. (1852). *A History of British Birds*. W.S. Orr, London.

MacKenzie, J. (1905). Notes on birds of St Kilda, compiled from memoranda of Rev. Neil MacKenzie. *Annals of Scottish Natural History*, pp.75-80 *and* 141-53.

Martin, M. (1698). *A Late Voyage to St. Kilda, the Remotest of All the Hebrides, or Western Isles of Scotland*. Gent, London.

Matthew, M. (1866). The Great Auk on Lundy Island. *Zoologist*, pp.100-1.

McClintock, F. (1860). The Great Auk. *Zoologist*, p.6981.

Mearns, B. & Mearns, R. (1988). *Biographies for Birdwatchers: The Lives of Those Commemorated in West Palearctic Bird Names*. Academic Press, London.

Mearns, B. & Mearns, R. (1992). *Audubon to Xántus: The Lives of Those Commemorated in North American Bird Names*. Academic Press, London.

Meldgaard, M. (1988). Isarukitsoq- den stumpvingede. *Grønland*, 36: 5-12.

Meldgaard, M. (1988). The Great Auk *Pinguinis impennis* in Greenland. *Historical Biology*, 1: 145-78.

Meyer, H. (1835-41). *Illustrations of British Birds*. London.

Meyer, H. (1852-57). *Coloured Illustrations of British Birds, and their Eggs*. London.

Michahelles, K. (1833). Zur Geschichte des *Alca impennis*. *Isis*, pp.648-51.

Milne, J. (1875). Relics of the Great Auk. *The Field* (March 27, April 3 and 10).

Milne-Edwards, M. & Oustalet, M. (1893). Notice sur quelques espèces d'oiseaux actuellement éteintes qui se trouvent représentées dans les collections du Muséum d'Histoire Naturelle. *Volume Commémoratif du Centenaire de la Fondation du Muséum d'Histoire Naturelle*.

Montagu, G. (1813). *Supplement to the Ornithological Dictionary*. London.

Montevecchi, W. & Kirk, D. (1996). Great Auk. In *The Birds of North America*, no.260 (A. Poole and F. Gill, eds.). The Academy of Natural Sciences, Philadelphia and The American Ornithologists' Union, Washington.

Montevecchi, W. & Tuck, L. (1987). *Newfoundland Birds: Exploitation, Study, Conservation*. Nuttall Ornithological Club, Cambridge, Mass.

Morris, Rev. F. (1851-7). *A History of British Birds*. London.

Morton, G. (1872). The Great Auk. *Zoologist*, p.3338.

Mowat, F. (1984). *Sea of Slaughter*. McClelland Stewart, Toronto.

Mullens, W. (1917). Some museums of old London. *Museums' Journal* (October).

Mullens, W. (1921). Notes on the Great Auk. *British Birds*, 15 (5): 98-108.

Murray, J. (1968). *The Newfoundland Journal of Aaron Thomas, 1794*. London.

BIBLIOGRAPHY

Naumann, J. (1822-60). *Naturgeschichte der Vögel Deutschlands.* Leipzig and Stuttgart.

Naumann, J. (1896-1905). *Naturgeschichte der Vögel Mittel-Europas.* Gera-Untermhaus.

Neill, P. 1814. [Report of Great Auks visiting Orkney]. *Scots Magazine* (March), p.167.

Nettleship, D. & Birkhead, T. (1985). *The Atlantic Alcidae.* Academic Press, London.

Neufeldt, I. (1978). Extinct birds in the collection of the Zoological Institute of the Academy of Sciences of the U.S.S.R. *Proceedings of the Zoological Institute of the Academy of Sciences of the USSR*, 76, pp.101-110.

Newman, E. (1868). The Great Auk. *Zoologist*, p.1483.

Newton, A. (1861). Abstract of Mr. J. Wolley's researches in Iceland respecting the Gare-fowl or Great Auk. *Ibis*, pp.374-99

Newton, A. (1865). The Gare-fowl and its Historians. *Natural History Review*, 467-88.

Newton, A. (1866). On some old or little known works which mention *A. impennis*. *Ibis*, pp.223-4.

Newton, A. (1870). On Existing Remains of the Gare-fowl. *Ibis*, pp.256-261.

Newton, A. (1879). Gare-fowl. *Encyclopædia Brittanica*, 9th edition, vol.10, pp.78-80.

Newton, A. (1885). Mr. Grieve on the Gare-fowl. *Nature* (October 8), pp.545-6.

Newton, A. (1891). Notes on some old museums. *Annual Report of the Museums Association*, pp.28-46.

Newton, A. (1896). *A Dictionary of Birds.* Adam and Charles Black, London.

Newton, A. (1898). On the Orcadian home of the Gare-fowl. *Ibis*, pp.587-92.

Newton, A. (1905). *See* Wolley and Newton (1864-1907).

Nicholson, Fr. (1907). *Alca impennis. Memoirs and Proceedings of the Manchester Literary and Philosophical Society*, 51: XXVII-XXVIII.

Olson, S. (1977). A Great Auk *Pinguinis* from the Pliocene of North Carolina. *Proceedings of the Biological Society of Washington*, pp.690-7.

Olson, S., Swift, C. & Mokhiber, C. (1979). An attempt to determine the prey of the great auk. *Auk*, pp.790-2.

Orton, J. (1870). The Great Auk. *American Naturalist*, pp.539-42.

Owen, R. (1864). On the skeleton of the Gare-fowl and the probability of its being an extinct species. *Proceedings of the Zoological Society of London*, p.258.

Owen, R. (1865). Description of the skeleton of the Great Auk or Gare-fowl. *Transactions of the Zoological Society of London*, V: 317-35.

Owen, R. (1879). *Memoirs on the Extinct Wingless Birds of New Zealand with an appendix on those of England, Australia, Newfoundland, Mauritius and Rodriguez.* John van Voorst, London.

Palmer, R. (1949). Maine Birds. *Bulletin of the Museum of Comparative Zoology*, vol 102 [entire volume].

Parkin, T. (1894). *The Great Auk or Garefowl.* Published by the Hastings and St. Leonards Natural History Society, Hastings [pamphlet].

Parkin, T. (1911). The Great Auk. A Record of Sales of Birds and Eggs by Public Auction in Great Britain, 1806-1910. *Hastings and East Sussex Naturalist*, extra paper to pt.6 of vol.1: pp.1-36.

Parrot, C. (1895). [Letter to the editor]. *Ibis*, p.165.

Pässler, W. (1860). Die Eier der *Alca impennis* in deutschen Sammlungen. *Journal für Ornithologie*, pp.58-60.

Pelzeln, A. von (1864). The Gare Fowl or Great Auk. *Annals and Magazine of Natural History*, p.393.

Pennant, T. (1761-6). *The British Zoology.* London.

Petersen, A. (1995). Brot úr sögu Geirfuglsins. *Náttúrufræðingurinn*, 65 (1-2): 53-66.

Pile, S. (1986). How they scotched the great auk story. *The Sunday Times* (May 11), p.7.

Potts, T. (1870). Notes on the Egg of *Alca impennis* in the collection of the writer. *Transactions and Proceedings of the New Zealand Institute*, 3: 109-10.

Preyer, W. (1862). Der Brillenalk in Europäischen Sammlungen. *Journal für Ornithologie*, pp.77-9.

Preyer, W. (1862). Ueber *Plautus impennis* Brunn. *Journal für Ornithologie*, pp.110-24 *and* pp.337-56 [also published separately by Emmerling, Heidelberg].

Rapine, J. (1930). A propos d'un nouvel œuf d'*Alca impennis. L'Oiseau et la Revue Française d'Ornithologie*, pp.190-1.

Ree, V. (1984). Geirfuglen - utryddet i 140 år. *Norsk Natur*, pp.164-5.

Reeks, H. (1869). Notes on the Zoology of Newfoundland. *Zoologist*, pp.1854-6.

Reichenbach, H. (1850). *Novitiae ad Synopsin Avium Neueste Entdekungen und Nachträge zur Vervollstandigung der Classe der Vogel Natatores: oder Schwimmvogel.* Dresden.

Reinhardt, J. (1839). Om Gejerfuglens forekomst paa Island. *Kröyers Naturhistorisk Tidskrift, Kjobenhavn*, pp.532-5.

Reinhardt, J. (1861). List of the Birds hitherto observed in Greenland. *Ibis*, p.15.

Roberts, A. (1861). Eggs of the Great Auk. *Zoologist*, p.7438.

Roberts, A. (1861). Skins and Eggs of the Great Auk. *Zoologist*, p.7553.

Rosenberg, H. von (1874). De Reuzenalk. *Jaarboekje van het Koninklijk Zoologisch Genootschap Natura Artis Magistra, 1874*, pp.143-6.

Rothschild, W. (1907). *Extinct Birds*. Hutchinson, London.

Rothschild, M. (1983). *Dear Lord Rothschild*. Hutchinson, London.

Rowley, G. (1869). The Skins of *Alca impennis*. *Zoologist*, p.1645.

Rowley, G. (1872). *A paper upon the Gare Fowl*. Published by the author, Brighton [pamphlet].

Rowley, P. (1995). *Chronicles of the Rowleys*. Huntingdon Local History Society, Huntingdon.

Salomonsen, F. (1944-5). Gejrfuglen, et hundredaars Minde. *Dyr i Natur og Museum*, pp.99-110.

Salomonsen, F. (1950-1). *The Birds of Greenland*. Munksgaard, Copenhagen.

Saunders, H. (1869). Notes on the Ornithology of Italy and Spain. *Ibis*, pp.391-403.

Saunders, H. (1891). Notes on Birds observed in Switzerland. *Ibis*, pp.158-9.

Scherdlin, P. (1926). A propos du Grand Pingouin du Musée zoologique de Strasbourg. *Bulletin de l'Association Philomathique d'Alsace et de Lorraine*, tome VII, Fasc. 1 (1925): 10-17.

Schinz, H. (1830-6). *Naturgeschichte und Abbildungen der Vögel*. Leipzig.

Sclater, P., ed. (1888). The Turati Collection. *Ibis*, p.150.

Seebohm, H. (1896). *Coloured Figures of the Eggs of British Birds*. Pawson and Brailsford, Sheffield.

Seebohm, H. (1882-5). *A History of British Birds*. London.

Selby, P. (1821-34). *Illustrations of British Ornithology*. Edinburgh.

Sellar, J. (1706). *The English Pilot*. London.

Sélys-Longchamps, Baron E. (1870). Birds observed in Italian Museums. *Ibis*, pp.449-55.

Seton, G. (1878). *St Kilda - Past and Present*. William Blackwood, Edinburgh.

Sharpe, R. (1896-7). *A Hand-book to the Birds of Great Britain*. Lloyd's Natural History, London.

Sharpe, R. (1898). *Sketch-book of British Birds*. London.

Shaw, G. & Nodder, F. (1789-1813). *The Naturalists' Miscellany* (vol.XI). London.

Sheppard, J. & Whitear, W. (1826). Catalogue of Norfolk and Suffolk Birds. *Transactions of the Linnaean Society of London*, 15: 61.

Sheppard, T. (1922). An Unrecorded egg of the Great Auk. *Natural History, London*, p.254.

Skene, W., ed. (1871-2). *John of Fordun: Scotichronicon*. Edinburgh.

Smith, J. (1879-80). Notice of the remains of the Great Auk, or Gare-fowl, found in Caithness, with notes on its ocurrence in Scotland and of its early history *and* Additional Notes. *Proceedings of the Society of Antiquaries of Scotland*, 13: 76-105 *and* 14: 436-444.

Spink & Son Ltd. (c.1970). *Romantic Histories of Some Extinct Birds and Their Eggs*. Spink, London [pamphlet].

Steenstrup, J. (1855). Et Bidrag til *Geirfuglens, Alca impennis*, Naturhistorie, og særligt til Kundskaben om dens tidligere Udbredningskreds. *Videnskabelige Meddelelser fra den Naturhistoriske Forening i Kjöbenhavn*, pp.33-118 [French translation: *Bullétin de la Société Ornithologique Suisse*, tome II, pt.1: 5-70].

Storer, R. (1960). Evolution of Diving Birds. *Proceedings of the 12th International Ornithological Congress*, pp.694-707.

Strassen, O. (1910). Der Riesenalk. *Berichte Senckenbergischen Naturforschenden Geseltschaf im Frankfurt am Main*, 41: 184-190.

Strauch, J. (1985). The phylogeny of the Alcidae. *Auk*, 102: 520-39.

BIBLIOGRAPHY

Stresemann, E. (1954). Ausgestorbene und Aussterbende Vogelarten, Vertreten im Zoologischen Museum zu Berlin. *Mittheilungen aus dem Zoologischen Museum in Berlin*, pp.38-53.

Sundevall, C. & Kinberg, J. (1856-86). *Svenska Foglarna*. Stockholm.

Thayer, J. (1905). The Purchase of a Great Auk for the Thayer Museum at Lancaster, Mass. *Auk*, pp.300-2.

Thayer, J. (1912). Great Auk Eggs in the Thayer Museum. *Auk*, pp.208-9.

Thienemann, F. (1845-56). *Einhundert Tafeln Colorirter Abbildungen von Vogeleiern zur Fortpflanzungs Geschichte der Gesammten Vogel*. Dresden.

Thompson, W. (1849-56). *Natural History of Ireland*. Reeve, Benham and Reeve, London.

Ticehurst, N. (1908). Bird bones obtained by excavating an ancient mound known as the rocks of Ayree, near the Bay of Ayr, Orkney. *Nature* (London), p.467.

Tomkinson, P. & Tomkinson, J. (1966). Eggs of the Great Auk. *Bulletin of the British Museum (Natural History) Historical Series*, 3 (4): 95-128.

Tuck, J. (1975). *The Archaeology of Saglek Bay, Labrador*. National Museums of Canada, Museum of Man, Mercury Series no.32, Ottawa.

Tuck, J. (1976). *Ancient People of Port au Choix*. Newfoundland Social and Economic Studies no.17, St. John's.

Ussher, R. (1897). The discovery of bones of the Great Auk in County Waterford. *Irish Naturalist*, 6: 208.

Ussher, R. (1899). The Great Auk, once an Irish Bird. *Irish Naturalist*, 8: 1-3.

Ussher, R. (1902). Great Auk in County Clare. *Irish Naturalist*, 11: 188.

Ussher, R. & Warren, R. (1900). *The Birds of Ireland*. Gurney and Jackson, London.

Violani, C. (1974). Ecologia di un' estinzione: l'Alca impenne. *Bolletin de Museo Civico Storia Naturale, Venezia*, 25: 49-60.

Violani, C. (1974). L'Alca impenne (*Alca impennis*) Nelle Collezioni Italiane. *Natura- Rivista di Scienze Naturali, Milano*, pp.13-24.

Violani, C. *et al.* (1984). Uccelli Estinti e Rari nei Musei Naturalistici. *Revista Italiana di Ornitologie*, p.144.

Vouga, A. (1868). [*In* Extraits de Procés Verbaux]. *Bullétin de la Société Ornithologique Suisse*, tome II, pt. 1, p.113-4.

Walcott, J. (1789). *Synopsis of British Birds*. Published by the author, London.

Walker, P. (1883). The Great Auk, formerly eaten in Lent. *Zoologist*, p.38

Wallis, J. (1769). *The Natural History and Antiquities of Northumberland*. London.

Walter, J. (1828-41). *Nodisk Ornithologie*. Copenhagen.

Werner, J. & Temminck, C. (1826-42). *Atlas des Oiseaux d'Europe*. Paris.

Westerman, G. (1880). [Obituary of G.A. Frank]. *Zoologischer Anzeiger* (Leipzig), 54: 216.

Wheeler, R. (1992). Tell Them! *Island Journal*, 10: 42-5.

Whitbourne, R. (1622). *A Discourse and Discovery of Newfoundland*. Felix Kinston, London.

Whitear, Rev. W. (1880-1). Extracts from the Calendar of the Rev. William Whitear (1809 to 1826). *Transactions of the Norfolk and Norwich Naturalists' Society*, 3 (2): 231-60.

Wiglesworth, J. (1903). St. Kilda and Its Birds. *Transactions of the Liverpool Biological Society*.

Williamson, K. (1939). The Great Auk in Man. *Journal of the Manx Museum*, 4: 168-72.

Williamson, K. (1941). Early drawings of the Great Auk and Gannet made in the Isle of Man in 1652. *Ibis*, 301-10.

Williamson, K. (1970). *The Atlantic Islands*. Routledge and Kegan Paul, London.

Williamson, K. & Boyd, J. (1960). *St Kilda Summer*. London.

Wilson, J. (1831). *Illustrations of British Zoology*. William Blackwood, Edinburgh.

Winge, H. (1898). Grønlands Fugle. *Meddelelser Grønland, 21 (1): 1-316.*

Winge, H. (1903). Om jordfundne Fugle fra Danmark. *Videnskabelige Meddelelser fra den Naturhistoriske Forening i Kjøbenhavn*, pp.31-109.

Witherby, H., Jourdain, F., Ticehurst, N. & Tucker, B. (1941). *The Handbook of British Birds*. Witherby, London.

Wollaston, A. (1921). *Life of Alfred Newton*. London.

Wolley, J. & Newton, A. (1864-1907). *Ootheca Wolleyana*. London.

Worm, O. (1655). *Museum Wormianum seu Historiae Rerum Rariorum*. Amsterdam.

Wyman, J. (1868). Remarks on finding bones of the Great Auk on Goose Island, Casco Bay, Maine. *Proceedings of the Boston Society for Natural History*, 11: 301-13.

Yarrell, W. (1843). *A History of British Birds*. Van Voorst, London.

INDEX

INDEX